Molecular Advances in Mechanism and Regulation of Lifespan and Aging

Molecular Advances in Mechanism and Regulation of Lifespan and Aging

Guest Editor

Mark McCormick

MDPI

Basel • Beijing • Wuhan • Barcelona • Belgrade • Novi Sad • Cluj • Manchester

Guest Editor
Mark McCormick
Department of Biochemistry
and Molecular Biology
University of New Mexico
Health Sciences Center
Albuquerque, NM
USA

Editorial Office
MDPI AG
Grosspeteranlage 5
4052 Basel, Switzerland

This is a reprint of the Special Issue, published open access by the journal *Biomolecules* (ISSN 2218-273X), freely accessible at: https://www.mdpi.com/journal/biomolecules/special_issues/HZ10VQKVD5.

For citation purposes, cite each article independently as indicated on the article page online and as indicated below:

Lastname, A.A.; Lastname, B.B. Article Title. *Journal Name* **Year**, *Volume Number*, Page Range.

ISBN 978-3-7258-7528-3 (Hbk)
ISBN 978-3-7258-7529-0 (PDF)
https://doi.org/10.3390/books978-3-7258-7529-0

Contents

About the Editor . . . . **vii**

Olivia C. Heath, Madison P. Otero, Alexander T. Achusim, Dorothy S. Karr, Antonio W. Leger and Mark A. McCormick
Novel tRNA Synthetase Inhibitors Increase Healthspan, Lifespan, and Autophagic Flux in *C. elegans*
Reprinted from: *Biomolecules* **2026**, *16*, 73, https://doi.org/10.3390/biom16010073 . . . . **1**

Emily Barr, Maude Comtois-Cabana, Andressa Coope, Sylvana M. Coté, Michael S. Kobor, Chaini Konwar, et al.
Early-Life Adversity and Epigenetic Aging: Findings from a 17-Year Longitudinal Study
Reprinted from: *Biomolecules* **2025**, *15*, 887, https://doi.org/10.3390/biom15060887 . . . . **16**

Timur Saliev and Prim B. Singh
Targeting Senescence: A Review of Senolytics and Senomorphics in Anti-Aging Interventions
Reprinted from: *Biomolecules* **2025**, *15*, 860, https://doi.org/10.3390/biom15060860 . . . . **35**

Mario García-Domínguez
Pathological and Inflammatory Consequences of Aging
Reprinted from: *Biomolecules* **2025**, *15*, 404, https://doi.org/10.3390/biom15030404 . . . . **54**

Supaya Wenuganen, Kenneth G. Walton, Frederick T. Travis, Tobias Stalder, R. Keith Wallace, Meera Srivastava and John Fagan
Possible Anti-Aging and Anti-Stress Effects of Long-Term Transcendental Meditation Practice: Differences in Gene Expression, EEG Correlates of Cognitive Function, and Hair Steroids
Reprinted from: *Biomolecules* **2025**, *15*, 317, https://doi.org/10.3390/biom15030317 . . . . **87**

Nicholas Pontillo and Yang Lyu
Perception and Longevity Control in Invertebrate Model Organisms—A Mini-Review of Recent Advances
Reprinted from: *Biomolecules* **2025**, *15*, 187, https://doi.org/10.3390/biom15020187 . . . . **115**

Azra Hasan and Brian E. Staveley
Bcl-2 Orthologues, *Buffy* and *Debcl*, Can Suppress *Drp1*-Dependent Age-Related Phenotypes in Drosophila
Reprinted from: *Biomolecules* **2024**, *14*, 1089, https://doi.org/10.3390/biom14091089 . . . . **128**

Bradford T. Hull, Kayla M. Miller, Caroline Corban, Grant Backer, Susan Sheehan, Ron Korstanje and George L. Sutphin
3-Hydroxyanthranilic Acid Delays Paralysis in *Caenorhabditis elegans* Models of Amyloid-Beta and Polyglutamine Proteotoxicity
Reprinted from: *Biomolecules* **2024**, *14*, 599, https://doi.org/10.3390/biom14050599 . . . . **144**

About the Editor

Mark McCormick

Mark McCormick is an Associate Professor of Biochemistry and Molecular Biology in the School of Medicine at the University of New Mexico Health Sciences Center. Dr. McCormick has studied the biology of aging through his PhD, postdoctoral, and faculty career. He is a Fellow and current Board Member of the American Aging Association, and the current Vice-Chair of the Biological Sciences Section of the Gerontological Society of America.

biomolecules

MDPI

Article

Novel tRNA Synthetase Inhibitors Increase Healthspan, Lifespan, and Autophagic Flux in *C. elegans*

Olivia C. Heath [1], Madison P. Otero [1], Alexander T. Achusim [1], Dorothy S. Karr [1], Antonio W. Leger [1] and Mark A. McCormick [1,2,*]

[1] Department of Biochemistry and Molecular Biology, School of Medicine, University of New Mexico Health Sciences Center, Albuquerque, NM 87131, USA; oliviach@salud.unm.edu (O.C.H.)
[2] Autophagy, Inflammation, and Metabolism Center of Biomedical Research Excellence, Albuquerque, NM 87131, USA
* Correspondence: mmccormick@salud.unm.edu

Abstract

We previously demonstrated that the tRNA synthetase inhibitors mupirocin and borrelidin extend lifespan in *C. elegans* and *S. cerevisiae* and that tRNA synthetase inhibition enhances autophagy in mammalian cells. In this study, we identify four additional tRNA synthetase inhibitors, REP8839, REP3123, LysRS-In-2, and halofuginone, that extend both healthspan and lifespan in *C. elegans*. These compounds also trigger a significant upregulation of autophagy, specifically at their lifespan-extending doses. These phenotypes partially depend on the conserved transcription factor ATF-4. Our findings further establish tRNA synthetase inhibition as a conserved mechanism for promoting increased lifespan and now healthspan, with potential implications for therapeutic interventions targeting age-related decline in humans.

Keywords: Gcn4; ATF-4; ATF4; tRNA synthetase; autophagy; healthspan

1. Introduction

Aging has a profound impact on individuals, societies, and economies worldwide. As the proportion of older adults grows rapidly, so does the economic burden that comes with inevitable age-related diseases. The need for interventions designed to delay the onset of these age-related diseases has never been greater.

C. elegans has long been used as a model for aging research [1–4]. They have a relatively short lifespan of two to three weeks, making them ideal for studying the many factors that influence aging. Additionally, among the over 18,000 *C. elegans* protein sequences, at least 83% have human homologs [5]. This makes *C. elegans* a fast and efficient model organism to study aging and age-related diseases.

Macroautophagy, henceforth referred to as autophagy, is a cellular process that, in part, can act to break down damaged, dysfunctional, or otherwise unwanted components. Autophagy is crucial for maintaining proteostasis and is a necessary system for cellular survival under stressful conditions. Autophagic efficiency declines during aging, leading to the buildup of damaged proteins and organelles, as well as other nonviable cellular debris.

The amino acid response (AAR) pathway is a highly conserved mechanism that ultimately leads to the increased translation of Gcn4 (in yeast), ATF-4 (in worms), and ATF4 (in mammals). We have previously shown that activation of this pathway through the chemical inhibition of tRNA synthetases (tRS) can extend lifespan in both worms and yeast [6]. tRNA synthetases play a critical role in attaching amino acids to their corresponding tRNA, a process necessary for translation. When tRS enzymes are inhibited, uncharged tRNA accumulates. This response is sensed by the uncharged tRNA sensor, general control non-depressible kinase 2 (Gcn2 in yeast, GCN-2 in worms, and GCN2 in mammals).

The accumulation of uncharged tRNA prompts GCN2 to phosphorylate eukaryotic initiation factor 2 (eIF2) [7], which inhibits translation re-initiation until an eIF2 phosphatase restores eIF2's function [8]. This delay in re-initiation leads to a global reduction in protein synthesis. Still, it paradoxically enhances the translation of certain genes, including the transcription factor ATF4 [9] (or its orthologs like Gcn4 and ATF-4), which regulates hundreds of target genes. ATF4 and its orthologs are controlled at the translational level by upstream open reading frames [10], a mode of regulation that is evolutionarily conserved. This suggests that ATF4 and its counterparts function as a conserved stress response system for translational disruption. However, the full downstream effects of tRS inhibition on this system remain largely undefined. Previous research in our lab has demonstrated that tRS inhibitors activate the AAR and extend lifespan in worms [6] and yeast, a process that is entirely dependent on Gcn4/ATF-4. This finding suggests a strong link between this conserved transcription factor and the aging process.

2. Materials and Methods

2.1. *C. elegans* Strains

All worm strains were maintained at 20 °C on NGM plates (Fisher FB0875714, Thermo-Fisher, Waltham, MA, USA) containing *E. coli* strain OP-50 as a food source [11]. Wild-type N2 worms, *atf-4* deletion worms (RB790 *atf-4(ok576)*), and autophagy reporter worms (MAH215 sqIs11 [*lgg-1*p::mCherry::GFP::*lgg-1* + *rol-6*]), as well as *E. coli* strain OP-50, were obtained directly from the *Caenorhabditis* Genetics Center (https://cgc.umn.edu/), URL accessed on January 15 2025, Minneapolis, MN, USA which is funded by NIH Office of Research Infrastructure Programs (P40 OD010440).

2.2. *C. elegans* Lifespan Experiments

Lifespan analysis was conducted at 20 °C unless otherwise stated [12]. Specifically, all lifespans were carried out on 6 cm plates (T3306, Tritech Research, Los Angeles, CA, USA) containing standard NGM media [11] using on-plate UV-killed bacterial food (*E. coli* strain OP-50 unless otherwise specified) that was first plated at 50 μL per lifespan plate in the center of each plate, allowed to grow for 72 h at 25 °C, and then treated to 3 × 9999 μm Joules ×100 using a UV Stratalinker 2400 (Stratagene, San Diego, CA, USA) on uncovered plates. In all lifespans, 0.1 mM FUDR (5-fluorodeoxyuridine) (518265, Bio-World, Dublin, OH, USA) was added to plates at late L4 stage in order to suppress development of progeny. All drug treatments were added to plates after UV treatment and before addition of eggs and allowed to equilibrate overnight to ensure even diffusion through the agar. Reported concentrations represent the final agar concentrations. Each day, all worms were scored for movement using a 6.7×–45× Trinocular Zoom Stereo Microscope with attached incandescent reflected transillumination base (ZM-2T-EB, AmScope, Irvine, CA, USA), and any dead worms were removed using a pick made from a 5.75 inch plain soda-lime glass Pasteur pipette handle (Thermo-Fisher, Waltham, MA, USA) and a platinum wire tip (Tritech Research, Los Angeles, CA, USA). Bagged, missing (crawled off), or exploded worms were noted and censored daily. Mean lifespan (in days) and the number of animals analyzed (n) are reported in parentheses in the figure legends. As a reference for baseline

 https://doi.org/10.3390/biom16010073

viability, *atf-4(ok576)* showed a lifespan comparable to wild type under vehicle conditions in our assays (see Figures 1 and 2 legends for mean and n).

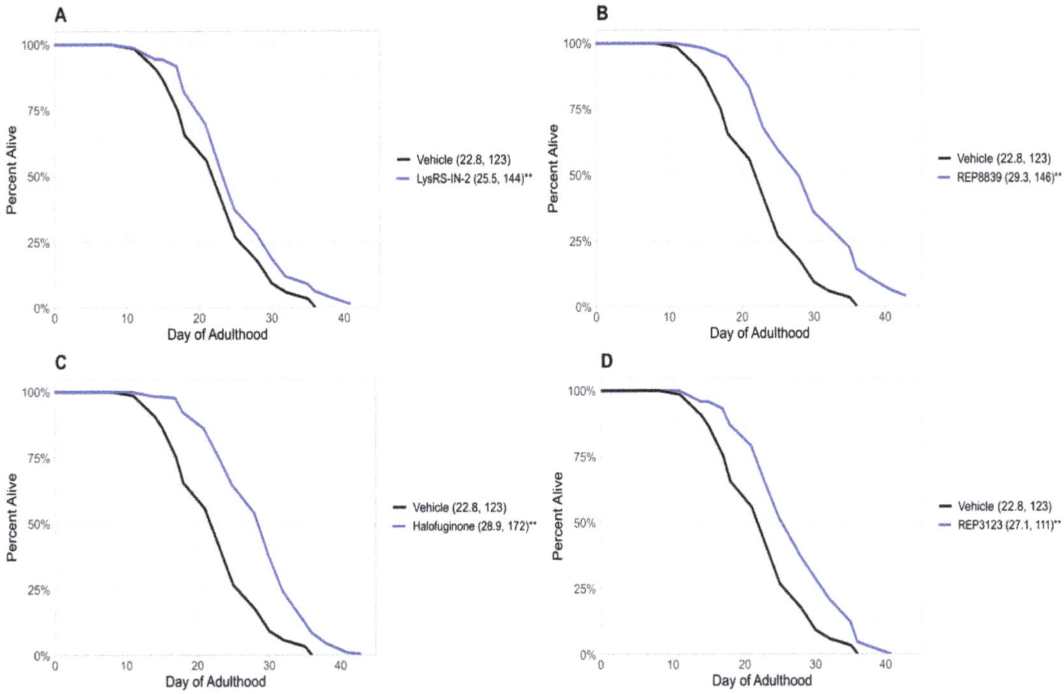

Figure 1. Novel tRNA synthetase inhibitors increase lifespan in wild-type *C. elegans*. (**A**) A total of 10 µM LysRS-IN-2 increases lifespan in wild-type *C. elegans* (blue) compared to the vehicle-treated group (black). (**B**) A total of 60 µM REP8839 increases lifespan in wild-type *C. elegans* (blue) compared to the vehicle-treated group (black). (**C**) A total of 1 µM halofuginone increases lifespan in wild-type *C. elegans* (blue) compared to the vehicle-treated group (black). (**D**) A total of 40 µM REP3123 increases lifespan in wild-type *C. elegans* (blue) compared to the vehicle-treated group (black). Vehicle = DMSO. ** $p < 0.001$. Values in parentheses indicate mean lifespan (days) and number of animals analyzed. p values were calculated using the log-rank (Mantel–Cox) test.

2.3. tRNA Synthetase Inhibitor Drug Concentrations

All tRNA synthetase inhibitor concentrations were selected based on pilot lifespan experiments and literature precedent, where available. For each compound, several concentrations were screened to identify the lowest dose that reproducibly extended lifespan without visible toxicity (e.g., developmental delay, death, or growth defects). The concentrations reported in this study represent those optimal, non-toxic doses. Compounds showing no further benefit or exhibiting toxicity at higher concentrations were not pursued further.

2.4. Autophagy Assays

MAH215 *C. elegans* were maintained at 20 °C on NGM plates containing *E. coli* strain OP-50 as a food source [11]. tRS-treated nematodes were grown on plates with their respective drug treatments. For imaging, *C. elegans* were mounted on 1% agarose pads and paralyzed with 10 mM levamisole in M9 prior to fluorescence imaging, following established protocols [13] (Fisher, Waltham MA, USA, AC187870100). Imaging was performed on an Olympus iX83 Yokogawa spinning disc confocal microscope (Olympus, Center Valley, PA, USA) in the UNM Comprehensive Cancer Center Fluorescence Microscopy and Cell

https://doi.org/10.3390/biom16010073

Imaging Shared Resource, which is supported by UNM Comprehensive Cancer Center Support Grant NCI P30CA118100. Wild-type and *atf-4(ok576)* animals were assayed concurrently under identical conditions. Vehicle-only controls for *atf-4(ok576)* are shown in Supplementary Figure S1. Autophagy flux measurements were conducted using animals derived from a single synchronized cohort per experiment. Following synchronization by egg lay, individual animals were imaged and quantified on the same day for each strain and condition, with 13–16 animals analyzed per strain × condition × day. These measurements reflect multiple individual animals within one biological experiment.

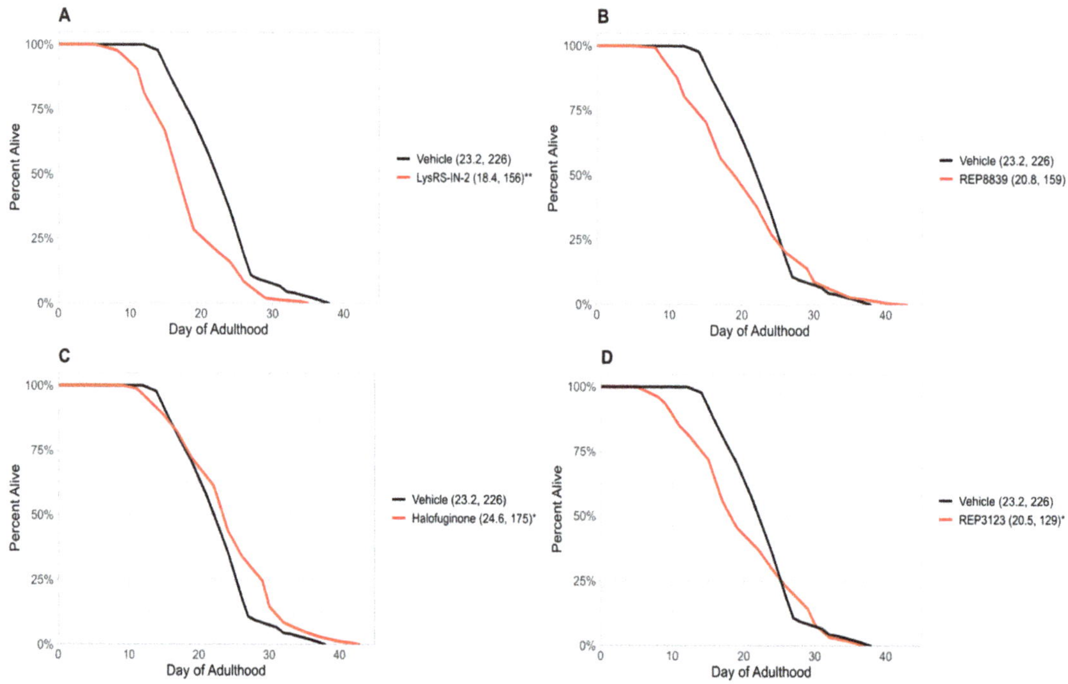

Figure 2. tRNA synthetase inhibitors do not increase lifespan in *atf-4(ok576) C. elegans*. (**A**) A total of 10 μM LysRS-IN-2 does not increase lifespan in *atf-4(ok576) C. elegans* (red) compared to the vehicle-treated group (black). (**B**) A total of 60 μM REP8839 does not increase lifespan in *atf-4(ok576) C. elegans* (red) compared to the vehicle-treated group (black). (**C**) A total of 1 μM halofuginone does not increase lifespan in *atf-4(ok576) C. elegans* (red) compared to the vehicle-treated group (black). (**D**) A total of 40 μM REP3123 does not increase lifespan in *atf-4(ok576) C. elegans* (red) compared to the vehicle-treated group (black). Vehicle = DMSO. * $p < 0.01$, ** $p < 0.001$. Values in parentheses indicate mean lifespan (days) and number of animals analyzed. p values were calculated using the log-rank (Mantel–Cox) test.

2.5. *Healthspan/Thrashing Assays*

C. elegans were grown on NGM plates containing their respective drug treatment. For the thrashing assays, nematodes were placed on a glass slide (AmScope, Irvine, CA, USA) containing liquid M9 and were allowed to acclimate to the liquid for 1 min. Subsequently, body bends were counted manually under a microscope for 1 min. Thrashing assays were performed on animals obtained from a single synchronized population per experiment. Worms were synchronized by egg lay and maintained under identical conditions, and individual animals were scored on the same day for each strain and treatment. Approximately

https://doi.org/10.3390/biom16010073

8 individual animals were measured per strain × treatment × day. These measurements represent multiple animals assessed within a single biological experiment.

2.6. Quantification and Statistical Analysis

Lifespan curves were plotted using Kaplan–Meier [14] estimates using R 4.5.0 (R Foundation, Vienna, Austria). Statistical significance for lifespan analyses was determined using the log-rank (Mantel–Cox) test implemented via survival::survdiff() from package survival 3.7-0 in R 4.5.0 (R Foundation, Vienna, Austria). Thrashing and autophagy assays were analyzed using Student's *t*-test [15] with Bonferroni correction in R 4.5.0 (R Foundation, Vienna, Austria).

2.7. Materials

Halofuginone was purchased from Ambeed, Buffalo Grove, IL, USA [CAS No. 64924-67-0]. Borrelidin was purchased from BioViotica, San Diego, CA, USA (BVT-0098). LysRS-IN-2 was purchased from GLPBio, Montclair, CA, USA (GC65058). REP3123 and REP8839 were purchased from Axon Medchem, Reston, VA, USA (1704, 1705). Mupirocin was purchased from BOC Sciences, Shirley, NY, USA (B0084-056590). Borrelidin and mupirocin were included as positive controls for lifespan extension, based on prior studies. Because their effects on healthspan and autophagy had not been examined previously, they were also tested here to determine whether their lifespan-extending activity is accompanied by improvements in these related phenotypes.

3. Results

3.1. Novel tRNA Synthetase Inhibitors Extend Lifespan in C. elegans in an Atf-4-Dependent Manner

Previous studies from our group have demonstrated that treatment with tRNA synthetase inhibitors can extend lifespan in *C. elegans* and *S. cerevisiae* [6]. In the present study, we identify four novel tRNA synthetase inhibitors that also extend lifespan in wild-type *C. elegans* (Figure 1). These novel lifespan-extending inhibitors include two methionyl tRNA synthetase inhibitors, REP8839 (Figure 1B) and REP3123 (Figure 1D), as well as the lysyl tRNA synthetase inhibitor LysRS-In-2 (Figure 1A) and the prolyl tRNA synthetase inhibitor halofuginone (Figure 1C). Given that we have previously shown that these compounds greatly upregulate ATF4 translation in mammalian cells [16] and that previously identified tRNA synthetase inhibitors depend on *atf-4* for lifespan extension in *C. elegans* [6], we next used an *atf-4(ok576)* deletion strain to ask whether the observed lifespan extensions depend on *atf-4*. To assess this, we treated *atf-4(ok576) C. elegans* with the same tRNA synthetase inhibitors and found no significant increase in lifespan (Figure 2). Under our assay conditions, *atf-4(ok576)* exhibited a baseline lifespan comparable to the wild type, as indicated by the vehicle control curves (e.g., N2 mean ≈ 23.2 days vs. *atf-4* mean ≈ 22.8 days). This supports that loss of *atf-4* does not inherently reduce viability under baseline conditions. The results of Figure 2 indicate that the lifespan extension induced by these tRNA synthetase inhibitions is *atf-4*-dependent. For each of these tRNA synthetase inhibitors, the lifespan extension in wild-type animals was dose-dependent, and the largest effect size is shown in Figure 1. The doses used for each compound correspond to the lowest concentrations that consistently extended lifespan without visible toxicity, as determined by pilot dose-range assays.

These findings extend our prior work by broadening the claim that the entire class of drugs, tRNA synthetase inhibitors, elicits a pro-longevity effect. Notably, although the four compounds target distinct aminoacyl-tRNA synthetases, they all converge on a common downstream genetic requirement for *atf-4*, highlighting a conserved transcription factor as

https://doi.org/10.3390/biom16010073

a key mediator of lifespan extension. This finding is promising for the possibility of the use of these drugs in higher organisms with the goal of promoting longevity in humans.

Together, these data identify a new set of candidate small molecules that extend lifespan in an *atf-4*-dependent manner and provide further mechanistic support for tRNA synthetase inhibition as a viable, safe, and evolutionarily conserved longevity intervention.

3.2. tRNA Synthetase Inhibitors Extend Healthspan, Specifically Thrashing Activity, in C. elegans at the Same Doses That Extend Lifespan

Having shown that these four novel compounds can extend lifespan in *C. elegans*, we next asked whether they could also influence healthspan. For comparison, we included two previously characterized tRNA synthetase inhibitors—borrelidin and mupirocin—that we have previously shown to extend lifespan in *C. elegans* and yeast [6]. These compounds had not yet been tested for effects on healthspan or autophagy and thus served both as positive controls for lifespan extension and as comparators to evaluate whether previously known tRNA synthetase inhibitors similarly enhance functional and cellular maintenance phenotypes. Healthspan refers to the healthy, active period of life, underscoring a growing interest in interventions that do not solely delay mortality without delaying morbidity but that increase the healthy, productive period of life as well [17–19]. One way that healthspan can be measured in *C. elegans* is by quantifying thrashing rates. When placed into liquid, these animals thrash back and forth at a steady rate, and this rate decreases consistently with age in wild-type animals [20]. To evaluate whether tRNA synthetase inhibition improves healthspan, we used a thrashing assay to measure the frequency of body bends that *C. elegans* voluntarily makes when suspended in liquid.

Wild-type *C. elegans* were treated with tRNA synthetase inhibitors at the same concentrations used to extend lifespan (Figure 1), and their thrashing rate was assessed throughout adulthood. Treatment with some of these compounds significantly enhanced thrashing, particularly later in life (Figure 3). Similar to lifespan extension, this improvement in healthspan was partially dependent on *atf-4* (Figure 4). In particular, animals treated with the compounds maintained higher thrashing rates at day 7 and beyond, a stage when untreated control animals typically show marked declines in movement and responsiveness. This preservation of neuromuscular function suggests that the benefits of tRNA synthetase inhibition are not limited to longevity alone but are accompanied by an overall delay in functional decline.

Importantly, these effects were consistent across multiple different tRNA synthetase inhibitors, indicating that healthspan extension may be a generalizable consequence of tRNA synthetase inhibition rather than an off-target effect of any individual compound. Moreover, the observation that *atf-4(ok576)* mutants showed less improved thrashing behavior in response to treatment reinforces the notion that this transcription factor plays a central role, although it is not solely responsible, in the protective response elicited by these compounds. Together, these results provide strong evidence that tRNA synthetase inhibitors not only extend lifespan but also maintain functional capacity during aging, an important goal of lifespan-extending therapies.

3.3. tRNA Synthetase Inhibitors Upregulate Autophagy at the Same Doses That Extend Lifespan

The effect of tRNA synthetases on lifespan that we have observed is dependent on *atf-4* (Figure 2). ATF-4 is itself a transcription factor that is translationally upregulated during the accumulation of uncharged tRNA [8,10,21]. Previous work has shown that upregulation of autophagy has been linked to increased lifespan [22], and we have previously shown that tRNA synthetase inhibitor treatment causes ATF-4-dependent changes in transcripts relating to autophagy in both mouse embryonic fibroblasts [16] and *S. cerevisiae* [23]. We have also shown that these tRNA synthetase inhibitors cause partially ATF-4-dependent

https://doi.org/10.3390/biom16010073

increases in autophagic activity in mouse cells in vitro [16]. As a result, we wanted to ask whether autophagy was changed in *C. elegans* by lifespan-extending doses of tRNA synthetase inhibitors.

To further investigate this, we examined autophagy in *C. elegans* using a dual-color fluorescent reporter strain, MAH215, which expresses tandem-tagged GFP and mCherry fused to LGG-1, the *C. elegans* homolog of human LC3 [24]. We found that treatment with tRNA synthetase inhibitors leads to a substantial increase in autophagic activity in *C. elegans* at the same concentrations that extend lifespan (Figure 5). This increase in autophagy was dependent in part on the presence of *atf-4* (Figure 6), suggesting a potential link between ATF-4 upregulation, autophagy, and lifespan extension.

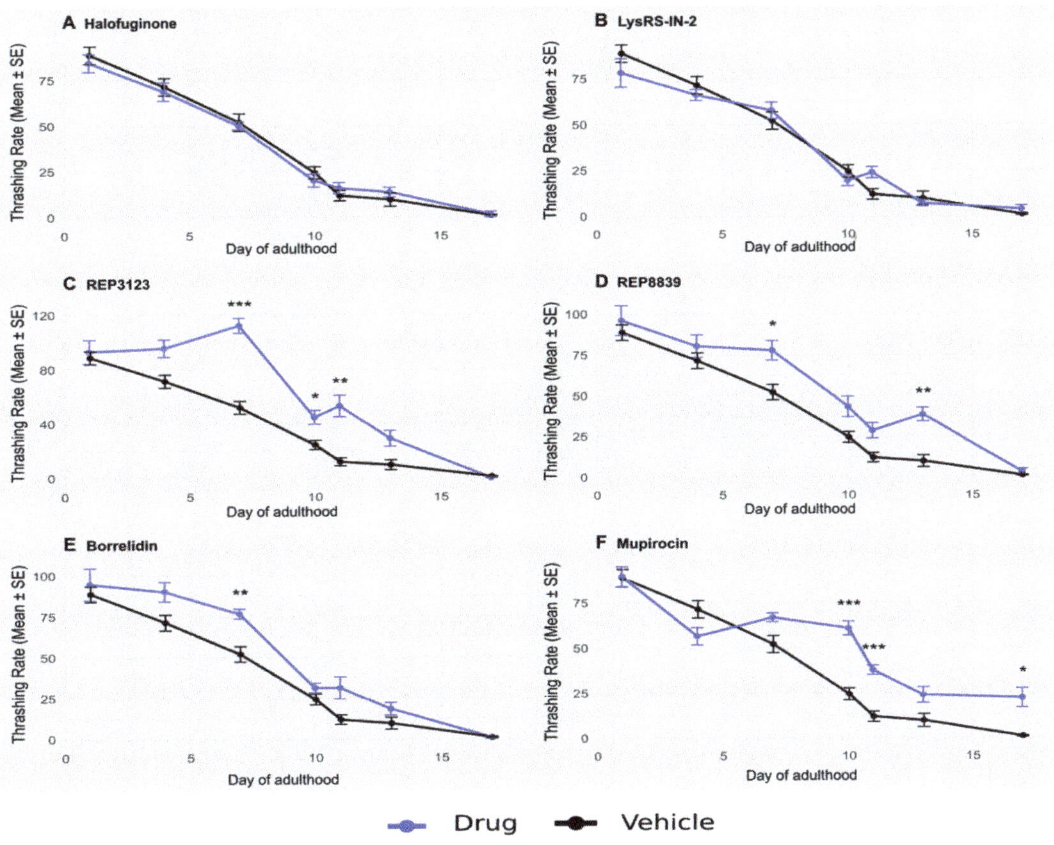

Figure 3. Thrashing in wild-type *C. elegans* is significantly improved upon treatment with some tRNA synthetase inhibitors. (**A**) A total of 1 μM halofuginone does not increase thrashing in wild-type *C. elegans* (blue) compared to the vehicle-treated group (black). (**B**) A total of 10 μM LysRS-IN-2 does not increase thrashing in wild-type *C. elegans* (blue) compared to the vehicle-treated group (black). (**C**) A total of 40 μM REP3123 significantly increases thrashing in wild-type *C. elegans* (blue) compared to the vehicle-treated group (black) on days 7, 10, and 11. (**D**) A total of 60 μM REP8839 significantly increases thrashing in wild-type *C. elegans* (blue) compared to the vehicle-treated group (black) on days 7 and 13. (**E**) A total of 75 μM borrelidin significantly increases thrashing in wild-type *C. elegans* (blue) compared to the vehicle-treated group (black) on day 7. (**F**) A total of 550 μM mupirocin significantly increases thrashing in wild-type *C. elegans* (blue) compared to the vehicle-treated group (black) on days 10, 11, and 17. * $p < 0.05$, ** $p < 0.01$, *** $p < 0.001$. p values calculated with Student's t-test with Bonferroni correction ($n = 7$ comparisons per drug panel).

 https://doi.org/10.3390/biom16010073

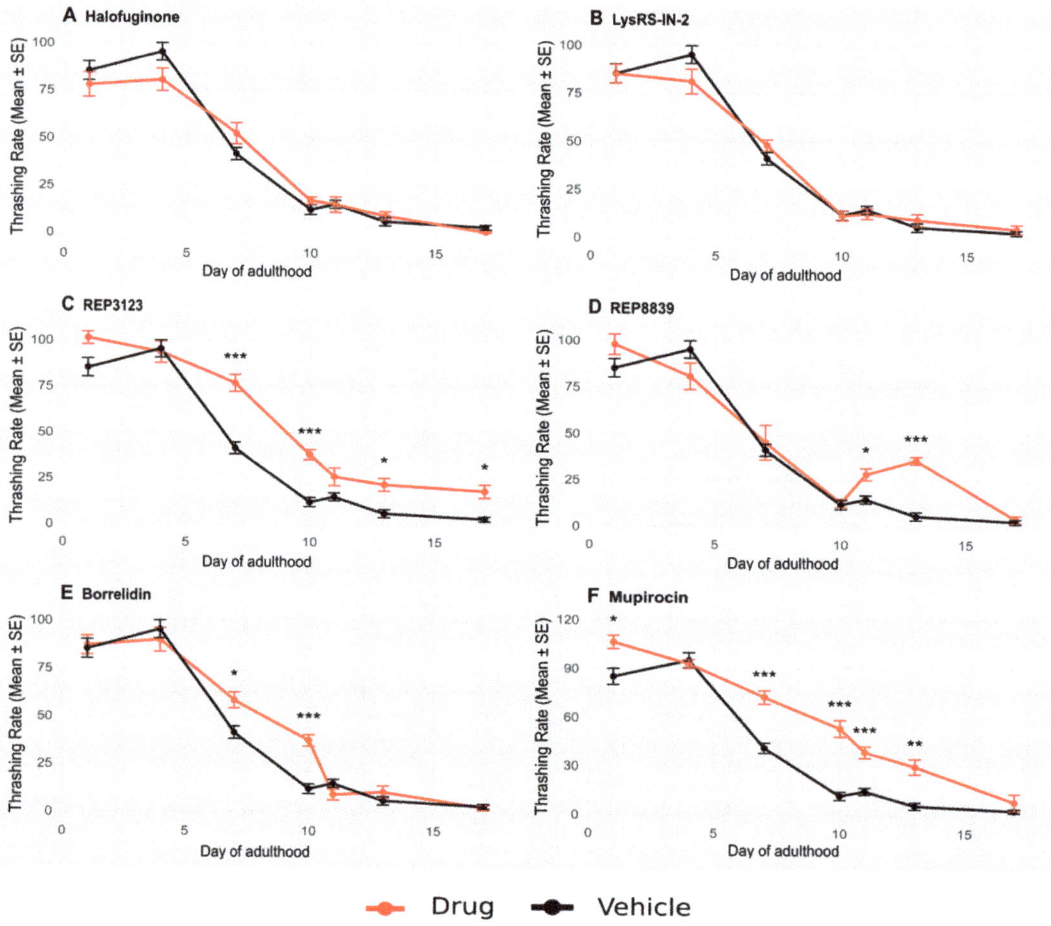

Figure 4. Thrashing in *atf-4(ok576) C. elegans* is significantly improved upon treatment with some tRNA synthetase inhibitors, although less so than the wild type. (**A**) A total of 1 μM halofuginone does not increase thrashing in *atf-4(ok576) C. elegans* (red) compared to the vehicle-treated group (black). (**B**) A total of 10 μM LysRS-IN-2 does not increase thrashing in *atf-4(ok576) C. elegans* (red) compared to the vehicle-treated group (black). (**C**) A total of 40 μM REP3123 significantly increases thrashing in *atf-4(ok576) C. elegans* (red) compared to the vehicle-treated group (black) on days 7, 10, 13, and 17. (**D**) A total of 60 μM REP8839 significantly increases thrashing in *atf-4(ok576) C. elegans* (red) compared to the vehicle-treated group (black) on days 11 and 13. (**E**) A total of 75 μM borrelidin significantly increases thrashing in *atf-4(ok576) C. elegans* (red) compared to the vehicle-treated group (black) on days 7 and 10. (**F**) A total of 550 μM mupirocin significantly increases thrashing in *atf-4(ok576) C. elegans* (red) compared to the vehicle-treated group (black) on days 1, 7, 10, 11, and 13. Vehicle = DMSO. * $p < 0.05$, ** $p < 0.01$, *** $p < 0.001$. p values calculated with Student's t-test with Bonferroni correction ($n = 7$ comparisons per drug panel).

Consistent with these findings, quantitative analysis of tagged LGG-1 puncta revealed a marked elevation in autophagolysosome formation in tRNA synthetase inhibitor-treated animals, particularly during mid-to-late adulthood when autophagic flux typically declines. The use of the dual GFP::mCherry tag allowed us to distinguish between autophagosomes and autolysosomes, and our data indicate enhanced autophagic flux rather than simple accumulation of autophagic intermediates. Importantly, for some of these compounds,

https://doi.org/10.3390/biom16010073

no increase in autophagic activity was observed in *atf-4(ok576)* mutants treated with the same doses, supporting a model in which ATF-4 acts upstream of autophagy induction under these conditions. Notably, not all tRNA synthetase inhibitors produced a significant increase in autophagic flux. This variability likely reflects compound-specific mechanisms, as each inhibitor targets a distinct aminoacyl-tRNA synthetase with unique cellular roles and stress response profiles. Such differences in target enzyme abundance, localization, or affinity could lead to varying levels of uncharged tRNA accumulation and downstream activation of the ATF-4 pathway. These findings suggest that the induction of autophagy by tRNA synthetase inhibitors is not universal but instead depends on the degree and nature of translational stress imposed by each compound.

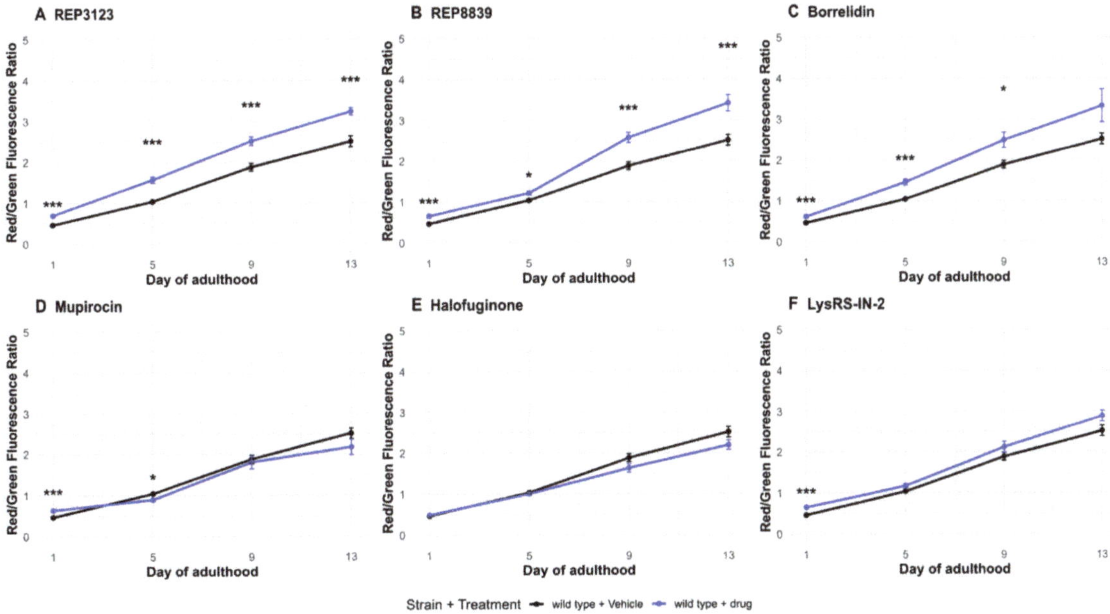

Figure 5. Autophagic flux increases following some tRNA synthetase inhibitor treatments. (**A**) A total of 40 µM REP3123 significantly increases the ratio of Red/Green fluorescence in wild-type drug-treated *C. elegans* (blue) compared to the vehicle-treated group (black) on all days tested (1, 5, 9, and 13). (**B**) A total of 60 µM REP8839 significantly increases the ratio of Red/Green fluorescence in wild-type drug-treated *C. elegans* (blue) compared to the vehicle-treated group (black) on all days tested (1, 5, 9, and 13). (**C**) A total of 75 µM borrelidin significantly increases the ratio of Red/Green fluorescence in wild-type drug-treated *C. elegans* (blue) compared to the vehicle-treated group (black) on days 1, 5, and 9. (**D**) A total of 550 µM mupirocin significantly increases the ratio of Red/Green fluorescence in wild-type drug-treated *C. elegans* (blue) compared to the vehicle-treated group (black) on day 1. (**E**) A total of 1 µM halofuginone does not significantly increase autophagic flux on any day tested. (**F**) A total of 10 µM Lys-RS-In-2 significantly increases the ratio of Red/Green fluorescence in wild-type drug-treated *C. elegans* (blue) compared to the vehicle-treated group (black) on day 1. Vehicle = DMSO. * $p < 0.05$, *** $p < 0.001$. p values calculated with Student's t-test with Bonferroni correction ($n = 4$ comparisons per timepoint).

Taken together, these results suggest that autophagy acts as one of several effector pathways downstream of ATF-4 in response to tRNA synthetase inhibition. However, not all compounds required ATF-4 for their beneficial effects, indicating that additional or parallel signaling pathways likely contribute to the observed improvements in lifespan and motility. The induction of autophagy likely contributes to improved proteostasis and

https://doi.org/10.3390/biom16010073

cellular maintenance during aging, thereby promoting both the increased lifespan and healthspan seen following treatments with these same concentrations of tRS inhibitors. Thus, tRNA synthetase inhibitors appear to elicit a conserved, adaptive response involving ATF-4 and autophagy that supports healthy aging.

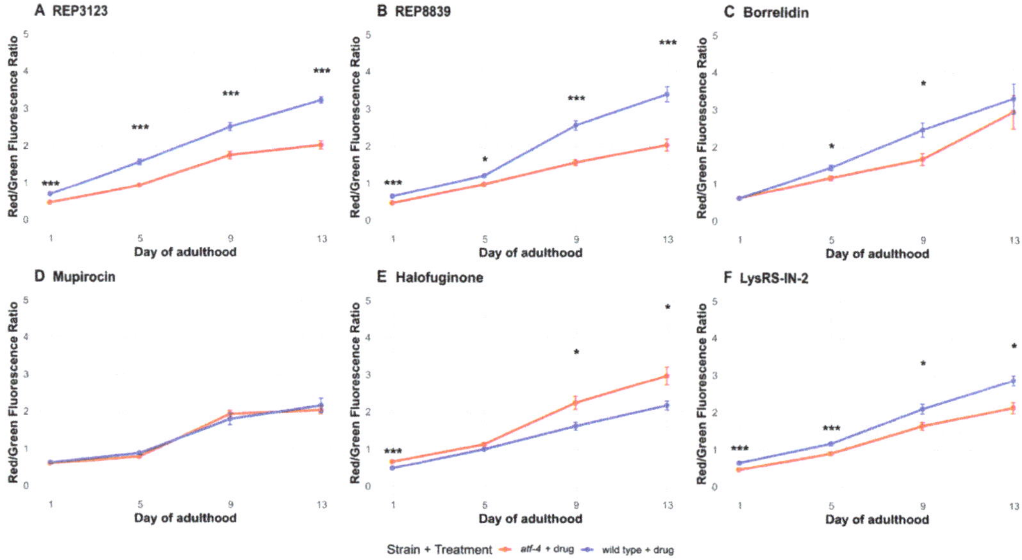

Figure 6. Autophagic flux increases following some tRNA synthetase inhibitor treatments partially depend on *atf-4(ok576)*. (**A**) A total of 40 μM REP3123 significantly increases the ratio of Red/Green fluorescence in wild-type drug-treated *C. elegans* (blue) compared to the *atf-4(ok576)* drug-treated group (red) on all days tested (1, 5, 9, and 13). (**B**) A total of 60 μM REP8839 significantly increases the ratio of Red/Green fluorescence in wild-type drug-treated *C. elegans* (blue) compared to the *atf-4(ok576)* drug-treated group (red) on all days tested (1, 5, 9, and 13). (**C**) A total of 75 μM borrelidin significantly increases the ratio of Red/Green fluorescence in wild-type drug-treated *C. elegans* (blue) compared to the *atf-4(ok576)* drug-treated group (red) on days 5 and 9. (**D**) A total of 550 μM mupirocin did not significantly increase the ratio of Red/Green fluorescence in wild-type drug-treated *C. elegans* (blue) compared to the *atf-4(ok576)* drug-treated group (red) on any day tested. (**E**) A total of 1 μM halofuginone significantly increases the ratio of Red/Green fluorescence in *atf-4(ok576)* drug-treated *C. elegans* (red) compared to the wild-type drug-treated group (blue) on days 1, 9, and 13. (**F**) A total of 10 μM Lys-RS-In-2 significantly increases the ratio of Red/Green fluorescence in wild-type drug-treated *C. elegans* (blue) compared to the *atf-4(ok576)* drug-treated group (red) on all days tested (1, 5, 9, and 13). Vehicle = DMSO. (mean thrashing, *n*) * $p < 0.05$, *** $p < 0.001$. p values calculated with Student's t-test with Bonferroni correction ($n = 4$ comparisons per timepoint). p values calculated with Student's t-test with Bonferroni correction ($n = 4$ comparisons per timepoint). The *atf-4(ok576)* + vehicle control was omitted here for clarity, but is shown in Supplementary Figure S1, where all conditions are displayed side by side.

4. Discussion

Our findings identify four novel tRNA synthetase inhibitors that can robustly extend lifespan in *C. elegans*. These inhibitors target different tRNA synthetases, consistent with our model in which the accumulation of many distinct uncharged tRNAs should lead to translational upregulation of ATF-4 and increased transcription of its targets. Combined with our work showing similar effects of these compounds on ATF4 in mammalian cells [16], and the work of others showing upregulation of ATF4 in long-lived mice [25,26], these results support the idea that tRNA synthetase inhibition may engage conserved stress

 https://doi.org/10.3390/biom16010073

response pathways relevant to aging. Notably, several of these compounds also improved healthspan in *C. elegans,* as assessed by age-dependent declines in thrashing behavior at the same doses that extended lifespan, raising the possibility that tRNA synthetase inhibition could delay functional decline during aging.

Two of the compounds identified in this study—REP3123 and REP8839—target methionyl tRNA synthetase, while LysRS-In-2 and halofuginone inhibit lysyl and prolyl-tRNA synthetases, respectively. All four significantly increase lifespan in wild-type *C. elegans* (Figure 1). This effect was completely abolished in *atf-4(ok576)* mutants (Figure 2), confirming that ATF-4 activation is necessary for the observed longevity benefits for the tested compounds. These data support a possible model in which tRNA synthetase inhibition engages an adaptive transcription factor response, ultimately leading to increased organism survival [8,21].

Halofuginone increased lifespan in the wild type, yet showed no wild-type autophagic flux increase and, in fact, showed a relative flux elevation in *atf-4(ok576)* animals at several timepoints. Given halofuginone's known tolerability limits and reported off-target effects [27,28], it is possible that this compound may trigger compensatory, ATF-4-independent stress responses at the tested maximally lifespan-extending dose.

Interestingly, some tRNA synthetase inhibitors appeared to reduce lifespan in *atf-4(ok576)* mutants, possibly suggesting increased sensitivity to translational stress in *atf-4(ok576)*. Loss of *atf-4* compromises the ability to upregulate stress-responsive and proteostasis-related genes, which may render animals less tolerant to inhibitors that partially suppress protein synthesis. Because the concentrations used were optimized for maximal lifespan extension in the wild type, these same doses may approach the toxicity threshold in the potentially more sensitive *atf-4* background. Thus, the apparent toxicity in *atf-4* mutants could reflect a combination of dose-dependent effects and reduced stress resilience rather than a fundamentally distinct mechanism of action.

The degree of *atf-4* dependence varied across the inhibitors tested in our autophagy and thrashing data. This variability could arise from several factors. Although all compounds ultimately limit aminoacyl-tRNA synthetase activity, they act on distinct enzymes that may differ in expression level, substrate specificity, or cellular localization. As a result, each inhibitor could produce a unique pattern of uncharged tRNA accumulation and thus distinct activation kinetics of the AAR pathway. Additionally, some inhibitors—such as halofuginone, which is near its tolerability limit—may engage other stress response pathways, including the unfolded protein response or oxidative stress signaling, and do so in an *atf-4*-independent manner. Differences in compound stability, uptake, or metabolism may also contribute to apparent differences in *atf-4* dependence. Together, these factors could explain the varying *atf-4* interactions with healthspan and autophagy phenotypes observed across this drug panel.

Although our results demonstrate that loss of *atf-4* suppresses lifespan and partially suppresses autophagy responses to tRNA synthetase inhibition, we did not directly measure ATF-4 activation in this study. Future work employing ATF-4 reporter strains or transcriptional analysis of known ATF-4 target genes will be essential to confirm pathway activation and to define the broader transcriptional program engaged by these compounds.

Previous work in MEFs showed that tRNA synthetase inhibition leads to increased ATF4 and also to ATF4-dependent increased autophagy [16]. Here, using the autophagy reporter strain MAH215, we demonstrate that autophagy is markedly increased in *C. elegans* following treatment with lifespan-extending doses of some of the same tRNA synthetase inhibitors (Figure 5). Importantly, this autophagic response is in some cases partially ATF-4-dependent (Figure 6), suggesting a possible mechanistic link between ATF-4 activation, autophagy, and longevity. These findings provide in vivo support for a model in which

https://doi.org/10.3390/biom16010073

ATF-4-driven enhancement of autophagy could be an effector of the longevity response to translational stress.

That said, the observation that only a subset of compounds robustly increased autophagic flux indicates that this pathway is not uniformly activated by all tRNA synthetase inhibitors at the lifespan-extending doses tested here. The extent of autophagy induction may depend on differences in enzyme target class, compound potency, or the specific stress response pathways engaged. Some inhibitors may primarily activate alternative adaptive mechanisms, such as unfolded protein or oxidative stress responses, rather than canonical ATF-4-driven autophagy.

While our data indicate that tRNA synthetase inhibitors enhance autophagic flux and that this response is partially *atf-4*-dependent, we cannot conclude any cause–effect relationship between increased autophagic flux and increased lifespan in these contexts based on these data alone. Directly testing this hypothesis by blocking autophagy is challenging in part because autophagy-deficient mutants, such as *bec-1*, *atg-7*, or *lgg-1*, exhibit shortened lifespans, developmental arrest, and compromised health even under control conditions [22,29], although chemical inhibitors such as bafilomycin could be utilized in future studies. Nevertheless, the parallel increases in autophagic flux and lifespan, and the loss of flux induction in the *atf-4(ok576)* background, are consistent with the possibility that autophagy might contribute functionally to the adaptive response elicited by tRNA synthetase inhibition.

Beyond lifespan, we also assessed healthspan through age-related declines in motility using the thrashing assay. Treatment with some tRNA synthetase inhibitors resulted in a sustained increase in thrashing behavior throughout aging, suggesting improved overall vitality in later life (Figure 3). Strikingly, as with lifespan and autophagy, these improvements seemed to be partially dependent on the presence of *atf-4* (Figure 4) in some of the drug treatments, further reinforcing its role as a central mediator of both longevity and functional health [30]. It should be noted that our healthspan analysis was based solely on thrashing behavior, which primarily reflects neuromuscular function. While this is a commonly used metric in *C. elegans* aging studies, other measures of healthspan, such as reproductive capacity, pharyngeal pumping, or stress resistance, may not always correlate directly with motility. Future work employing multiple independent healthspan metrics will be important to determine the extent to which the benefits of tRNA synthetase inhibition extend across physiological systems.

One limitation of this study is that thrashing and autophagic flux measurements were obtained from single synchronized cohorts per experiment, with multiple individual animals assessed within each strain × condition × day, rather than from independent biological replicates performed across separate experimental days. As a result, these datasets capture within-cohort variability but do not fully account for between-cohort variation. Replication across independent synchronized populations will be important in future work to confirm the robustness and generalizability of these healthspan and autophagy phenotypes.

Together, these results establish a clear and compelling link between tRNA synthetase inhibition, ATF-4 activation, and healthy aging. By extending these findings across multiple inhibitors and phenotypic readouts, we propose that tRNA synthetase inhibition acts as a broadly effective strategy to trigger conserved stress response pathways that enhance longevity and quality of life. Future work will be needed to explore the translational relevance of these findings in mammalian systems.

In particular, identifying specific ATF-4 target genes required for the observed longevity benefits may provide insights into the mechanisms by which this transcription factor influences aging. Additionally, the conservation of these effects across species

 https://doi.org/10.3390/biom16010073

raises the possibility of developing targeted therapeutics that mimic the effects of tRNA synthetase inhibition. Small molecules that selectively modulate the ATF-4 pathway or its downstream effectors could therefore represent a novel class of interventions aimed at promoting healthy aging. Finally, longitudinal studies in mammalian models will be crucial to determine whether similar improvements in healthspan and lifespan can be achieved in complex organisms and to evaluate the safety and efficacy of long-term modulation of this pathway.

5. Conclusions

In summary, our study identifies multiple tRNA synthetase inhibitors that extend both lifespan and healthspan in *C. elegans* through a conserved and partially ATF-4–dependent mechanism. These findings not only reinforce the role of ATF-4 as a critical mediator of the cellular response to translational stress but also highlight autophagy as a likely effector of this longevity pathway. By linking tRNA synthetase inhibition to increased vitality and extended lifespan, this work lays the foundation for exploring ATF-4 activation as a therapeutic strategy for promoting healthy aging in higher organisms.

Supplementary Materials: The following supporting information can be downloaded at: https://www.mdpi.com/article/10.3390/biom16010073/s1, Figure S1: Autophagic flux increases following some tRNA synthetase inhibitor treatments in a manner that is partially dependent on atf-4(ok576).; Table S1: Autophagic flux *p*-values. Vehicle = DMSO. *p* values calculated with Student's *t*-test with Bonferroni correction ($n = 4$ comparisons per timepoint).; Table S2: Thrashing *p*-values. Vehicle = DMSO. *p* values calculated with Student's *t*-test with Bonferroni correction ($n = 7$ comparisons per drug panel).; Table S3: Summary of mean lifespan values for *C. elegans* wild-type (N2) and *atf-4(ok576)* animals treated with tRNA synthetase inhibitors. Each dataset represents one independent experiment with approximately 100-250 worms per condition. Values indicate mean lifespan (days) and the number of animals analyzed (*n*). Statistical significance was determined using the log-rank (Mantel-Cox) test implemented in R. These data correspond to Figures 1 and 2 in the main text. The *atf-4* condition's significance indicates groups that were significantly shorter lived than the vehicle condition.

Author Contributions: Conceptualization, O.C.H. and M.A.M.; methodology, O.C.H., M.P.O., A.T.A., D.S.K., A.W.L. and M.A.M.; software, A.T.A.; validation, O.C.H. and M.A.M.; formal analysis, O.C.H. and M.A.M.; investigation, O.C.H., M.P.O., A.T.A., D.S.K., A.W.L. and M.A.M.; resources, M.A.M.; data curation, M.A.M.; writing—original draft preparation, O.C.H.; writing—review and editing, O.C.H. and M.A.M.; visualization, O.C.H. and M.A.M.; supervision, M.A.M.; project administration, M.A.M.; funding acquisition, M.A.M. All authors have read and agreed to the published version of the manuscript.

Funding: This research was funded by the National Institutes of Health grant P20GM121176, National Institutes of Health grant R01AG07077601, Longevity Impetus Grant, Glenn Foundation for Medical Research/American Federation for Aging Research Junior Investigator Grant, American Federation for Aging Research Reboot Award, University of New Mexico School of Medicine Research Allocation Committee New Investigator Award, and Japan Agency for Medical Research and Development/New York Academy of Sciences Longevity Interstellar Initiative Award to M.A.M., M.P.O. was supported by NIH 5T32GM144834. This work was supported in part by resources from the *Caenorhabditis* Genetics Center (https://cgc.umn.edu/), which is funded by NIH Office of Research Infrastructure Programs (P40 OD010440); the UNM Comprehensive Cancer Center Fluorescence Microscopy and Cell Imaging Shared Resource, which is funded by UNM Comprehensive Cancer Center Support Grant NCI P30CA118100; and the Autophagy, Inflammation, and Metabolism Center of Biomedical Research Excellence (AIM CoBRE) Core, which is funded by NIH P20GM121176.

Data Availability Statement: The original contributions presented in this study are included in the article/Supplementary Material. Further inquiries can be directed to the corresponding author.

https://doi.org/10.3390/biom16010073

Acknowledgments: We thank Jack Peterson and members of the McCormick lab and the Ponomarova lab at the University of New Mexico for the technical assistance and helpful discussions.

Conflicts of Interest: The authors declare no conflicts of interest.

Abbreviations

The following abbreviations are used in this manuscript:

ATF-4	Activating transcription factor 4
tRNA	Transfer RNA
TRS	tRNA synthetase
AAR	Amino acid stress response
eIF-2	Eukaryotic initiation factor 2
NGM	Nematode growth media
GCN2	General control nonderepressible 2
MEFs	Mouse embryonic fibroblasts
GCN4	General control nonderepressible 4
LC3	Microtubule-associated protein 1 light chain 3

References

1. Klass, M.R. A Method for the Isolation of Longevity Mutants in the Nematode *Caenorhabditis elegans* and Initial Results. *Mech. Ageing Dev.* **1983**, *22*, 279–286. [CrossRef]
2. Friedman, D.B.; Johnson, T.E. A Mutation in the Age-1 Gene in *Caenorhabditis elegans* Lengthens Life and Reduces Hermaphrodite Fertility. *Genetics* **1988**, *118*, 75–86. [CrossRef]
3. Kenyon, C.; Chang, J.; Gensch, E.; Rudner, A.; Tabtiang, R. A *C. elegans* Mutant That Lives Twice as Long as Wild Type. *Nature* **1993**, *366*, 461–464. [CrossRef] [PubMed]
4. Kimura, K.D.; Tissenbaum, H.A.; Liu, Y.; Ruvkun, G. Daf-2, an Insulin Receptor-like Gene That Regulates Longevity and Diapause in *Caenorhabditis elegans*. *Science* **1997**, *277*, 942–946. [CrossRef] [PubMed]
5. Lai, C.-H.; Chou, C.-Y.; Ch'ang, L.-Y.; Liu, C.-S.; Lin, W. Identification of Novel Human Genes Evolutionarily Conserved in *Caenorhabditis elegans* by Comparative Proteomics. *Genome Res.* **2000**, *10*, 703–713. [CrossRef] [PubMed]
6. Robbins, C.E.; Patel, B.; Sawyer, D.L.; Wilkinson, B.; Kennedy, B.K.; McCormick, M.A. Cytosolic and Mitochondrial tRNA Synthetase Inhibitors Increase Lifespan in a *GCN4/Atf-4*-Dependent Manner. *iScience* **2022**, *25*, 105410. [CrossRef]
7. Ramirez, M.; Wek, R.C.; De Aldana, C.R.; Jackson, B.M.; Freeman, B.; Hinnebusch, A.G. Mutations Activating the Yeast eIF-2a Kinase GCN2: Isolation of Alleles Altering the Domain Related to Histidyl-tRNA Synthetases. *Mol. Cell. Biol.* **1992**, *12*, 5801–5815. [CrossRef]
8. Hinnebusch, A.G.; Natarajan, K. Gcn4p, a Master Regulator of Gene Expression, Is Controlled at Multiple Levels by Diverse Signals of Starvation and Stress. *Eukaryot. Cell* **2002**, *1*, 22–32. [CrossRef]
9. Ameri, K.; Harris, A.L. Activating Transcription Factor 4. *Int. J. Biochem. Cell Biol.* **2008**, *40*, 14–21. [CrossRef]
10. Lu, P.D.; Harding, H.P.; Ron, D. Translation Reinitiation at Alternative Open Reading Frames Regulates Gene Expression in an Integrated Stress Response. *J. Cell Biol.* **2004**, *167*, 27–33. [CrossRef]
11. Brenner, S. The Genetics of *Caenorhabditis elegans*. *Genetics* **1974**, *77*, 71–94. [CrossRef] [PubMed]
12. Hsin, H.; Kenyon, C. Signals from the Reproductive System Regulate the Lifespan of *C. elegans*. *Nature* **1999**, *399*, 362–366. [CrossRef]
13. Meneely, P.M.; Dahlberg, C.L.; Rose, J.K. Working with Worms: *Caenorhabditis elegans* as a Model Organism. *Curr. Protoc. Essent. Lab. Tech.* **2019**, *19*, e35. [CrossRef]
14. Kaplan, E.L.; Meier, P. Nonparametric Estimation from Incomplete Observations. *J Amer Stat. Assn* **1958**, *53*, 457–481. [CrossRef]
15. Student, A. The Probably Error of a Mean. *Biometrika* **1908**, *6*, 1–25. [CrossRef]
16. Mariner, B.L.; Rodriguez, A.S.; Heath, O.C.; McCormick, M.A. Induction of Proteasomal Activity in Mammalian Cells by Lifespan-Extending tRNA Synthetase Inhibitors. *GeroScience* **2024**, *46*, 1755–1773. [CrossRef]
17. Cinader, B. Aging, Evolution and Individual Health Span: Introduction. *Genome* **1989**, *31*, 361–367. [CrossRef] [PubMed]
18. Globerson, A.; Barzilai, N. The Voyage to Healthy Longevity: From Experimental Models to the Ultimate Goal. *Mech. Ageing Dev.* **2005**, *126*, 225–229. [CrossRef]
19. Carter, C.S.; Hofer, T.; Seo, A.Y.; Leeuwenburgh, C. Molecular Mechanisms of Life- and Health-Span Extension: Role of Calorie Restriction and Exercise Intervention. *Appl. Physiol. Nutr. Metab. Physiol. Appl. Nutr. Metab.* **2007**, *32*, 954–966. [CrossRef] [PubMed]

 https://doi.org/10.3390/biom16010073

20. Koopman, M.; Seinstra, R.I.; Nollen, E.A.A. *C. elegans* as a Model for Synucleinopathies and Other Neurodegenerative Diseases: Tools and Techniques. *Methods Mol. Biol. Clifton NJ* **2019**, *1948*, 93–112. [CrossRef]
21. Hinnebusch, A.G. Translational Regulation of GCN4 and the General Amino Acid Control of Yeast*. *Annu. Rev. Microbiol.* **2005**, *59*, 407–450. [CrossRef]
22. Hansen, M.; Chandra, A.; Mitic, L.L.; Onken, B.; Driscoll, M.; Kenyon, C. A Role for Autophagy in the Extension of Lifespan by Dietary Restriction in *C. elegans*. *PLoS Genet.* **2008**, *4*, e24. [CrossRef] [PubMed]
23. Mariner, B.L.; Felker, D.P.; Cantergiani, R.J.; Peterson, J.; McCormick, M.A. Multiomics of GCN4-Dependent Replicative Lifespan Extension Models Reveals Gcn4 as a Regulator of Protein Turnover in Yeast. *Int. J. Mol. Sci.* **2023**, *24*, 16163. [CrossRef]
24. Chang, J.T.; Kumsta, C.; Hellman, A.B.; Adams, L.M.; Hansen, M. Spatiotemporal Regulation of Autophagy during *Caenorhabditis elegans* Aging. *ELife* **2017**, *6*, e18459. [CrossRef]
25. Li, W.; Li, X.; Miller, R.A. ATF4 Activity: A Common Feature Shared by Many Kinds of Slow-Aging Mice. *Aging Cell* **2014**, *13*, 1012–1018. [CrossRef] [PubMed]
26. Li, W.; Miller, R.A. Elevated ATF4 Function in Fibroblasts and Liver of Slow-Aging Mutant Mice. *J. Gerontol. A. Biol. Sci. Med. Sci.* **2015**, *70*, 263–272. [CrossRef]
27. EFSA Panel on Additives and Products or Substances used in Animal Feed (FEEDAP); Bampidis, V.; Azimonti, G.; Bastos, M.d.L.; Christensen, H.; Dusemund, B.; Fašmon Durjava, M.; Kouba, M.; López-Alonso, M.; López Puente, S.; et al. Safety and Efficacy of STENOROL® (Halofuginone Hydrobromide) as a Feed Additive for Chickens for Fattening and Turkeys. *EFSA J.* **2020**, *18*, e06169. [CrossRef] [PubMed]
28. Pines, M.; Spector, I. Halofuginone—The Multifaceted Molecule. *Molecules* **2015**, *20*, 573–594. [CrossRef]
29. Meléndez, A.; Tallóczy, Z.; Seaman, M.; Eskelinen, E.-L.; Hall, D.H.; Levine, B. Autophagy Genes Are Essential for Dauer Development and Life-Span Extension in *C. elegans*. *Science* **2003**, *301*, 1387–1391. [CrossRef]
30. Rousakis, A.; Vlassis, A.; Vlanti, A.; Patera, S.; Thireos, G.; Syntichaki, P. The General Control Nonderepressible-2 Kinase Mediates Stress Response and Longevity Induced by Target of Rapamycin Inactivation in *Caenorhabditis elegans*. *Aging Cell* **2013**, *12*, 742–751. [CrossRef]

 https://doi.org/10.3390/biom16010073

biomolecules

MDPI

Article

Early-Life Adversity and Epigenetic Aging: Findings from a 17-Year Longitudinal Study

Emily Barr [1,2,3], Maude Comtois-Cabana [1,2,3,4], Andressa Coope [1], Sylvana M. Coté [5], Michael S. Kobor [2,3,6], Chaini Konwar [2,3], Sonia Lupien [4], Marie-Claude Geoffroy [7,8], Michel Boivin [9], Nadine Provençal [1,2], Nicole L. A. Catherine [1], Jessica K. Dennis [2,3,6,*] and Isabelle Ouellet-Morin [10,11,*]

1 Faculty of Health Sciences, Simon Fraser University, Burnaby, BC V5A 1S6, Canada; ebarr@sfu.ca (E.B.); maude.comtois-cabana@umontreal.ca (M.C.-C.); andressa_coope@sfu.ca (A.C.); nadine_provencal@sfu.ca (N.P.); nicole_catherine@sfu.ca (N.L.A.C.)
2 BC Children's Hospital Research Institute, Vancouver, BC V5Z 4H4, Canada; msk@cmmt.ubc.ca (M.S.K.); ckonwar@bcchr.ca (C.K.)
3 Centre for Molecular Medicine and Therapeutics, BC Children's Hospital, Vancouver, BC V5Z 4H4, Canada
4 Department of Psychology, University of Montreal, Montreal, QC H3C 3J7, Canada; sonia.lupien@umontreal.ca
5 Department of Social and Preventive Medicine, University of Montreal, Montreal, QC H3N 1X9, Canada; sylvana.cote.1@umontreal.ca
6 Department of Medical Genetics, University of British Columbia, Vancouver, BC V6T 1Z3, Canada
7 Department of Educational & Counselling Psychology, McGill University, Montreal, QC H3A 1X1, Canada; marie-claude.geoffroy@mcgill.ca
8 McGill Group for Suicide Studies, McGill University, Montreal, QC H3A 0G4, Canada
9 School of Psychology, Laval University, Quebec City, QC G1V 0A6, Canada; michel.boivin@psy.ulaval.ca
10 School of Criminology, University of Montreal, Montreal, QC H3T 1J4, Canada
11 Research Centre of the Montreal Mental Health University Institute, Montreal, QC H1N 3M5, Canada
* Correspondence: jessica.dennis@ubc.ca (J.K.D.); isabelle.ouellet-morin@umontreal.ca (I.O.-M.); Tel.: +1-514-343-6191 (I.O.-M.)

Abstract: Youth exposed to early-life adversity (ELA) are at greater risk for poorer physical and mental health outcomes in adolescence and adulthood. Although the biological mechanisms underlying these associations remain elusive, DNA methylation (DNAm) has emerged as a potential pathway. DNAm-based measures of epigenetic age have been associated with ELA, indicating accelerated aging. According to the stress sensitization hypothesis, prenatal adversity may further heighten sensitivity to subsequent stressors in childhood and adolescence. This study examined the associations between ELA and six epigenetic aging measures, considering both the timing of adversity and the participant's sex. Data were drawn from the Quebec Longitudinal Study of Child Development, with two cumulative indices of ELA derived from prospectively collected data: the Perinatal Adversity and the Child and Adolescent Adversity indices. Higher Perinatal Adversity scores were associated with accelerated DunedinPACE scores. No significant associations were found between ELA and the other epigenetic clocks, nor did we find support for the stress sensitization hypothesis—though a sex-specific trend emerged among girls. The findings suggest that DunedinPACE may be more sensitive to variations in ELA than other clocks. Future research should systematically investigate sex-dimorphic associations between ELA and epigenetic aging, with particular attention to the impact of perinatal adversity.

Keywords: aging; DNA methylation; epigenetic aging; early-life adversity; stress sensitization

1. Introduction

Early-life adversity (ELA) encompasses a range of potentially detrimental experiences, traumas, or stressors that occur during childhood and adolescence. Common examples of ELA include household dysfunction, family socioeconomic hardship, harsh parenting, and neighborhood violence [1,2]. Importantly, ELA is a cumulative concept at its core; it encompasses multiple forms of experiences, severity, and extent of exposure, occurring within and outside the family environment, and arising at any point from the prenatal period to the end of adolescence [3]. Longitudinal studies have shown that ELA robustly predicts physical, socioemotional, and behavioral problems later in life [4–6]. Despite ELA's known risk for health and socioemotional functioning, the mechanisms underlying these associations remain poorly understood.

A common hypothesis suggests that ELA may induce persistent changes in physical and mental health by altering neurobiological stress systems designed to help the child cope with unpredictable, uncontrollable, and threatening environments [3,4,7]. For instance, studies have shown that children exposed to different levels of ELA have a disrupted pattern of hypothalamic–pituitary–adrenal (HPA) axis activity in both basal conditions (e.g., circadian rhythm [8,9]) and stressful conditions (e.g., magnitude and length of activation and recovery [10–12]). However, the strength and direction of these associations vary across studies. This inconsistency may stem from differences in the types of ELA considered, as well as a lack of attention to the co-occurring and cumulative effects of multiple adverse experiences [13]. Simultaneously accounting for multiple indicators has proven to be a promising approach. For example, Ouellet-Morin et al. [14] created a cumulative index of ELA that spanned several stressors within and beyond the family environment, measured from infancy to mid-adolescence. This index included eight indicators such as socioeconomic status, harsh parenting, and peer victimization, and was found to be non-linearly associated with hair cortisol concentration measured in late adolescence. Notably, only one indicator, neighborhood dangerousness, predicted hair cortisol concentration on its own, providing further support that cumulative indices of ELA better predict changes in stress systems than independent indicators.

The mechanisms by which ELA affects later health are complex and involve several biological processes [15]. Epigenetic alterations, particularly DNA methylation (DNAm), have been proposed to take part in the associations between ELA, HPA axis activity, and subsequent health [16,17]. DNAm refers to the addition of a methyl group to specific sites in the genome where cytosine and guanine meet along the phosphate backbone, called CpG sites. However, these CpG sites can gain or lose a methyl group, potentially activating or inactivating genes and altering gene function across the life course. DNAm can both be affected by and induce changes in HPA axis activity, creating recursive loops that may perpetuate and exacerbate the impact of ELA on later functioning [18,19]. This emphasizes DNAm as a critical focus for understanding how adversity becomes embedded and influences vulnerability to later stress [4,16,20]. Clarifying whether DNAm serves as a mediating pathway between ELA and later health may inform targeted prevention strategies for individuals with a history of ELA [21–23].

Most studies to date have examined the association between ELA and DNAm using a candidate gene approach, focusing on DNAm near or within genes previously associated with ELA and/or relevant health outcomes [17]. Although this approach may be valuable, the greater part of the epigenome is overlooked, thus missing the opportunity to identify new loci of interest. Moreover, previous studies targeting the methylation of genes implicated in the regulation of stress, emotions, and behaviors, such as NR3C1, FKBP5, and SLC6A4, have yielded inconsistent findings with ELA experiences [17]. The potential partial overlap between candidate genes and other yet unidentified genes may impact the reproducibility of findings drawn from candidate genes vs. epigenome-wide approaches [24]. Conversely, epigenome-wide association studies comprise close to 1 million CpG sites depending on the array, exponentially increasing the number of tests and the possibility of false-positive and -negative findings (if the correction criteria are overly liberal or conservative, respectively). Adopting a more biologically informed approach, many studies elected to use polyepigenetic scores [17]. These scores encompass DNAm information across several CpG sites of interest collated into single scores, such as epigenetic clocks [25,26]. This approach enables the examination of multiple sites that may be more uniformly affected by ELA, extending beyond previously identified candidate genes, while reducing the number of tests by avoiding a full epigenome-wide analysis.

Epigenetic clocks estimate aging based on DNAm patterns, which can deviate from their chronological age. Epigenetic ages that exceed an individual's chronological age have been linked with poorer health outcomes later in life, ranging from cognitive functioning [27] to high risk of cancer [28]. The first generation of epigenetic clocks, including the Horvath Pan-tissue Clock, focused on CpG sites that related solely to chronological age [25,29]. These clocks were then improved to encompass various tissues, such as the Skin and Blood Clock, and age ranges, such as with the Pediatric Buccal (PedBE) Clock [26,30]. The PedBE enabled researchers to shift the focus from aging-related decline to measuring developmental progress, offering a new perspective on what epigenetic clocks can capture. The focus then shifted from optimizing epigenetic clocks for chronological age to capturing the processes of biological aging and mortality more acutely, leveraging the information from biomarkers known to change with age and smoking, begetting a second generation of epigenetic clocks such as the PhenoAge and GrimAge Clocks [31,32]. To appropriately analyze first- and second-generation clocks, an epigenetic age acceleration variable ought to be derived to indicate the extent to which an individual's epigenetic aging is accelerated or decelerated as compared to their chronological age [33]. Belsky et al. [34] adopted another approach by focusing on biomarkers that assess the integrity of physiological systems over time, both within and between individuals, resulting in DunedinPoAm, which was subsequently refined into DunedinPACE [35]. DunedinPACE reflects whether an individual is aging quickly or slowly, and requires no transformation into epigenetic age acceleration, thereby streamlining the analysis process and heralding a third generation of epigenetic aging measures.

Epigenetic clocks have been, since the first generation, associated with numerous mental health outcomes linked to ELA, as well as experiences of ELA themselves. The original Horvath Pan-tissue Clock, for example, has been associated with psychiatric disorders [36–39], as well as experiences of childhood maltreatment [40,41] and socioeconomic disadvantage [42]. This trend has continued through the second generation, with GrimAge, for example, also being associated with psychiatric disorders [37,43], as well as experiences of ELA encompassing both violence [44] and socioeconomic disadvantage [45]. DunedinPACE has shown similar associations with these ELA indicators [43,46]. However, the inconsistent results are reported for all three generations of clocks [17]. We, and others [47], propose that the heterogeneity in these findings may partly stem from variations in the types of ELA examined—often in isolation—and differences in the timing of exposure (see Lupien et al. [48] for a similar assertion regarding HPA axis activity). Comparisons across studies focusing on different developmental periods, such as the prenatal period and adolescence, may yield misleading conclusions. Moreover, few studies have examined these associations using cumulative indices derived from prospectively and repeatedly collected indicators of adversity [37,49], which are expected to reveal more robust deviations in epigenetic clocks [17].

While ELA occurring during the pre- (before birth), peri- (around the time of birth), and postnatal (a few months after birth) periods—collectively referred to as the perinatal period—is often considered to exert the most enduring impact on neurobiological systems, increasing vulnerability for later problems [18,50,51], many studies overlooked their unique effects alongside later adversity [17]. Furthermore, emerging evidence suggests that ELA may increase sensitivity to future stressors through lasting neurobiological changes, as described by the stress sensitization hypothesis [52–54]. However, determining whether additive or synergic contributions better describe the associations between ELA and DNAm requires repeated measures across distinct periods of development (e.g., perinatal, childhood, and adolescence). To our knowledge, while this hypothesis has been explored in relation to DNAm and outcomes such as bipolar disorder and depression [52], it has not yet been tested with epigenetic clocks.

Finally, while sex differences are often reported for both ELA and epigenetic clocks, and the participant's sex is typically included as a confounder in analyses, few studies have tested whether the direction or magnitude of these associations differed by sex [55]. This contrasts with recent evidence suggesting sex-specific differences between ELA and epigenetic aging, where adult males often show greater acceleration in epigenetic age compared to females who experienced similar adversity [56,57].

This study aimed to investigate the association between cumulative indices of ELA and six measures of epigenetic aging, each with its own strengths for use within our dataset. The specific objectives were threefold: (1) to test whether cumulative indices of ELA derived during the perinatal period and during childhood and adolescence (from 5 months to 15 years of age) were associated with the measures of epigenetic aging ascertained at 17 years of age; (2) to test the stress sensitization hypothesis, whereby the exposure to perinatal ELA exacerbated the strength of the associations between childhood and adolescent ELA and the measures of epigenetic aging; and (3) to examine whether these associations varied between males and females.

2. Materials and Methods

2.1. Cohort Description

The Quebec Longitudinal Study of Child Development (QLSCD) was initiated by the Institut de la Statistique du Quebec (ISQ) and included 2120 singleton children born between 1997 and 1998 in Quebec, Canada. The cohort has been described in detail elsewhere [58]. In brief, participants were recruited from all administrative regions of the province of Quebec, excluding First Nations' Territories and Northern Quebec due to differences in institutional services and socio-cultural context. Families were invited to participate if the pregnancy lasted between 24 and 42 weeks, and if parents could speak French or English. The resulting participants were representative of the general population. Participants were followed annually or biennially from 5 months to the present. At age 17, 1150 youths still in contact with the survey and who lived in the province of Quebec were invited to provide a saliva sample for the measurement of DNA methylation. Ultimately, 773 (67.2%) of the invited youths provided a saliva sample. The QLSCD protocol was approved by the ISQ and the University of Montreal ethics committees, and informed consent forms (and the participating child's assent when over 10 years of age) were obtained at each data collection.

2.2. Measures

2.2.1. Perinatal Adversity Index

Information about adverse experiences during perinatal development was provided by parents via questionnaires when the participating children were 5 months old or retrieved from hospital records (birth weight and gestational time). Seven indicators of adversity were included, based on the cumulative index for adversity previously designed and implemented in the Maternal Adversity, Vulnerability, and Neurodevelopment (MAVAN) cohort [59]. These indicators were (i) low birth weight, (ii) pre-term birth, (iii) low socioeconomic status (SES), (iv) maternal smoking during pregnancy, (v) hospitalization in the first 5 months of life, (vi) maternal depression at 5 months postpartum, and (vii) poor family function. Due to the general low-risk nature of this population-based cohort, all items were dichotomized to reflect the presence (1) or absence of (0) a given adverse experience. (i) Low birth weight was recorded as <2500 g [60]. (ii) Pre-term birth was recorded as <37 weeks. (iii) Low SES was recorded as a <CAD 30,000 gross household annual income. (iv) Maternal smoking during pregnancy (yes/no) was derived from the maternal questionnaire item "Did you smoke during pregnancy?". (v) Hospitalization in the first five months of life (yes/no) was derived from a parental report. (vi) Postnatal maternal depression (yes/no) was assessed with a shortened (13 items) version of the Center for Epidemiologic Studies Depression Scale (CES-D), which focused on depressive symptoms in the previous week. Mothers were scored on 16 items indicating clinically relevant depressive symptomatology (e.g., "I was bothered by things that usually don't bother me") on a Likert scale varying from 0 (rarely or none of the time (less than one day)) to 3 (most or all of the time (5–7 days)) [61]. (vii) Poor family functioning (yes/no) was assessed using the Family Functioning FNC Survey (Q1A-Q1M) [58]. Both parents independently completed the survey, which measured marital strain, as well as a low sense of cohesiveness and closeness in the first 5 months postpartum. The survey included 13 items measured on a 4-point Likert scale (e.g., "We express feelings to each other"; 1 (strongly agree) to 4 (strongly disagree)). If either parent had a score above the 90th percentile, the family was considered to have poor family functioning. The seven binary indicators were summed to derive the cumulative Perinatal Adversity index. Only three indicators had missing data (maternal smoking during pregnancy: n = 2; low SES: n = 1; and family function: n = 2). The missing data was replaced with the indicator's most frequent response (0). Although the cumulative index theoretically ranged from 0 to 7, very few participants had scores above 3. We therefore collapsed our scores above 2 into a single group of 2+ to ensure the distribution was not too positively skewed for the analyses (0: 44.4%; 1: 32.0%; 2: 23.6%).

2.2.2. Child and Adolescent Adversity Index

We derived the Child and Adolescent Adversity index as previously reported [14]. Briefly, the cumulative index included information from eight distinct indicators of socioeconomic and psychosocial adversity, all previously associated with poor health [22]. Indicators were collected prospectively and repeatedly from 5 months to 15 years of age. The cumulative index included the following indicators: (i) young motherhood, (ii) single-parent household, (iii) low SES, (iv) harsh and coercive parenting, (v) maternal depression, (vi) maternal alcohol consumption, (vii) peer victimization, and (viii) dangerous neighborhoods. (i) Young motherhood was collected at 5 months old and recorded as present if the mother was younger than 21 years of age when she had her first child, regardless of whether the first child was the child included in the QLSCD. (ii) Single-parent household was determined from maternal reports at any point during the child's preschool years, which were collected up to six times between 5 months and 5 years. (iii) Low SES was reported by the parents, and this data was collected up to 11 times between 5 months and 15 years, which incorporated information about low family income (<CAD 30,000 per year), mothers' and fathers' education, and occupational prestige. (iv) Harsh and coercive parenting practices were assessed by maternal reports at up to 10 occasions between 2.5 and 15 years.

These reports included 4–8 items, dependent on the child's age, and referred to power-assertive behaviors manifested toward the child (i.e., "I have been angry with my child when he was particularly fussy", from 0 ("Not at all what I did") to 10 "(Exactly what I did")). (v) Maternal Depression was reported by mothers 7 times, between 5 months and 13 years, according to the same questionnaire described for the Perinatal Adversity index. (vi) Maternal alcohol consumption was reported by mothers through questionnaires completed at up to 6 time points between 5 months and 15 years on an 8-point Likert scale ranging from "never" to "every day". (vii) Peer victimization was reported by the participating youth at seven time points from age 6 to 15, utilizing the Self-Report Victimization Scale [62]. This scale includes seven items, encompassing the perceived occurrence of physical, verbal, relational, and cybervictimization, and was measured on a 3-point Likert scale from "never" to "often or very often". (viii) Neighborhood dangerousness was reported by mothers at up to eight time points between 5 months and 15 years to measure a lack of security and cohesion, using five items from the Simcha-Fagan Neighborhood Questionnaire (i.e., "Around here, when there is a problem, neighbors get together to find a solution.").

As previously reported [14], group-based growth mixture models used all available data from each indicator collected over time (excluding young motherhood) to estimate stable patterns of adversity exposure. The best fitting model for each indicator was selected based on fit and parsimony indices (e.g., Bayesian Information Criterion (BIC), Lo–Mendell–Rubin likelihood ratio test (LMR-LRT), and entropy estimates). Class membership then designated groups exposed to low (0), moderate (1), or high (2) levels of adversity according to each indicator. These scores were then summed across all eight indicators to create the Child and Adolescent Adversity index, varying between 0 and 13. As very few individuals scored 12 or higher, we collapsed our distribution between 0 and 11 to avoid the disproportionate weight of the higher scores in the regression analyses using all available data. Only individuals with complete data for the Child and Adolescent Adversity index (n = 696; 99.6%) were included for subsequent analyses.

2.3. DNA Methylation Quantification

Participants provided saliva samples using the Oragene Self-Collection Kit (DNA Genotek, Ottawa, ON, Canada). DNA purification was performed following the Beckman Coulter Genomics Laboratory protocol for the manual purification of DNA from 500 µL of Oragene/saliva samples using the Agencourt DNAdvance Kit (Beckman Coulter, Brea, CA, USA). DNA was quantified primarily using absorbance measured on the NanoDrop 8000 Spectrophotometer (Thermo Fisher Scientific, Waltham, MA, USA), and a subset of samples was tested with fluorescence (pico green) on the Qubit dsDNA BR Assay kit (Thermo Fisher Scientific, Waltham, MA, USA) to ensure the absorbance was in an acceptable range. Finally, bisulfite conversion was performed with the EZ-96 DNA Methylation Kit (Zymo Research, cat. No. D5004, Irvine, CA, USA), following the manufacturer's protocol. Bisulfite DNA was quantified again using the absorbance method (for RNA), and 250 ng of bisulfite DNA from each sample was extracted with the EZ-96 DNA Methylation Kit (Zymo Research, Irvine, CA, USA) and then run on the Infinium MethylationEPIC v1.0 BeadChip (Illumina, San Diego CA, USA).

DNA methylation beta values were extracted and processed using R software (version 3.6.2), as well as with the minfi R package, version 1.54.1 [63]. Poor quality samples were initially flagged based on the ratio of the log2 unmethylated to methylated intensity of the control probes at a default cutoff of 10.5 and further confirmed by inspecting control probe intensities in the quality control report from minfi. To adjust for the different intensities of the two different probe types on the array, we used Noob normalization for dye-bias and background correction (the preprocessNoob command from the minfi package, version 1.54.1). We performed a sex check to confirm concordance between reported and inferred sex, based on XX/XY intensities, and removed mismatches. Further, probe filtering was performed based on published annotations [64,65] to remove probes that cross-react or cross-hybridize to multiple autosomal and sex probes, as well as probes that had a detection *p*-value greater than 0.01 in 25% of samples, indicating that probes were either too sensitive or not sensitive enough. Batch correction was not performed, as our normalization method effectively removed the most prominent batch effects. Overall, after probe and sample quality checks, 817,658 probes and 696 samples were retained for subsequent analyses.

Cell Type Estimation

As DNA methylation is cell type-specific, and saliva comprises multiple cell types including neutrophils, monocytes, and epithelial cells, we used DNA methylation-based cell type deconvolution in the R package EpiDISH version 2.24.0 [66] to infer the saliva cell types (Supplemental Figure S1).

2.4. Epigenetic Age Acceleration

To account for the different strengths and limitations of different epigenetic clocks, including their target age ranges, cell types, and generations, we measured epigenetic aging in the QLSCD with six different clocks. We used R (version 4.3.1) and adjusted the publicly available code to calculate all epigenetic age measures but one, the Horvath Pan-tissue Clock, which we calculated by submitting our data to an online portal. We included three measures of first-generation clocks, which were designed to measure chronological age: (1) Horvath Pan-tissue Clock [25]; (2) the Skin and Blood Clock [26]; and (3) the PedBE Clock [30]. We also included two measures of second-generation clocks, which improved epigenetic age derivation by incorporating phenotypic measures of aging: (4) PhenoAge [31] and (5) GrimAge [32]. Finally, we investigated a third-generation measure of epigenetic aging, (6) DunedinPACE [35], which expands on both prior generations by incorporating comprehensive measures of changes in physiological biomarkers for organ-system integrity over time. The first five clocks were regressed against chronological age to derive a measure of epigenetic age acceleration; this step was unnecessary for DunedinPACE, as it directly provides a measure of the "pace of aging" instead of an epigenetic age. For each clock and DunedinPACE, we also removed the confounding effect of cell type composition by deriving the residuals from regressions; therefore, epigenetic age acceleration was calculated with the regression against both chronological age and cell type. We used this approach rather than including cell type proportions as covariates due to high correlation with cell type across the epigenetic clocks, and a low covariation between cell type and ELA. In addition, the GrimAge Clock includes sex differences in its derivation, resulting in epigenetic ages heavily skewed by sex. As such, sex was also regressed from the GrimAge variable along with both chronological age and cell type, and this clock was excluded from testing the moderating role of sex (Supplemental Figure S2).

Assessment of Epigenetic Age Measure Performance

Pearson's correlation coefficients (r) between our measures of epigenetic age and participant chronological age were measured for all five clocks, excluding DunedinPACE, as it does not measure age in years. Correlations were also measured between all predicted measures of epigenetic age. Given the homogeneity of chronological age in our cohort (mean = 17.22; SD = 0.27), we expected weaker correlations between each epigenetic age and chronological age compared to the correlations between the epigenetic aging measures. We calculated the mean absolute error and maximum absolute error (MAE and MaxAE) for each measure of epigenetic age and chronological age using the mae and maxae commands from the mlr3measures package [67].

2.5. Covariates

Race/ethnicity was self-reported by the parents. Individuals were considered non-White if at least one parent selected any option for race/ethnicity other than "White". As our cohort was overwhelmingly White (93%), we dichotomized race/ethnicity as "White" vs. "non-White". Although this dichotomization masks potentially important differences across non-White race/ethnicity people, it ensured that we maintained sufficient participants in each group. Information about tobacco, cannabis, and alcohol consumption within the year prior to saliva sample collection was provided by participants at age 17 on a 4-point Likert scale ("never" to "daily"). BMI at age 17 was also self-reported. Age at the time of sample collection was also included as a covariate in our final models of epigenetic age acceleration because even slight differences in age may confound the adversity–epigenetic age associations [33] (Supplemental Table S1).

2.6. Statistical Analyses

All statistical analyses were performed in R 4.3.1, and linear regressions used the base linear model (lm) function. Statistical significance was initially defined as a p-value < 0.05, and we subsequently applied the false discovery rate (FDR) < 0.10 for multiple testing correction. First, we examined the associations between our six measures of epigenetic age with Perinatal Adversity as well as with Child and Adolescent Adversity in unadjusted and adjusted models, with the latter including all covariates. Second, we tested the unique contribution of each index of ELA, and the hypothesized sensitization influence of Perinatal Adversity on the association between Child and Adolescent Adversity and the measures of epigenetic clocks. Finally, we examined the moderating influence of sex in these associations by including an interaction term between sex and each adversity index separately and together (i.e., triple interaction). Significant interactions were decomposed by the moderator (sex and/or Perinatal Adversity) to depict their simple slopes. Effect sizes were measured using standardized betas and Cohen's f^2 for full models (where $f > 0.02$ indicates a small effect, $f > 0.15$ is a medium effect, and $f > 0.35$ is a large effect) (Supplemental Table S2).

3. Results

3.1. Exposure to Adversity Was Low Overall and Differed by Sex

Of the 773 participants who provided a saliva sample for DNAm quantification, 39 did not have enough DNA for bisulfite conversion, 35 were dropped due to poor probe intensity in the array data, and 3 were dropped due to incomplete adversity data for both Perinatal and Child and Adolescent Adversity. The analyses were carried out on 696 participants (385 females, 55%). The prevalence of Perinatal Adversity in the cohort was low and similar across sexes (Table 1).

In contrast, a greater proportion of the cohort experienced some types of Child and Adolescent Adversity, although, on average, the overall score was low to moderate (Table 2). Compared to females, males experienced higher levels of peer victimization, lower family SES, and harsher and more coercive parenting. This difference translated into a statistically significantly higher cumulative score in males compared to females.

The two indicators of adversity are moderately and significantly correlated, according to Kendall's correlation ($\tau = 0.28$).

Table 1. Summary of each indicator included in the Perinatal Adversity index according to the total sample and by sex in the QLSCD [a].

	Total (n = 696)	Female (n = 385)	Male (n = 311)	Mean Differences
	n (%) or Mean (SD)	n (%) or Mean (SD)	n (%) or Mean (SD)	Chi-Square Test (*p*-Value) or *t*-test (*p*-Value)
Specific Indicators				
Low Birth Weight (<2500 g)	26 (3.7%)	14 (3.6%)	12 (3.9%)	6.8×10^{-31} (1.00)
Pre-Term Birth (<37 weeks)	92 (13.2%)	44 (11.4%)	48 (15.4%)	2.07 (0.15)
Low SES (<CAD 30,000)	156 (22.4%)	89 (23.1%)	67 (21.5%)	0.16 (0.69)
Maternal Smoking (Yes)	147 (21.1%)	82 (21.3%)	65 (20.9%)	0.001 (0.97)
Hospital Admissions (First 5 Months)	80 (11.5%)	39 (10.1%)	41 (13.2%)	1.29 (0.26)
Maternal Depression (CESD > 16)	31 (4.5%)	20 (5.2%)	11 (3.5%)	0.76 (0.39)
Poor Family Functioning	61 (8.8%)	35 (9.1%)	32 (8.4%)	0.04 (0.84)
Total Score				
Perinatal Adversity (Range 0–2)	0.74 (0.79)	0.73 (0.79)	0.77 (0.78)	0.73 (0.46)

Notes: There were no mean differences by sex for any indicator included in Perinatal Adversity. [a] Data were compiled from the final master file of the Québec Longitudinal Study of Child Development (1998–2015), ©Gouvernement du Québec, Institut de la statistique du Québec.

Table 2. Summary of each indicator included in the Child and Adolescent Adversity variable for the total sample and by sex in the QLSCD [a].

	Total (n = 696)	Female (n = 385)	Male (n = 311)	Mean Differences
	n (%) or Mean (SD)	n (%) or Mean (SD)	n (%) or Mean (SD)	Chi-Square Test (*p*-Value) or *t*-test (*p*-Value)
Specific Indicators				
Young Motherhood Yes	133 (19.1%)	69 (17.9%)	64 (20.6%)	0.62 (0.43)
Single-Parent Household Yes	90 (12.9%)	46 (11.9%)	44 (14.1%)	0.56 (0.46)
Socioeconomic Status (Low vs.) Medium High	283 (40.7%) 174 (25.0%)	141 (36.6%) 106 (27.5%)	142 (45.7%) 68 (21.9%)	6.23 (**0.044**)
Coercive Parenting (Low vs.) Medium High	327 (47.0%) 84 (12.1%)	173 (44.9%) 36 (9.35%)	154 (49.5%) 48 (15.4%)	10.82 (**0.004**)
Maternal Depression (Low vs.) Medium High	285 (40.9%) 76 (10.9%)	152 (39.5%) 42 (10.9%)	133 (42.7%) 34 (10.9%)	0.85 (0.66)
Maternal Alcohol Consumption (Low vs.) Yes	143 (20.5%)	76 (19.7%)	67 (21.5%)	0.24 (0.62)
Peer Victimization (Low vs.) Medium High	390 (56.0%) 84 (12.1%)	199 (51.7%) 36 (9.35%)	191 (61.4%) 48 (15.4%)	21.66 (**<0.001**)
Dangerous Neighborhood (Low vs.) Medium High	561 (80.6%) 42 (6.03%)	306 (79.5%) 27 (7.01%)	255 (82.0%) 15 (4.82%)	1.52 (0.47)
Total Score				
Child and Adolescent Adversity (Range 0–11)	4.97 (2.30)	4.76 (2.31)	5.23 (2.26)	2.72 (**0.007**)

Notes: Mean differences by sex with a $p < 0.05$ are marked in bold. [a] Data were compiled from the final master file of the Québec Longitudinal Study of Child Development (1998–2015), ©Gouvernement du Québec, Institut de la statistique du Québec.

3.2. Measures of Epigenetic Age Among Adolescents

As most measures of epigenetic aging are trained on individuals in adulthood, and more specifically, middle-aged individuals, we examined their association with chronological age in our adolescent cohort. Although we report the correlation estimates between each clock and chronological age, the limited age range (16.8–18.1 years) of the present sample greatly limited this otherwise widely used validation technique. Unsurprisingly, all measures showed low correlation with chronological age, with only the Skin and Blood Clock (r = 0.09) and GrimAge (r = 0.08) reaching significance. Comparing correlations between clocks, DunedinPACE and PhenoAge had the highest correlation (r = 0.67), followed by DunedinPACE and GrimAge (r = 0.27). The rest of the clocks were weakly correlated (r > 0.15), except for the Pediatric Clock, which did not significantly correlate with any of our measures or chronological age. We also included mean and maximum errors, which were low for all clocks (MAE < 5.5 years of error) except for GrimAge, which had an MAE > 10 years (Supplemental Table S3). This indicated that, while there was limited overlap between the clocks, the error in clock calculations was small across all but the GrimAge Clock.

3.3. Only Perinatal Adversity Is Associated with Epigenetic Age Acceleration

Regression analyses between measures of epigenetic age and adversity indices revealed that DunedinPACE was the only measure of epigenetic age that was significantly associated with the Perinatal Adversity index, even after adjusting for confounders (β = 0.079, *p*-value = 0.045, Cohen's f^2 = 0.098) (Table 3). This finding did not survive multiple testing correction, though.

Table 3. Adjusted associations between both adversity indices and measures of epigenetic age acceleration in the QLSCD [a].

Model	Horvath	Pediatric Clock	Skin and Blood Clock	PhenoAge	GrimAge	DunedinPACE
	Estimate (*p*val)	Estimate (*p*val)	Estimate (*p*val)	Estimate (*p*val)	Estimate (*p*val)	Estimate (*p*val)
Perinatal Adversity	−0.00025 (0.99)	0.024 (0.55)	0.020 (0.62)	0.023 (0.57)	0.0033 (0.94)	**0.079 (0.045)**
Child and Adolescent Adversity	−0.035 (0.41)	0.056 (0.17)	−0.0016 (0.97)	0.0037 (0.93)	0.029 (0.50)	0.064 (0.12)

Notes: Bold values are significant at $p < 0.05$. All clocks, excluding DunedinPACE, were adjusted to give epigenetic age acceleration by taking the residuals of the calculated age regressed on chronological age. All age measures were additionally regressed against cell type proportions before multivariable modeling. The GrimAge Clock was also adjusted for sex before multivariable modeling. All models were controlled for self-reported race/ethnicity, tobacco smoking, cannabis smoking, alcohol consumption, BMI, age, and sex. Estimate refers to standardized regression coefficients. [a] Data were compiled from the final master file of the Québec Longitudinal Study of Child Development (1998–2015), ©Gouvernement du Québec, Institut de la statistique du Québec.

For the Child and Adolescent Adversity index, the unadjusted models indicated associations with two measures of epigenetic aging at a trend level: PedBE (β = 0.065, *p*-value = 0.085, Cohen's f^2 = 0.0043) and DunedinPACE (β = 0.066, *p*-value = 0.083, Cohen's f^2 = 0.0043). However, none of these measures of epigenetic age were significantly associated with the Child and Adolescent Adversity index after controlling for the covariates (Table 3).

We performed forward regression to more precisely describe the relative proportion of variance in each measure of epigenetic age explained by the adversity indices and each confounder (i.e., race/ethnicity, tobacco, cannabis, and alcohol consumption, BMI, age, and sex) (see Figure 1). DunedinPACE had the highest explained variance (12.0%), as well as the most variance explained by adversity indices (1%). In addition, DunedinPACE was the only measure of epigenetic age with variance explained by chronological age (3%) in this otherwise age-homogeneous sample. As expected, tobacco, cannabis, and alcohol consumption, BMI, and sex each contributed to the explained variance across the clocks, even though the magnitude of these effects varied.

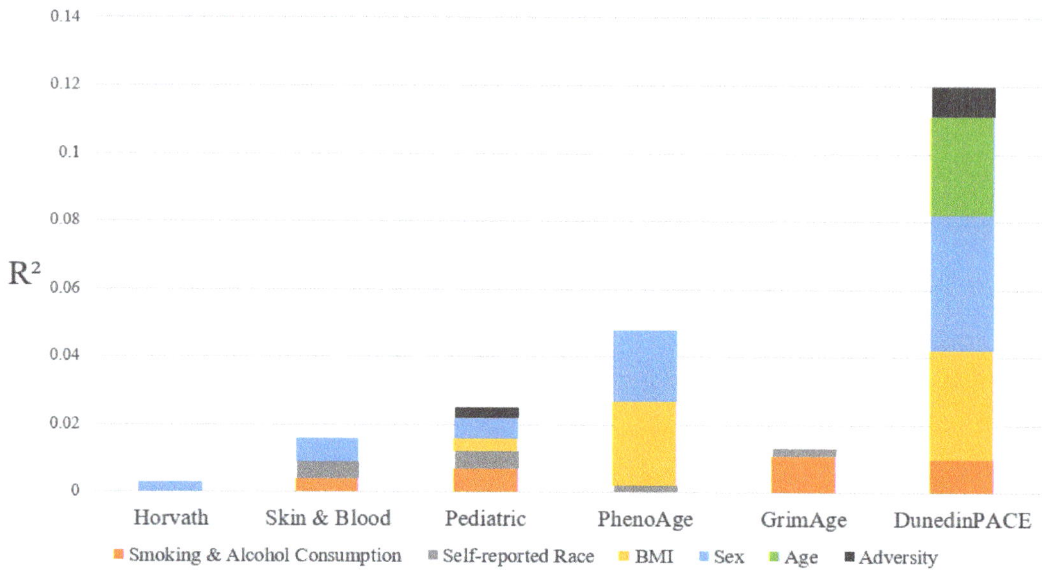

Figure 1. Portion of variance in epigenetic age measures explained by confounders and adversity indices in the QLSCD [a]. Notes: Smoking & Alcohol Consumption includes tobacco, cannabis, and alcohol consumption. Adversity includes both Perinatal Adversity and Child and Adolescent Adversity indexes. All clocks, excluding DunedinPACE, were adjusted to give epigenetic age acceleration by taking the residuals of the calculated age regressed on chronological age. All age measures were additionally regressed against cell type proportions before multivariable modeling. The GrimAge Clock was also adjusted for sex before multivariable modeling. [a] Data were compiled from the final master file of the Québec Longitudinal Study of Child Development (1998–2015), ©Gouvernement du Québec, Institut de la statistique du Québec.

3.4. Limited Evidence for the Stress Sensitization Hypothesis

Before formally testing the stress sensitization hypothesis, we first included both Perinatal (β = 0.065, *p*-value = 0.12) and Child and Adolescent Adversity (β = 0.045, *p*-value = 0.29) indices (additive model, Cohen's $f^2 = 0.10$) to evaluate their unique contributions. We found no main significant associations, including for DunedinPACE, suggesting that both adversity indices were weakly associated with a common portion of DunedinPACE variance (Table 4). We then tested the stress sensitization hypothesis by adding an interaction term to the two adversity indices. A significant association between the interaction term and epigenetic age acceleration would lend support to this hypothesis. However, none of the interactions were significant.

3.5. Epigenetic Age Differs by Sex, but Evidence of Sex Moderation Is Limited

Significant sex differences emerged in Welch's two-sample *t*-tests for two clocks, PhenoAge ($t = -4.56$, $p = < 0.001$) and DunedinPACE ($t = -5.79$, $p = < 0.001$), whereby males had a lower PhenoAge and DunedinPACE than females. Regression analyses including interaction terms between sex and each measure of ELA did not detect sex-dimorphic associations for the Perinatal or Child and Adolescent Adversity indices alone. However, a trend for a triple interaction emerged for DunedinPACE (β = 0.073, *p*-value = 0.072, Cohen's $f^2 = 0.10$), suggesting that a distinct pattern of stress sensitization may be present between males and females (Table 5).

Table 4. Associations between adversity variables and epigenetic age acceleration, testing adversity timing in the QLSCD [a].

Model	Horvath	Pediatric Clock	Skin and Blood Clock	PhenoAge	GrimAge	DunedinPACE
	Estimate (*pval*)	Estimate (*pval*)	Estimate (*pval*)	Estimate (*pval*)	Estimate (*pval*)	Estimate (*pval*)
Perinatal Adversity + Child and Adolescent Adversity						
Perinatal Adversity	0.011 (0.79)	0.0070 (0.87)	0.022 (0.88)	0.024 (0.58)	−0.0050 (0.91)	0.065 (0.12)
Child and Adolescent Adversity	−0.039 (0.38)	0.055 (0.21)	−0.0067 (0.61)	−0.0022 (0.96)	0.027 (0.55)	0.045 (0.29)
Perinatal Adversity*Child and Adolescent Adversity						
Perinatal Adversity*Child and Adolescent Adversity	−0.029 (0.48)	0.0077 (0.85)	−0.020 (0.62)	−0.054 (0.18)	0.010 (0.80)	0.027 (0.49)

Notes: All clocks, excluding DunedinPACE, were adjusted to give epigenetic age acceleration by taking the residuals of the calculated age regressed on chronological age. All age measures were additionally regressed against cell type proportions before multivariable modeling. The GrimAge Clock was also adjusted for sex before multivariable modeling. All models were controlled for self-reported race/ethnicity, tobacco smoking, cannabis smoking, alcohol consumption, BMI, age, and sex. Estimate refers to standardized regression coefficients. Models used * to denote an interaction term. [a] Data were compiled from the final master file of the Québec Longitudinal Study of Child Development (1998–2015), ©Gouvernement du Québec, Institut de la statistique du Québec.

Table 5. Associations between adversity variables and epigenetic age acceleration, testing sex moderation in the QLSCD [a].

Model	Horvath	Pediatric Clock	Skin and Blood Clock	PhenoAge	GrimAge	DunedinPACE
	Estimate (*pval*)	Estimate (*pval*)	Estimate (*pval*)	Estimate (*pval*)	Estimate (*pval*)	Estimate (*pval*)
Perinatal Adversity*Sex	−0.049 (0.23)	−0.002 (0.96)	−0.021 (0.60)	0.056 (0.15)	-	0.018 (0.64)
Child and Adolescent Adversity*Sex	0.0080 (0.85)	−0.0097 (0.81)	0.0077 (0.85)	0.065 (0.11)	-	0.033 (0.40)
Perinatal Adversity*Child and Adolescent Adversity*Sex	0.05 (0.22)	−0.049 (0.24)	−0.015 (0.71)	0.051 (0.22)	-	0.073 (0.072)

Notes: All clocks, excluding DunedinPACE, were adjusted to give epigenetic age acceleration by taking the residuals of the calculated age regressed on chronological age. All age measures were additionally regressed against cell type proportions before multivariable modeling. The GrimAge Clock was also adjusted for sex before multivariable modeling, and as such was removed from sex moderation analyses. All models were controlled for self-reported race/ethnicity, tobacco smoking, cannabis smoking, alcohol consumption, BMI, and age. Estimate refers to standardized regression coefficients. In all models, sex = females. Models used * to denote an interaction term [a] Data were compiled from the final master file of the Québec Longitudinal Study of Child Development (1998–2015), ©Gouvernement du Québec, Institut de la statistique du Québec.

To explore this tentative signal, we tested whether distinct patterns of stress sensitization could have been hidden (i.e., canceled out) for males and females. No robust indication of significant interactions between Perinatal Adversity and Child and Adolescent Adversity emerged for males ($\beta = -0.068$, p-value = 0.28) or females ($\beta = 0.081$, p-value = 0.13). Figure 2 illustrates the simple slopes between Child and Adolescent Adversity and the DunedinPACE scores according to ± 1 standard deviation of the mean of the Perinatal Adversity, for males and females separately. Notably, however, opposite patterns of moderation were noted: females exposed to higher levels of Perinatal Adversity tended to have higher DunedinPACE scores in association with higher levels of Child and Adolescent Adversity ($\beta = 0.15$, $p = 0.042$), whereas no associations were detected at or below the sample's mean of Perinatal Adversity (mean: $\beta = 0.06$, $p = 0.81$; -1SD: $\beta = -0.02$, $p = 0.25$). None of the simple slopes were significant in males. As post hoc analyses were conducted based on a trend for significance, interpretations of the sex-moderated stress sensitization effect should be approached with caution.

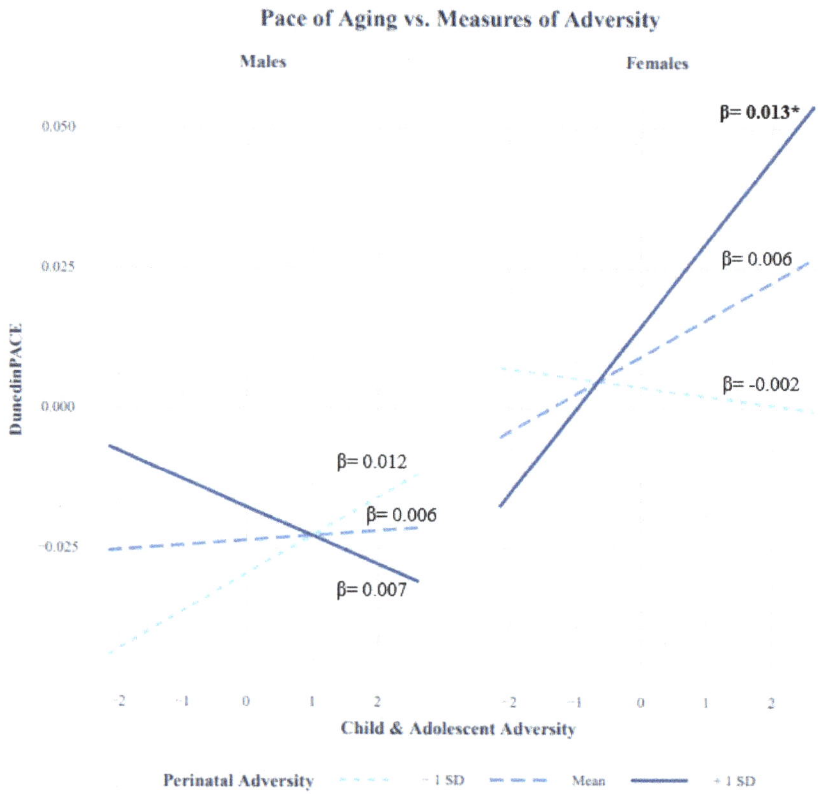

Figure 2. Model-predicted associations between Child and Adolescent Adversity and DunedinPACE for male (**left**) and female (**right**) participants with Perinatal Adversity scores at ±1 SD above the mean in the QLSCD [a]. Bold values with * are significant at $p < 0.05$. Perinatal Adversity and Child and Adolescent Adversity indices were mean-centered to investigate simple slopes. [a] Data were compiled from the final master file of the Québec Longitudinal Study of Child Development (1998–2015), ©Gouvernement du Québec, Institut de la statistique du Québec.

4. Discussion

This study examined the association between two cumulative indices of ELA and six measures of epigenetic aging in a cohort of 17-year-old youth. This allowed us to investigate whether Prenatal vs. Child and Adolescent adversity is differently related to epigenetic aging. We also formally tested the stress sensitization hypothesis and whether these associations differed by sex. Our findings provided limited evidence for these hypotheses. However, a few isolated findings suggested that the DunedinPACE measure may be more sensitive than the other epigenetic clocks at detecting such effects.

DunedinPACE was the only measure of epigenetic aging that captured, albeit inconsistently, variations in ELA in our cohort of adolescents. Furthermore, it was the only measure that captured chronological age as a measure that influenced variance in our data. This distinct pattern of findings may echo the differences in how each measure of epigenetic aging was derived. Unlike other epigenetic clocks, DunedinPACE was built from data collected repeatedly in the same individuals to capture physiological changes both within and between individuals. In contrast, other epigenetic clocks were derived from inter-individual differences in chronological age or changes in biomarkers at a single time point [35]. DunedinPACE may thus be better suited to capture deviations induced by ELA over time, especially earlier in development, given its focus on physiological systems related to accelerated aging. Also, both our sample and the cohort from which the DunedinPACE measure was derived are population-based, with relatively few individuals exposed to very high levels of adversity. In community-based samples like ours, variation in socioeconomic deprivation has been associated with DunedinPACE [34,35,68,69]. Raffington et al. [46], for instance, have reported an association between a composite measure of socioeconomic disadvantage—including a combination of household income, parental education, occupational prestige, and measures of neighborhood disadvantage—and DunedinPoAm, the precursor to DunedinPACE. Consistent with our findings, they found no associations between their measures of deprivation and the Horvath Pan-tissue Clock, the Hannum Clock, PedBE, PhenoAge, and GrimAge. Our null findings for Child and Adolescent Adversity and these other measures of epigenetic age are thus in line with much of the literature focused on community-based cohorts [41,42,49,70]. In contrast, associations with ELA in other epigenetic clocks appear to be more robustly detected in at-risk samples, comprising those exposed to greater levels of adversity [47,71] or those specifically associated with changes in epigenetic age over time [72].

Contrary to our expectations, our findings did not offer robust support for the stress sensitization hypothesis. We did, however, find that female participants who experienced higher levels of Perinatal Adversity had higher DunedinPACE scores when they had also experienced higher levels of Child and Adolescent Adversity than their peers with lower or no Perinatal Adversity. This could suggest that females may be prone to be more negatively affected by re-exposure to adversity. Although the stress sensitization hypothesis has been scarcely examined in relation to measures of DNAm, there is support for this sex-specific finding in both stress sensitization research and studies investigating overall stress [73,74]. A 2017 meta-analysis by Bunea et al. [53] found that women with a history of ELA required less exposure to recent stress, as measured by cortisol levels, to trigger post-traumatic stress disorder compared to their male counterparts. Stroud (2020) [75] provided an in-depth review of Post's model of stress sensitization, summarizing findings across the literature and discussing the model's strengths and limitations. Notably, sex emerged as a key factor influencing stress sensitization outcomes, with sex-specific effects observed depending on pubertal stage.

Sex differences in methylation are also well documented, pointing to higher epigenetic age in males in comparison to females [55,76,77]. However, we found higher measures of epigenetic aging in females than in males using both PhenoAGE and DunedinPACE, both of which derive epigenetic age from physiological biomarkers of aging. A possible explanation for this difference in our study compared to the existing literature is the age of our participants (17 years) compared to previous studies conducted with adult samples [31]. Both genetic and epigenetic mechanisms interact with hormone activity, changing during puberty. At 17 years of age, neither males nor females have typically reached full pubertal maturation; however, females generally tend to be more advanced in the pubertal process than males [78]. The PedBE Clock included few individuals over age 15 and shows increasing prediction error with age, limiting its accuracy in adolescence, as acknowledged by its developers [30]. Its good performance at age 10 in a subsample of our cohort [70] further supports these age-related limitations in its design. It remains uncertain whether measuring epigenetic aging a decade later (e.g., in one's twenties) may provide additional insights, but it would enable the analysis of intra-individual changes in epigenetic aging over time and document, if present, any shift in mean levels between males and females [71,76,79]. Nonetheless, these preliminary findings underscore the importance of systematically investigating and reporting sex differences in epigenetic aging in future studies.

Our study had several limitations. First, we measured DNAm in saliva, and while each measure of epigenetic aging has been adjusted for use in saliva by controlling for cell type proportions, three measures were originally derived for use in blood and may not be optimized for use in other tissues [31,32,35]. Second, our study sample was 17 years old at the time of DNAm collection, and each measure of epigenetic aging used in this study was derived and optimized for use in age groups younger than (PedBE, optimized for children prior to adolescence) or older than (remaining epigenetic aging measures) our sample. Third, our cohort included youth primarily of self-reported White race/ethnicity, limiting the generalizability of our results. Fourth, levels of exposure to ELA were relatively low in our cohort; most parents had completed a high school diploma or higher, and the average household income exceeded the national poverty cutoff. Given that exposure to threat and abuse has more frequently been associated with epigenetic age acceleration than experiences of neglect and deprivation, it is possible that distinct findings could be uncovered in higher-risk cohorts [40,46,47].

5. Conclusions

This study sought to expand upon the existing literature investigating the relationship between ELA and epigenetic aging by incorporating numerous clocks and two cumulative measures of ELA and by accounting for the timing of ELA and the potential impact of sex differences in adolescence. Strengths of our study include our use of comprehensive, prospectively, and repeatedly collected indicators of adversity. Future studies would benefit from examining these associations in higher-risk samples and according to a dimensional approach of adversity, such as exposure to socioeconomic vs. relational adversity or threat vs. abuse experiences of maltreatment (1). Overall, exposure to ELA was low in our cohort, and we found minimal evidence of associations between ELA and epigenetic aging. Nonetheless, of all the measures of epigenetic aging tested, DunedinPACE captured the most variation in ELA, and we found some evidence that female children exposed to more adversity from before birth to adolescence experienced greater epigenetic aging measured by DunedinPACE. Future studies should systematically examine sex differences in DNAm in response to ELA.

Supplementary Materials: The following supporting information can be downloaded at https://www.mdpi.com/article/10.3390/biom15060887/s1: Figure S1: Cell type proportions predicted from epigenetic data in the QLSCD; Figure S2: Epigenetic age acceleration and DunedinPACE pace of aging in female and male adolescents in the QLSCD; Table S1: Summary of each covariate used in all models and their mean differences by sex in the QLSCD; Table S2: Cohen's f^2 for all models and measure of epigenetic age in the QLSCD; Table S3: Pearson's correlation (r) with chronological age, mean absolute error, and maximum absolute error of the Horvath, Pediatric, Skin and Blood, PhenoAge, and GrimAge epigenetic clocks in the QLSCD.

Author Contributions: I.O.-M., S.M.C., S.L., M.-C.G., N.P. and M.B. contributed to the conception and design of the study. I.O.-M., E.B., M.C.-C. and J.K.D. contributed to the design of the study methodology. E.B., A.C. and M.C.-C. extracted DNA methylation data from the QLSCD salivary DNA samples. M.S.K. provided lab materials and the array on which the samples were run. E.B. conducted all statistical analysis and coding, and C.K. advised on coding. E.B., I.O.-M., J.K.D. and M.C.-C. reviewed all analyses and results. E.B. wrote the original draft of the manuscript. E.B. created all of the visualizations included in the manuscript. I.O.-M., J.K.D. and N.L.A.C. provided supervision and mentorship over the research project. I.O.-M. and N.P. acquired and provided funding for the completion of this project. All authors have read and agreed to the published version of the manuscript.

Funding: The Québec Longitudinal Study of Child Development was supported by funding from the Ministère de la Santé et des Services sociaux, Ministère de la Famille, Ministère de l'Éducation, and Ministère de l'Enseignement supérieur (Québec ministries), the Lucie and André Chagnon Foundation, the Institut de recherche Robert-Sauvé en santé et en sécurité du travail, the Research Centre of the Sainte-Justine University Hospital, the Ministère de l'Emploi et de la Solidarité sociale, and the Institut de la statistique du Québec. IO-M received funding through CIHR grant MOP-133501. JKD is a Michael Smith Health Research BC Scholar. NP was funded through the Jacobs Foundation, which supported AC and EB salaries.

Institutional Review Board Statement: The study was conducted in accordance with the Declaration of Helsinki, and approved by the Research Ethics Committee of Université de Montréal (2014-3257—CERAS-2014-15-136-P).

Informed Consent Statement: Informed consent was obtained from all subjects involved in the study.

Data Availability Statement: Access to the data is restricted in accordance with the regulations governing data privacy in the province of Quebec (Canada), as enforced by the Institut de la statistique du Québec. As per these regulations, survey participants were assured that raw data would remain strictly confidential and would not be disclosed or shared outside a secure environment.

Conflicts of Interest: The authors declare no conflicts of interest.

Abbreviations

The following abbreviations are used in this manuscript:

ELA	Early-Life Adversity
DNAm	DNA Methylation

References

1. McLaughlin, K.A. Future Directions in Childhood Adversity and Youth Psychopathology. *J. Clin. Child Adolesc. Psychol.* **2016**, *45*, 361–382. [CrossRef]
2. Afifi, T.O. Considerations for expanding the definition of ACEs. In *Adverse Childhood Experiences*; Elsevier: Amsterdam, The Netherlands, 2020; pp. 35–44.
3. Gunnar, M.R. Early adversity, stress, and neurobehavioral development. *Dev. Psychopathol.* **2020**, *32*, 1555–1562. [CrossRef]
4. Boyce, W.T.; Levitt, P.; Martinez, F.D.; McEwen, B.S.; Shonkoff, J.P. Genes, Environments, and Time: The Biology of Adversity and Resilience. *Pediatrics* **2021**, *147*, e20201651. [CrossRef]
5. Arseneault, L. Annual Research Review: The persistent and pervasive impact of being bullied in childhood and adolescence: Implications for policy and practice. *J. Child Psychol. Psychiatry* **2018**, *59*, 405–421. [CrossRef]
6. Forte, A.; Orri, M.; Galera, C.; Pompili, M.; Turecki, G.; Boivin, M.; Tremblay, R.E.; Côté, S.M. Developmental trajectories of childhood symptoms of hyperactivity/inattention and suicidal behavior during adolescence. *Eur. Child Adolesc. Psychiatry* **2020**, *29*, 145–151. [CrossRef]
7. Hertzman, C. Putting the concept of biological embedding in historical perspective. *Proc. Natl. Acad. Sci. USA* **2012**, *109* (Suppl. S2), 17160–17167. [CrossRef]
8. Engert, V.; Efanov, S.I.; Dedovic, K.; Dagher, A.; Pruessner, J.C. Increased cortisol awakening response and afternoon/evening cortisol output in healthy young adults with low early life parental care. *Psychopharmacology* **2011**, *214*, 261–268. [CrossRef]
9. Trickett, P.K.; Noll, J.G.; Susman, E.J.; Shenk, C.E.; Putnam, F.W. Attenuation of cortisol across development for victims of sexual abuse. *Tissue Eng.* **2007**, *23*, 165–175. [CrossRef]
10. Gunnar, M.R.; Vazquez, D.M. Low cortisol and a flattening of expected daytime rhythm: Potential indices of risk in human development. *Dev. Psychopathol.* **2001**, *13*, 515–538. [CrossRef]
11. Adam, E.K.; Quinn, M.E.; Tavernier, R.; McQuillan, M.T.; Dahlke, K.A.; Gilbert, K.E. Diurnal cortisol slopes and mental and physical health outcomes: A systematic review and meta-analysis. *Psychoneuroendocrinology* **2017**, *83*, 25–41. [CrossRef]
12. Young, E.S.; Farrell, A.K.; Carlson, E.A.; Englund, M.M.; Miller, G.E.; Gunnar, M.R.; Roisman, G.I.; Simpson, J.A. The Dual Impact of Early and Concurrent Life Stress on Adults' Diurnal Cortisol Patterns: A Prospective Study. *Psychol. Sci.* **2019**, *30*, 739–747. [CrossRef]

13. Tarullo, A.R.; Tuladhar, C.T.; Kao, K.; Drury, E.B.; Meyer, J. Cortisol and socioeconomic status in early childhood: A multidimensional assessment. *Dev. Psychopathol.* **2020**, *32*, 1876–1887. [CrossRef]
14. Ouellet-Morin, I.; Cantave, C.; Lupien, S.; Geoffroy, M.C.; Brendgen, M.; Vitaro, F.; Tremblay, R.E.; Boivin, M.; Côté, S.M. Cumulative exposure to socioeconomic and psychosocial adversity and hair cortisol concentration: A longitudinal study from 5 months to 17 years of age. *Psychoneuroendocrinology* **2021**, *126*, 105153. [CrossRef]
15. Ioannidis, K.; Askelund, A.D.; Kievit, R.A.; van Harmelen, A.L. The complex neurobiology of resilient functioning after childhood maltreatment. *BMC Med.* **2020**, *18*, 32. [CrossRef]
16. Weder, N.; Zhang, H.; Jensen, K.; Yang, B.Z.; Simen, A.; Jackowski, A.; Lipschitz, D.; Douglas-Palumberi, H.; Ge, M.; Perepletchikova, F.; et al. Child Abuse, Depression, and Methylation in Genes Involved With Stress, Neural Plasticity, and Brain Circuitry. *J. Am. Acad. Child Adolesc. Psychiatry* **2014**, *53*, 417–424. [CrossRef]
17. Parade, S.H.; Huffhines, L.; Daniels, T.E.; Stroud, L.R.; Nugent, N.R.; Tyrka, A.R. A systematic review of childhood maltreatment and DNA methylation: Candidate gene and epigenome-wide approaches. *Transl. Psychiatry* **2021**, *11*, 134. [CrossRef]
18. Provençal, N.; Binder, E.B. The effects of early life stress on the epigenome: From the womb to adulthood and even before. *Exp. Neurol.* **2015**, *268*, 10–20. [CrossRef]
19. Wiechmann, T.; Röh, S.; Sauer, S.; Czamara, D.; Arloth, J.; Ködel, M.; Beintner, M.; Knop, L.; Menke, A.; Binder, E.B.; et al. Identification of dynamic glucocorticoid-induced methylation changes at the FKBP5 locus. *Clin. Epigenet.* **2019**, *11*, 83. [CrossRef]
20. McEwen, B.S. Allostasis and the Epigenetics of Brain and Body Health Over the Life Course. *JAMA Psychiatry* **2017**, *74*, 551. [CrossRef]
21. Martin, D.I.K.; Cropley, J.E.; Suter, C.M. Epigenetics in disease: Leader or follower? *Epigenetics* **2011**, *6*, 843–848. [CrossRef]
22. Oh, D.L.; Jerman, P.; Silvério Marques, S.; Koita, K.; Purewal Boparai, S.K.; Burke Harris, N.; Bucci, M. Systematic review of pediatric health outcomes associated with childhood adversity. *BMC Pediatr.* **2018**, *18*, 83. [CrossRef]
23. Suarez, A.; Lahti, J.; Lahti-Pulkkinen, M.; Girchenko, P.; Czamara, D.; Arloth, J.; Malmberg, A.L.; Hämäläinen, E.; Kajantie, E.; Laivuori, H.; et al. A polyepigenetic glucocorticoid exposure score at birth and childhood mental and behavioral disorders. *Neurobiol. Stress* **2020**, *13*, 100275. [CrossRef]
24. Bick, J.; Naumova, O.; Hunter, S.; Barbot, B.; Lee, M.; Luthar, S.S.; Raefski, A.; Grigorenko, E.L. Childhood adversity and DNA methylation of genes involved in the hypothalamus–pituitary–adrenal axis and immune system: Whole-genome and candidate-gene associations. *Dev. Psychopathol.* **2012**, *24*, 1417–1425. [CrossRef]
25. Horvath, S. DNA methylation age of human tissues and cell types. *Genome Biol.* **2013**, *14*, 3156. [CrossRef]
26. Horvath, S.; Oshima, J.; Martin, G.M.; Lu, A.T.; Quach, A.; Cohen, H.; Felton, S.; Matsuyama, M.; Lowe, D.; Kabacik, S.; et al. Epigenetic clock for skin and blood cells applied to Hutchinson Gilford Progeria Syndrome and ex vivo studies. *Aging* **2018**, *10*, 1758–1775. [CrossRef]
27. Jain, P.; Binder, A.M.; Chen, B.; Parada, H.; Gallo, L.C.; Alcaraz, J.; Horvath, S.; Bhatti, P.; Whitsel, E.A.; Jordahl, K.; et al. Analysis of Epigenetic Age Acceleration and Healthy Longevity Among Older US Women. *JAMA Netw. Open* **2022**, *5*, e2223285. [CrossRef]
28. Johnstone, S.E.; Gladyshev, V.N.; Aryee, M.J.; Bernstein, B.E. Epigenetic clocks, aging, and cancer. *Science* **2022**, *378*, 1276–1277. [CrossRef]
29. Hannum, G.; Guinney, J.; Zhao, L.; Zhang, L.; Hughes, G.; Sadda, S.; Klotzle, B.; Bibikova, M.; Fan, J.B.; Gao, Y.; et al. Genome-wide Methylation Profiles Reveal Quantitative Views of Human Aging Rates. *Mol. Cell* **2013**, *49*, 359–367. [CrossRef]
30. McEwen, L.M.; O'Donnell, K.J.; McGill, M.G.; Edgar, R.D.; Jones, M.J.; MacIsaac, J.L.; Lin, D.T.S.; Ramadori, K.; Morin, A.; Gladish, N.; et al. The PedBE clock accurately estimates DNA methylation age in pediatric buccal cells. *Proc. Natl. Acad. Sci. USA* **2020**, *117*, 23329–23335. [CrossRef]
31. Levine, M.E.; Lu, A.T.; Quach, A.; Chen, B.H.; Assimes, T.L.; Bandinelli, S.; Hou, L.; Baccarelli, A.A.; Stewart, J.D.; Li, Y.; et al. An epigenetic biomarker of aging for lifespan and healthspan. *Aging* **2018**, *10*, 573–591. [CrossRef]
32. Lu, A.T.; Quach, A.; Wilson, J.G.; Reiner, A.P.; Aviv, A.; Raj, K.; Hou, L.; Baccarelli, A.A.; Li, Y.; Stewart, J.D.; et al. DNA methylation GrimAge strongly predicts lifespan and healthspan. *Aging* **2019**, *11*, 303–327. [CrossRef] [PubMed]
33. Krieger, N.; Chen, J.T.; Testa, C.; Diez Roux, A.; Tilling, K.; Watkins, S.; Simpkin, A.J.; Suderman, M.; Davey Smith, G.; De Vivo, I.; et al. Use of Correct and Incorrect Methods of Accounting for Age in Studies of Epigenetic Accelerated Aging: Implications and Recommendations for Best Practices. *Am. J. Epidemiol.* **2023**, *192*, 800–811. [CrossRef] [PubMed]
34. Belsky, D.W.; Caspi, A.; Arseneault, L.; Baccarelli, A.; Corcoran, D.L.; Gao, X.; Hannon, E.; Harrington, H.L.; Rasmussen, L.J.; Houts, R.; et al. Quantification of the pace of biological aging in humans through a blood test, the DunedinPoAm DNA methylation algorithm. *Elife* **2020**, *9*, e54870. [CrossRef] [PubMed]
35. Belsky, D.W.; Caspi, A.; Corcoran, D.L.; Sugden, K.; Poulton, R.; Arseneault, L.; Baccarelli, A.; Chamarti, K.; Gao, X.; Hannon, E.; et al. DunedinPACE, a DNA methylation biomarker of the pace of aging. *eLife* **2022**, *11*, e73420. [CrossRef]
36. Wolf, E.J.; Logue, M.W.; Morrison, F.G.; Wilcox, E.S.; Stone, A.; Schichman, S.A.; McGlinchey, R.E.; Milberg, W.P.; Miller, M.W. Posttraumatic psychopathology and the pace of the epigenetic clock: A longitudinal investigation. *Psychol. Med.* **2019**, *49*, 791–800. [CrossRef]

37. Mehta, D.; Bruenig, D.; Pierce, J.; Sathyanarayanan, A.; Stringfellow, R.; Miller, O.; Mullens, A.B.; Shakespeare-Finch, J. Recalibrating the epigenetic clock after exposure to trauma: The role of risk and protective psychosocial factors. *J. Psychiatr. Res.* **2022**, *149*, 374–381. [CrossRef]
38. Han, L.K.M.; Aghajani, M.; Clark, S.L.; Chan, R.F.; Hattab, M.W.; Shabalin, A.A.; Zhao, M.; Kumar, G.; Xie, L.Y.; Jansen, R.; et al. Epigenetic Aging in Major Depressive Disorder. *Am. J. Psychiatry* **2018**, *175*, 774–782. [CrossRef]
39. Cerveira de Baumont, A.; Hoffmann, M.S.; Bortoluzzi, A.; Fries, G.R.; Lavandoski, P.; Grun, L.K.; Guimarães, L.S.; Guma, F.T.; Salum, G.A.; Barbé-Tuana, F.M.; et al. Telomere length and epigenetic age acceleration in adolescents with anxiety disorders. *Sci. Rep.* **2021**, *11*, 7716. [CrossRef]
40. Jovanovic, T.; Vance, L.A.; Cross, D.; Knight, A.K.; Kilaru, V.; Michopoulos, V.; Klengel, T.; Smith, A.K. Exposure to Violence Accelerates Epigenetic Aging in Children. *Sci. Rep.* **2017**, *7*, 8962. [CrossRef]
41. Lawn, R.B.; Anderson, E.L.; Suderman, M.; Simpkin, A.J.; Gaunt, T.R.; Teschendorff, A.E.; Widschwendter, M.; Hardy, R.; Kuh, D.; Relton, C.L.; et al. Psychosocial adversity and socioeconomic position during childhood and epigenetic age: Analysis of two prospective cohort studies. *Hum. Mol. Genet.* **2018**, *27*, 1301–1308. [CrossRef]
42. Austin, M.K.; Chen, E.; Ross, K.M.; McEwen, L.M.; Maclsaac, J.L.; Kobor, M.S.; Miller, G.E. Early-life socioeconomic disadvantage, not current, predicts accelerated epigenetic aging of monocytes. *Psychoneuroendocrinology* **2018**, *97*, 131–134. [CrossRef] [PubMed]
43. Yusupov, N.; Dieckmann, L.; Erhart, M.; Sauer, S.; Rex-Haffner, M.; Kopf-Beck, J.; Brückl, T.M.; Czamara, D.; Binder, E.B. Transdiagnostic evaluation of epigenetic age acceleration and burden of psychiatric disorders. *Neuropsychopharmacology* **2023**, *48*, 1409–1417. [CrossRef] [PubMed]
44. Hamlat, E.J.; Prather, A.A.; Horvath, S.; Belsky, J.; Epel, E.S. Early life adversity, pubertal timing, and epigenetic age acceleration in adulthood. *Dev. Psychobiol.* **2021**, *63*, 890–902. [CrossRef] [PubMed]
45. Simons, R.L.; Lei, M.K.; Klopach, E.; Berg, M.; Zhang, Y.; Beach, S.S.R. (Re)Setting Epigenetic Clocks: An Important Avenue Whereby Social Conditions Become Biologically Embedded across the Life Course. *J. Health Soc. Behav.* **2021**, *62*, 436–453. [CrossRef] [PubMed]
46. Raffington, L.; Belsky, D.W.; Kothari, M.; Malanchini, M.; Tucker-Drob, E.M.; Harden, K.P. Socioeconomic Disadvantage and the Pace of Biological Aging in Children. *Pediatrics* **2021**, *147*, e2020024406. [CrossRef]
47. Sumner, J.A.; Colich, N.L.; Uddin, M.; Armstrong, D.; McLaughlin, K.A. Early Experiences of Threat, but Not Deprivation, Are Associated With Accelerated Biological Aging in Children and Adolescents. *Biol. Psychiatry* **2019**, *85*, 268–278. [CrossRef]
48. Lupien, S.J.; McEwen, B.S.; Gunnar, M.R.; Heim, C. Effects of stress throughout the lifespan on the brain, behaviour and cognition. *Nat. Rev. Neurosci.* **2009**, *10*, 434–445. [CrossRef]
49. Marini, S.; Davis, K.A.; Soare, T.W.; Zhu, Y.; Suderman, M.J.; Simpkin, A.J.; Smith, A.D.; Wolf, E.J.; Relton, C.L.; Dunn, E.C. Adversity exposure during sensitive periods predicts accelerated epigenetic aging in children. *Psychoneuroendocrinology* **2020**, *113*, 104484. [CrossRef]
50. Holbrook, B.D. The effects of nicotine on human fetal development. *Birth Defects Res. C Embryo Today* **2016**, *108*, 181–192. [CrossRef]
51. Martins, J.; Czamara, D.; Sauer, S.; Rex-Haffner, M.; Dittrich, K.; Dörr, P.; de Punder, K.; Overfeld, J.; Knop, A.; Dammering, F.; et al. Childhood adversity correlates with stable changes in DNA methylation trajectories in children and converges with epigenetic signatures of prenatal stress. *Neurobiol. Stress* **2021**, *15*, 100336. [CrossRef]
52. Post, R.M. Epigenetic basis of sensitization to stress, affective episodes, and stimulants: Implications for illness progression and prevention. *Bipolar. Disord.* **2016**, *18*, 315–324. [CrossRef] [PubMed]
53. Bunea, I.M.; Szentágotai-Tătar, A.; Miu, A.C. Early-life adversity and cortisol response to social stress: A meta-analysis. *Transl. Psychiatry* **2017**, *7*, 1274. [CrossRef] [PubMed]
54. Young, E.S.; Doom, J.R.; Farrell, A.K.; Carlson, E.A.; Englund, M.M.; Miller, G.E.; Gunnar, M.R.; Roisman, G.I.; Simpson, J.A. Life stress and cortisol reactivity: An exploratory analysis of the effects of stress exposure across life on HPA-axis functioning. *Dev. Psychopathol.* **2021**, *33*, 301–312. [CrossRef]
55. McCrory, C.; Fiorito, G.; McLoughlin, S.; Polidoro, S.; Cheallaigh, C.N.; Bourke, N.; Karisola, P.; Alenius, H.; Vineis, P.; Layte, R.; et al. Epigenetic Clocks and Allostatic Load Reveal Potential Sex-Specific Drivers of Biological Aging. *J. Gerontol. Ser. A* **2019**, *75*, 495–503. [CrossRef]
56. Suarez, A.; Lahti, J.; Czamara, D.; Lahti-Pulkkinen, M.; Knight, A.K.; Girchenko, P.; Hämäläinen, E.; Kajantie, E.; Lipsanen, J.; Laivuori, H.; et al. The Epigenetic Clock at Birth: Associations With Maternal Antenatal Depression and Child Psychiatric Problems. *J. Am. Acad. Child Adolesc. Psychiatry* **2018**, *57*, 321–328. [CrossRef]
57. Engelbrecht, H.R.; Merrill, S.M.; Gladish, N.; MacIsaac, J.L.; Lin, D.T.S.; Ecker, S.; Chrysohoou, C.A.; Pes, G.M.; Kobor, M.S.; Rehkopf, D.H. Sex differences in epigenetic age in Mediterranean high longevity regions. *Front. Aging* **2022**, *3*, 1007098. [CrossRef] [PubMed]
58. Orri, M.; Boivin, M.; Chen, C.; Ahun, M.N.; Geoffroy, M.C.; Ouellet-Morin, I.; Tremblay, R.E.; Côté, S.M. Cohort Profile: Quebec Longitudinal Study of Child Development (QLSCD). *Soc. Psychiatry Psychiatr. Epidemiol.* **2021**, *56*, 883–894. [CrossRef]

59. Silveira, P.P.; Pokhvisneva, I.; Parent, C.; Cai, S.; Rema, A.S.S.; Broekman, B.F.P.; Rifkin-Graboi, A.; Pluess, M.; O'Donnell, K.J.; Meaney, M.J. Cumulative prenatal exposure to adversity reveals associations with a broad range of neurodevelopmental outcomes that are moderated by a novel, biologically informed polygenetic score based on the serotonin transporter solute carrier family C6, member 4. *Dev. Psychopathol.* **2017**, *29*, 1601–1617. [CrossRef]
60. Hughes, M.M.; Black, R.E.; Katz, J. 2500-g Low Birth Weight Cutoff: History and Implications for Future Research and Policy. *Matern. Child Health J.* **2017**, *21*, 283–289. [CrossRef]
61. Radloff, L.S. The CES-D Scale. *Appl. Psychol. Meas.* **1977**, *1*, 385–401. [CrossRef]
62. Ladd, G.W.; Kochenderfer-Ladd, B. Identifying victims of peer aggression from early to middle childhood: Analysis of cross-informant data for concordance, estimation of relational adjustment, prevalence of victimization, and characteristics of identified victims. *Psychol. Assess* **2002**, *14*, 74–96. [CrossRef] [PubMed]
63. Aryee, M.J.; Jaffe, A.E.; Corrada-Bravo, H.; Ladd-Acosta, C.; Feinberg, A.P.; Hansen, K.D.; Irizarry, R.A. Minfi: A flexible and comprehensive Bioconductor package for the analysis of Infinium DNA methylation microarrays. *Bioinformatics* **2014**, *30*, 1363–1369. [CrossRef] [PubMed]
64. Chen, Y.; Lemire, M.; Choufani, S.; Butcher, D.T.; Grafodatskaya, D.; Zanke, B.W.; Gallinger, S.; Hudson, T.J.; Weksberg, R. Discovery of cross-reactive probes and polymorphic CpGs in the Illumina Infinium HumanMethylation450 microarray. *Epigenetics* **2013**, *8*, 203–209. [CrossRef]
65. Pidsley, R.; Zotenko, E.; Peters, T.J.; Lawrence, M.G.; Risbridger, G.P.; Molloy, P.; Van Djik, S.; Muhlhausler, B.; Stirzaker, C.; Clark, S.J. Critical evaluation of the Illumina MethylationEPIC BeadChip microarray for whole-genome DNA methylation profiling. *Genome Biol.* **2016**, *17*, 208. [CrossRef] [PubMed]
66. Zheng, S.C.; Breeze, C.E.; Beck, S.; Teschendorff, A.E. Identification of differentially methylated cell types in epigenome-wide association studies. *Nat. Methods* **2018**, *15*, 1059–1066. [CrossRef]
67. Lang, M.; Binder, M.; Richter, J.; Schratz, P.; Pfisterer, F.; Coors, S.; Au, Q.; Casalicchio, G.; Kotthoff, L.; Bischl, B.; et al. mlr3: A modern object-oriented machine learning framework in R. *J. Open Source Softw.* **2019**, *4*, 1903. [CrossRef]
68. McCrory, C.; Fiorito, G.; O'Halloran, A.M.; Polidoro, S.; Vineis, P.; Kenny, R.A. Early life adversity and age acceleration at mid-life and older ages indexed using the next-generation GrimAge and Pace of Aging epigenetic clocks. *Psychoneuroendocrinology* **2022**, *137*, 105643. [CrossRef] [PubMed]
69. Schrempft, S.; Stringhini, S. Socioeconomic inequalities in the Pace of Aging. *Aging* **2023**, *15*, 1706–1707. [CrossRef]
70. Perret, L.C.; Geoffroy, M.C.; Barr, E.; Parnet, F.; Provencal, N.; Boivin, M.; O'Donnell, K.J.; Suderman, M.; Power, C.; Turecki, G.; et al. Associations between epigenetic aging and childhood peer victimization, depression, and suicidal ideation in adolescence and adulthood: A study of two population-based samples. *Front. Cell Dev. Biol.* **2023**, *10*, 1051556. [CrossRef]
71. Colich, N.L.; Rosen, M.L.; Williams, E.S.; McLaughlin, K.A. Biological aging in childhood and adolescence following experiences of threat and deprivation: A systematic review and meta-analysis. *Psychol. Bull.* **2020**, *146*, 721–764. [CrossRef]
72. Copeland, W.E.; Shanahan, L.; McGinnis, E.W.; Aberg, K.A.; van den Oord, E.J.C.G. Early adversities accelerate epigenetic aging into adulthood: A 10-year, within-subject analysis. *J. Child Psychol. Psychiatry* **2022**, *63*, 1308–1315. [CrossRef] [PubMed]
73. Yaroslavsky, I.; Bush, A.H.; France, C.M. Emotion regulation deficits mediate childhood sexual abuse effects on stress sensitization and depression outcomes. *Dev. Psychopathol.* **2022**, *34*, 157–170. [CrossRef]
74. Luo, J.; Willroth, E.C. Values and stress: Examining the relations between values and general and domain-specific stress in two longitudinal studies. *J. Pers. Soc. Psychol.* **2025**, *128*, 410–437. [CrossRef]
75. Stroud, C.B. The Stress Sensitization Model. In *The Oxford Handbook of Stress and Mental Health*; Harkness, K.L., Hayden, E.P., Eds.; Oxford University Press: Oxford, UK, 2020; pp. 348–370.
76. McCarthy, M.M.; Nugent, B.M. At the frontier of epigenetics of brain sex differences. *Front. Behav. Neurosci.* **2015**, *9*, 221. [CrossRef] [PubMed]
77. Kundakovic, M.; Lim, S.; Gudsnuk, K.; Champagne, F.A. Sex-specific and strain-dependent effects of early life adversity on behavioral and epigenetic outcomes. *Front. Psychiatry* **2013**, *4*, 78. [CrossRef] [PubMed]
78. Aylwin, C.F.; Toro, C.A.; Shirtcliff, E.; Lomniczi, A. Emerging Genetic and Epigenetic Mechanisms Underlying Pubertal Maturation in Adolescence. *J. Res. Adolesc.* **2019**, *29*, 54–79. [CrossRef]
79. Almstrup, K.; Lindhardt Johansen, M.; Busch, A.S.; Hagen, C.P.; Nielsen, J.E.; Petersen, J.H.; Juul, A. Pubertal development in healthy children is mirrored by DNA methylation patterns in peripheral blood. *Sci. Rep.* **2016**, *6*, 28657. [CrossRef]

biomolecules

MDPI

Review

Targeting Senescence: A Review of Senolytics and Senomorphics in Anti-Aging Interventions

Timur Saliev [1,*] and Prim B. Singh [2]

1 Institute of Fundamental and Applied Medical Research, S.D. Asfendiyarov Kazakh National Medical University, Tole Bi Street 94, Almaty 050000, Kazakhstan
2 School of Medicine, Nazarbayev University, Astana 010000, Kazakhstan; prim.singh@nu.edu.kz
* Correspondence: tim.saliev@gmail.com

Abstract: Cellular senescence is a fundamental mechanism in aging, marked by irreversible growth arrest and diverse functional changes, including, but not limited to, the development of a senescence-associated secretory phenotype (SASP). While transient senescence contributes to beneficial processes such as tissue repair and tumor suppression, the persistent accumulation of senescent cells is implicated in tissue dysfunction, chronic inflammation, and age-related diseases. Notably, the SASP can exert both pro-inflammatory and immunosuppressive effects, depending on cell type, tissue context, and temporal dynamics, particularly in early stages where it may be profibrotic and immunomodulatory. Recent advances in senotherapeutics have led to two principal strategies for targeting senescent cells: senolytics, which selectively induce their apoptosis, and senomorphics, which modulate deleterious aspects of the senescence phenotype, including the SASP, without removing the cells. This review critically examines the molecular mechanisms, therapeutic agents, and clinical potential of both approaches in the context of anti-aging interventions. We discuss major classes of senolytics, such as tyrosine kinase inhibitors, BCL-2 family inhibitors, and natural polyphenols, alongside senomorphics including mTOR and JAK inhibitors, rapalogs, and epigenetic modulators. Additionally, we explore the biological heterogeneity of senescent cells, challenges in developing specific biomarkers, and the dualistic role of senescence in physiological versus pathological states. The review also highlights emerging tools, such as targeted delivery systems, multi-omics integration, and AI-assisted drug discovery, which are advancing precision geroscience and shaping future anti-aging strategies.

Keywords: cellular senescence; senolytics; senomorphics; aging; senescence-associated secretory phenotype (SASP); anti-aging therapy; longevity

1. Introduction

Aging is an inevitable biological process characterized by a progressive decline in physiological integrity, leading to impaired function and increased vulnerability to death. Among the various cellular and molecular mechanisms that contribute to aging, cellular senescence has emerged as a fundamental driver of age-related dysfunction and chronic disease [1,2]. Initially described as a mechanism to prevent the proliferation of damaged cells, senescence involves a state of stable cell cycle arrest coupled with profound changes in gene expression, morphology, and function. While this process plays a critical role in tissue remodeling, wound healing, and tumor suppression, the chronic accumulation of senescent cells contributes to a pro-inflammatory tissue environment and promotes aging-related pathologies.

Senescent cells secrete a variety of bioactive molecules collectively known as the senescence-associated secretory phenotype (SASP), which includes pro-inflammatory cytokines, chemokines, growth factors, and matrix-remodeling enzymes [3,4]. These factors have a dual nature: although they can be beneficial in short-term physiological contexts, such as promoting tissue repair or halting the proliferation of precancerous cells, their long-term persistence fosters a deleterious milieu that disrupts tissue homeostasis, impairs stem cell function, and promotes chronic inflammation and fibrosis [5]. The SASP is now recognized as a key mediator of the "inflammaging" process, a low-grade, chronic inflammation associated with aging, and has been implicated in a broad spectrum of diseases, including osteoarthritis, atherosclerosis, neurodegeneration, diabetes, and cancer [6,7].

In recent years, the concept of targeting cellular senescence has gained traction as a promising therapeutic strategy to delay aging and extend health span. Two major classes of anti-senescence therapies have emerged: senolytics, which selectively induce apoptosis in senescent cells, and senomorphics, which suppress the harmful effects of the SASP without killing the cells [8,9]. These interventions aim to reduce the burden of senescent cells and modulate their secretory activity, thereby alleviating tissue dysfunction and enhancing resilience against stress and disease. Encouragingly, preclinical studies in mouse models have demonstrated that clearing senescent cells can improve tissue function, delay the onset of age-related diseases, and even extend lifespan [10]. These findings have spurred a surge in translational research, with multiple clinical trials now underway to evaluate the safety and efficacy of senescence-targeting agents in humans [11–13].

Senescent cells are key contributors to aging, marked by irreversible growth arrest and often the secretion of pro-inflammatory factors known as the senescence-associated secretory phenotype (SASP). While transient senescence aids in tissue repair and tumor suppression, chronic accumulation of senescent cells promotes tissue dysfunction, inflammation, and age-related decline. Kondratyeva et al. (2025) demonstrated that treatment with growth differentiation factor 11 (GDF11) partially reversed the senescent phenotype in mesenchymal stem cells (MSCs), improving viability and angiogenic function [14]. This highlights the impact of senescence on stem cell aging and the potential of rejuvenation therapies.

In the context of skin aging, Wyles et al. (2024) [15] showed that topical platelet-derived exosomes reduced senescence markers ($p16^{INK4a}$, $p21^{CIP1}$) and SASP components, improving skin structure, supporting the idea that local clearance or modulation of senescent cells can restore tissue function. At the systemic level, Nishizawa et al. [16] found that the transcription factor BACH1 suppresses senescence through ferroptosis-induced secretion of FGF21, which improved metabolism and extended lifespan in progeria models. This points to a non-cell-autonomous role of senescent cells in aging regulation. These studies affirm that senescent cells are not passive bystanders but active drivers of aging. They impair tissue regeneration, induce chronic inflammation, and influence systemic health. Therapeutic strategies that target senescent cells, via senolytics, senomorphics, or rejuvenation signals like GDF11 and FGF21, hold promise for mitigating age-related decline and extending health span. Emerging evidence strongly supports the therapeutic potential of eliminating or modulating senescent cells to enhance healthspan and delay age-related decline. Senescent cells accumulate with age and contribute to chronic inflammation, tissue dysfunction, and age-related diseases. Poblocka et al. (2021) developed a targeted antibody–drug conjugate (ADC) that selectively eliminates senescent cells by binding to β2-microglobulin (B2M), a surface marker enriched in senescence [17]. The ADC delivers a cytotoxic payload specifically to senescent cells, reducing their burden without affecting healthy tissue. This approach offers a promising, selective senolytic strategy. In contrast, Zumerle et al. (2024) used a natural polyphenol extract [18], Haenkenium (HK), from Salvia haenkei to reduce senescence markers and improve physical condition in aged mice. One active compound, luteolin, modulates the p16–CDK6 interaction, suppressing senescence-associated pathways without killing the cells. This senomorphic strategy demonstrates how modulating, rather than clearing, senescent cells can improve healthspan. Dorronsoro et al. [19] showed that extracellular vesicles (EVs) from young mesenchymal stem cells reduced cellular senescence and extended health span in mice, offering a regenerative, cell-free alternative with low risk and high translational potential. Together, these studies highlight that both senolytics and senomorphics can mitigate the harmful effects of senescent cells, offering viable strategies to improve tissue function and extend health span [20]. Despite the promise of these interventions, several challenges remain. The heterogeneity of senescent cell populations, the context-dependent effects of senescence, and the potential off-target consequences of eliminating or altering senescent cells raise concerns about the safety and specificity of current approaches [21]. Moreover, the lack of robust biomarkers to identify and monitor senescent cells in vivo complicates both research and clinical application [22]. A deeper understanding of the molecular mechanisms governing senescence, improved methods for detecting senescent cells, and the development of next-generation therapeutics with greater selectivity and minimal side effects will be critical to unlocking the full potential of this field [21,23].

As we continue to explore the complex biology of cellular senescence and refine our strategies for intervening in this process, targeting senescence holds significant promise not only for mitigating the effects of aging but also for transforming the way we approach the prevention and treatment of age-related diseases [24]. This review provides an overview of current strategies and advances in senolytics and senomorphics, examines their therapeutic potential and limitations, and discusses future directions in this rapidly evolving field of geroscience.

2. Senolytics: Eliminating Senescent Cells

Senolytics are a rapidly advancing class of therapeutic agents designed to selectively eliminate senescent cells, which accumulate with age and contribute to chronic inflammation, tissue degeneration, and age-related diseases [25,26]. Cellular senescence is a state of irreversible cell cycle arrest triggered by various stressors such as telomere shortening, DNA damage, mitochondrial dysfunction, and oncogene activation [27]. While transient senescence has beneficial roles in development, wound healing, and tumor suppression, the chronic persistence of senescent cells in aged tissues has been implicated in the pathogenesis of numerous conditions, including osteoarthritis, cardiovascular disease, pulmonary fibrosis, and neurodegeneration.

Senescent cells evade immune clearance by activating anti-apoptotic signaling pathways, often termed senescent cell anti-apoptotic pathways (SCAPs), which differentiate them from their healthy counterparts [28]. Senolytic agents exploit this vulnerability by disabling these pro-survival mechanisms, thereby triggering apoptosis selectively in senescent cells while sparing non-senescent tissue [29,30]. The first generation of senolytic therapies emerged with the identification of dasatinib, a tyrosine kinase inhibitor, and quercetin, a flavonoid that inhibits PI3K and other kinases [31] (Table 1). These two compounds, used in combination (D + Q), were found to clear senescent pre-adipocytes and endothelial cells in mouse models [32]. In aged or diseased animals, intermittent administration of D + Q improved physical function, reduced senescent cell burden, and in some cases, extended lifespan. These findings laid the groundwork for clinical trials in humans, where early results suggest improvements in physical performance and reductions in senescence biomarkers in patients with idiopathic pulmonary fibrosis and diabetic kidney disease.

Table 1. Comparative Overview of Major Senolytic Classes.

Senolytic Class	Molecular Targets	Represent Agents	Mode of Action	Strengths	Limitations/Challenges	Ref.
Tyrosine Kinase Inhibitors	Src family kinases, Eph receptors	Dasatinib	Inhibits pro-survival tyrosine kinases upregulated in certain SnC types	Effective in senescent preadipocytes and progenitors	Cell-type specificity; potential for systemic toxicity	[33]
Flavonoid Polyphenols	PI3K/AKT, NF-κB, ROS pathways	Quercetin, Fisetin	Induces apoptosis via oxidative stress and suppression of anti-apoptotic signaling	Low toxicity; orally bioavailable; broad applicability	Variable potency; poor bioavailability in vivo	[34]
BCL-2 Family Inhibitors	BCL-2, BCL-xL, BCL-w	Navitoclax (ABT-263), ABT-737	Blocks anti-apoptotic proteins, sensitizing SnCs to apoptosis	Potent and broad-acting across senescent phenotypes	Thrombocytopenia due to BCL-xL inhibition in platelets	[35]
FOXO4-p53 Disruptors	FOXO4-p53 complex	FOXO4-DRI peptide	Disrupts nuclear retention of p53, restoring apoptotic signaling	Selective SnC clearance; rejuvenates aged tissues	Peptide delivery limitations; currently preclinical	[36]
HSP90 Inhibitors	Heat shock protein 90	17-DMAG, Geldanamycin derivatives	Destabilizes chaperone-dependent survival proteins in SnCs	Targets multiple stress response pathways	General cytotoxicity; lacks SnC specificity	[37]
CDK/p53 Pathway Modulators	MDM2-p53, CDK4/6	UBX0101, Nutlin-3	Modulates cell cycle regulators and apoptotic checkpoints	Targeted for local intra-articular applications	Short half-life; mixed clinical trial results	[38]
Natural Senolytics (Plant-Derived)	ROS generation, NF-κB, SASP factors	Piperlongumine, Curcumin analogues	Promotes redox imbalance and downregulates SASP-related pathways	Low-cost, multi-targeted; dietary sources	Low potency; unclear pharmacokinetics and dosing strategies	[39]

Building on these foundational discoveries, researchers have developed a diverse range of senolytic agents targeting different aspects of senescent cell biology. One key area of focus is the BCL-2 family of proteins, which are upregulated in senescent cells to resist apoptosis. Navitoclax (ABT-263), an inhibitor of BCL-2 and BCL-XL, has demonstrated potent senolytic activity across several tissues [35,40] (Table 1). However, it has been limited in clinical translation by dose-dependent thrombocytopenia, as platelets also rely on BCL-XL for survival [41]. Similar agents, such as ABT-737, have shown promise in preclinical settings but share similar safety concerns [42]. A structurally related compound, ABT-737, was an earlier-generation BCL-2 family inhibitor that similarly targets BCL-2, BCL-XL, and BCL-w. ABT-737 has shown promising senolytic effects in preclinical models, particularly in the clearance of senescent cells from irradiated tissues and in models of pulmonary fibrosis and atherosclerosis [43]. However, ABT-737, like navitoclax, has demonstrated hematologic toxicity in vivo, including thrombocytopenia and neutropenia, due to the on-target depletion of non-senescent cells that rely on BCL-2/XL signaling for survival. This safety profile has limited its direct clinical use and prompted efforts to either develop selective delivery systems (e.g., nanoparticle carriers or conjugation with senescence-targeted ligands) or to identify next-generation BCL-2 inhibitors with improved senescent cell specificity.

Furthermore, both ABT-263 and ABT-737 underscore a broader challenge in senolytic drug development: the molecular overlap between survival pathways in senescent and healthy cells, especially in high-turnover tissues. As a result, recent research is focusing on transient, intermittent dosing strategies, combination regimens with senomorphics, or targeting downstream effectors unique to senescence, to mitigate adverse effects.

Another strategy targets the interaction between FOXO4 and p53, which plays a role in maintaining the survival of senescent cells. The FOXO4-DRI peptide disrupts this interaction, reactivating p53-mediated apoptosis and eliminating senescent cells in various models [44]. HSP90 inhibitors, such as 17-DMAG, destabilize chaperone-mediated survival pathways in senescent cells, offering another route for clearance [45,46]. More recently, UBX0101 and related molecules have been designed to modulate the p53-MDM2 axis and CDK activity to selectively eliminate senescent chondrocytes in osteoarthritic joints, although clinical progress has been mixed [38].

In addition to synthetic compounds, several naturally occurring molecules have demonstrated senolytic potential. Fisetin, a dietary flavonoid found in fruits such as strawberries and apples, is particularly notable for its broad efficacy across different senescent cell types [47,48]. It reduces systemic inflammation, improves tissue homeostasis, and extends health span in mouse models. Fisetin is currently being tested in several human trials for its potential to alleviate age-associated dysfunction. Other natural agents like piperlongumine and synthetic curcumin derivatives are also under investigation, offering potentially safer and more accessible alternatives to synthetic drugs [49,50].

Despite these promising advances, the development of senolytics faces several key challenges. Many compounds lack cell-type specificity, raising concerns about off-target effects and toxicity in non-senescent tissues. Some, like navitoclax, induce significant side effects at effective doses. Moreover, the heterogeneity of senescent cells, both within and across tissues, means that a one-size-fits-all senolytic is unlikely to exist. Optimization of dosing schedules, such as intermittent "hit-and-run" regimens, is being explored to limit toxicity while maintaining efficacy. Another major limitation is the absence of robust, non-invasive biomarkers for senescent cell burden in humans, which impairs the ability to stratify patients or monitor treatment response. Furthermore, there is concern that simply removing senescent cells without addressing the underlying damage or inflammatory microenvironment may lead to their rapid re-accumulation.

Nonetheless, the translational momentum in this field is strong. Several senolytic compounds are in early-phase clinical trials for diseases including Alzheimer's disease, chronic kidney disease, frailty, and osteoarthritis [51]. New approaches are being investigated to improve targeting and minimize off-target effects, including the use of antibody-drug conjugates, nanoparticles, and tissue-specific prodrugs. In the future, senolytic therapies may be combined with senomorphics, regenerative interventions, or immune modulators to enhance both efficacy and tissue repair. As our understanding of cellular senescence deepens, senolytics stand at the forefront of a potential paradigm shift in aging and chronic disease treatment, with the promise of not only extending lifespan but improving health span through selective cellular rejuvenation.

While senolytics represent a promising class of interventions to delay aging and mitigate age-related diseases, their clinical development is complicated by the heterogeneous nature of cellular senescence across different tissues. Senescence is not a uniform process; it varies significantly depending on cell type, tissue context, and developmental stage [52]. For example, senescent fibroblasts in dermal tissue may exhibit distinct SASPs and survival pathways compared to senescent endothelial cells in vascular tissue or chondrocytes in the joint [53]. Likewise, the function and fate of senescent cells can differ: in some settings, they are immunosuppressive and pro-fibrotic, while in others, they may promote inflammation or contribute to tissue repair.

This context dependency has direct implications for senolytic therapies, which are often designed based on generalized senescence mechanisms, such as the upregulation of BCL-2 family proteins or the FOXO4-p53 axis. However, given the variability in survival pathways between senescent cell types, senolytic agents may be effective in one tissue but ineffective, or even harmful, in another. For example, navitoclax and ABT-737 demonstrate robust senolytic activity in certain cell populations but cause thrombocytopenia due to BCL-XL dependence in platelets, highlighting the risk of unintended toxicity in non-senescent cells [35,54].

To overcome these challenges, future senolytic strategies will likely need to adopt organ-specific or cell-type-specific targeting approaches. This may involve the use of tissue-specific promoters, targeted delivery vectors (e.g., nanoparticles or antibody–drug conjugates), or prodrug strategies activated by tissue-enriched enzymes. These precision delivery systems could enhance the selective clearance of senescent cells in target tissues while sparing others.

Moreover, age and disease stage may influence the composition and behavior of senescent cells, particularly in organs with active regeneration or developmental plasticity. For instance, the response to senolytics in aged skeletal muscle stem cells may differ markedly from that in senescent astrocytes in neurodegeneration. Tailoring therapeutic timing and delivery to the developmental stage, disease context, and regenerative capacity will be crucial for optimizing outcomes.

Additionally, combinatorial approaches, such as pairing senolytics with senomorphics, anti-inflammatory agents, or regenerative therapies, may help address the complex interplay between senescent cells and their microenvironment. These strategies could mitigate tissue-specific rebound effects or functional deficits that occur following widespread cell clearance.

3. Senomorphics: Modulating the SASP

While senolytics aim to eliminate senescent cells from aging tissues, an alternative therapeutic approach, known as senomorphics, seeks instead to modulate the phenotype and secretory behavior of senescent cells without necessarily inducing their apoptosis [55,56] (Figure 1).

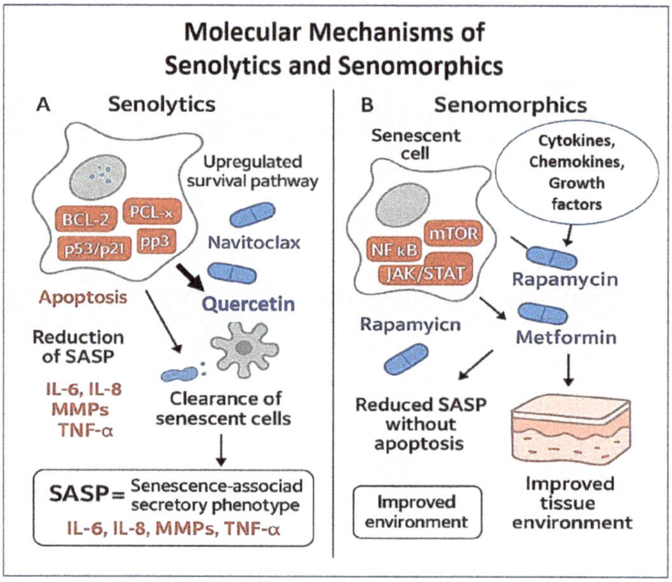

Figure 1. Molecular mechanisms of senolytics and senomorphics in anti-aging interventions.

This class of interventions is also frequently referred to as senostatics in the literature [57,58]. While the terms are often used interchangeably, some authors draw a subtle distinction: "senomorphics" emphasize the modulation of the senescent phenotype, particularly the suppression or alteration of the senescence-associated secretory phenotype (SASP), whereas "senostatics" highlight the stabilization of senescent cells in a non-harmful, quiescent state without promoting further senescence propagation. In practice, however, both terms describe agents that mitigate the detrimental effects of senescent cells without eliminating them, and they are commonly treated as synonyms in the current field.

The strategy of eliminating senescent cells from aging tissues holds great promise but may be inappropriate in certain physiological contexts, particularly in tissues where senescent cells play beneficial roles in processes such as wound healing, tissue remodeling, or immune surveillance. In such cases, senomorphics offer a more refined approach by targeting and suppressing the pro-inflammatory and tissue-damaging components of the senescence-associated secretory phenotype (SASP). The SASP comprises a complex and dynamic mixture of cytokines, chemokines, growth factors, matrix metalloproteinases (MMPs), and bioactive lipids secreted by senescent cells, which contribute to chronic inflammation and tissue dysfunction when left unchecked [13,59]. Persistent SASP signaling is now widely recognized as a key mediator of chronic inflammation in aging ("inflammaging"), as well as a driver of paracrine senescence, tumor promotion, and tissue degeneration.

The SASP is regulated by several stress-responsive transcription factors and intracellular signaling pathways, including NF-κB, C/EBPβ, mTOR, JAK/STAT, and p38 MAPK [60]. Targeting these pathways has led to the identification of various senomorphic agents capable of attenuating SASP expression and its downstream pathological effects (Table 2). Among the most studied are mTOR inhibitors such as rapamycin and its analogs (rapalogs), which can dampen SASP production by interfering with the translation of IL-1α, a key upstream driver of the SASP cascade [61,62]. In preclinical models, rapamycin has been shown to suppress senescence-related inflammation, improve stem cell function, and extend lifespan, without necessarily inducing cell death [63,64]. Its oral bioavailability and known safety profile from clinical use in transplantation and oncology make it one of the most promising senomorphic agents under evaluation for geroprotective interventions.

Another major class of senomorphics includes JAK inhibitors, such as ruxolitinib and baricitinib, which block downstream inflammatory signaling pathways induced by SASP components like IL-6 and IL-8 [65,66] (Table 2). In both mouse models and early human studies, JAK inhibition has reduced systemic inflammation, improved hematopoietic function, and restored tissue homeostasis in aged organs [67,68]. These agents are especially promising in chronic inflammatory conditions associated with clonal hematopoiesis and myeloid skewing. Importantly, senomorphics like JAK inhibitors offer a potentially safer long-term strategy compared to senolytics, particularly for chronic dosing in older adults, since they avoid the risks associated with cell ablation and regeneration imbalance [69,70].

Table 2. Overview of Major Senomorphic Agents in Aging and Disease Modulation.

Agent/Class	Primary Targets	Mechanism of Action	Current Status	Key Notes	Ref.
Rapamycin (Sirolimus)	mTORC1	Inhibits mTOR-mediated translation of SASP factors (e.g., IL-1α)	Multiple trials in aging (e.g., NCT02432287)	Shown to improve immune function, reduce inflammation	[71]
Ruxolitinib/ Baricitinib	JAK1/2	Blocks JAK-STAT signaling, suppressing IL-6/IL-8 mediated SASP amplification	Phase 2 trials for frailty, inflammation	Also used for myelofibrosis and rheumatoid arthritis	[72]
p38 MAPK Inhibitors	p38 MAPK	Reduces SASP via inhibition of upstream inflammatory signaling	Preclinical/early clinical	Reduces IL-6, TNF-α production in senescent cells	[73]
BET Inhibitors	BRD4, chromatin modifiers	Represses transcription of SASP-related genes by reducing enhancer/promoter activity	Preclinical	Emerging class with epigenetic modulation potential	[74]
NF-κB Inhibitors	NF-κB pathway (IKK complex)	Blocks transcription of pro-inflammatory cytokines central to the SASP	Mostly preclinical	Non-specific immunosuppression is a challenge	[75]
Glucocorticoids (e.g., Dexamethasone)	Glucocorticoid receptor/NF-κB	Represses SASP cytokine expression, suppresses general inflammation	Clinically approved; repurposing debated	Broad-spectrum effects; not ideal for long-term use	[76]

Table 2. *Cont.*

Agent/Class	Primary Targets	Mechanism of Action	Current Status	Key Notes	Ref.
Metformin	AMPK/NF-κB/mTOR	Indirectly suppresses SASP by activating AMPK, inhibiting mTOR, and dampening NF-κB	Widely used; TAME trial (NCT04245771) ongoing	Mild SASP modulation; favorable safety profile	[77]
Resveratrol	SIRT1/NF-κB	Activates SIRT1, inhibits NF-κB signaling and oxidative stress	Nutraceutical; limited clinical trials	Low potency; bioavailability limitations	[78]
HDAC Inhibitors	Histone deacetylases	Alters chromatin accessibility of inflammatory gene promoters	Preclinical/repurposing from oncology	Potential for selective SASP modulation	[79]

In addition, glucocorticoids, long used for their immunosuppressive effects, exhibit some senomorphic properties by downregulating SASP cytokine transcription through inhibition of NF-κB [80,81]. However, their broad-spectrum immunosuppression and significant side-effect profile limit their suitability for aging interventions. More targeted NF-κB inhibitors are under investigation, aiming to selectively block SASP-associated gene transcription while preserving beneficial immune function. Other experimental approaches include p38 MAPK inhibitors, which interfere with upstream signaling of SASP activation, and BET bromodomain inhibitors, which suppress SASP gene expression by modifying chromatin accessibility at pro-inflammatory loci [82].

Notably, the functional heterogeneity of the SASP presents both an opportunity and a challenge. While many SASP components are pro-inflammatory and deleterious, some elements are involved in tissue regeneration, immune surveillance, and fibrosis resolution. Therefore, future senomorphic strategies must aim for selective SASP modulation, suppressing harmful secretions while preserving or even enhancing beneficial paracrine signals [83]. Precision targeting of SASP regulators, possibly through AI-guided compound design or tissue-specific delivery systems, could offer a path forward in fine-tuning this complex phenotype [84].

By selectively modulating key components of the senescence-associated secretory phenotype, such as pro-inflammatory cytokines (e.g., IL-6, IL-1β), matrix metalloproteinases, and chemokines, it may be possible to preserve the beneficial functions of senescent cells, including tissue repair and tumor suppression, while minimizing their deleterious pro-aging effects. Advances in artificial intelligence and machine learning can accelerate the identification of small molecules or biologics that target specific SASP pathways, with greater predictive accuracy and reduced off-target toxicity. Simultaneously, innovations in nano-medicine and targeted delivery platforms, such as lipid nanoparticles, antibody-drug conjugates, or cell-penetrating peptides, are enabling the localization of senomorphic therapies to affected tissues or cell populations. This precision approach may help overcome the heterogeneity of senescent cell populations and the context-dependent nature of SASP, thereby reducing systemic side effects and improving therapeutic efficacy. Ultimately, such strategies could enable dynamic, tunable modulation of senescence-related inflammation, paving the way for more refined and personalized anti-aging interventions.

An important and increasingly recognized aspect of the senescence-associated secretory phenotype (SASP) is its time-dependent nature, which significantly influences its biological effects. The composition and functional consequences of the SASP evolve dynamically over time following the onset of senescence, and this temporal heterogeneity plays a crucial role in determining whether the SASP exerts beneficial or detrimental outcomes [85,86].

In the early stages of senescence, particularly during acute tissue responses or following oncogene activation, the SASP can exhibit pro-fibrotic and immunosuppressive characteristics. Early SASP components may include transforming growth factor-beta (TGF-β), vascular endothelial growth factor (VEGF), and various extracellular matrix (ECM) proteins, which promote fibrosis and tissue remodeling while concurrently suppressing certain immune responses [87]. In this context, the SASP may help maintain tissue integrity and dampen potentially excessive immune activation, an effect that can be protective in the short term. However, this transient immunosuppression may also create a local environment that allows for immune evasion, tissue scarring, or even tumor progression if senescent cells are not efficiently cleared [88].

As senescence becomes chronic and unresolved, the SASP transitions to a more pro-inflammatory profile, marked by the upregulation of interleukins such as IL-6, IL-8, and IL-1β, as well as matrix metalloproteinases (MMPs) [89]. These factors promote immune cell recruitment, stem cell dysfunction, and tissue degeneration. This shift contributes to the pathological remodeling of tissues, persistent inflammation, and the propagation of senescence to neighboring cells via paracrine signaling.

Understanding the temporal evolution of the SASP is therefore essential when designing therapeutic interventions. For instance, efforts to modulate the SASP (e.g., via senomorphics) should consider not only which SASP components to suppress but also when during the senescence timeline such modulation will be most effective or least disruptive to physiological repair mechanisms [90]. Therapeutic strategies that blunt early pro-fibrotic or immunosuppressive SASP elements without impairing beneficial acute functions may offer the most precise and context-sensitive benefits.

Senomorphics may also complement senolytic therapy in combinatorial regimens. For instance, transient senomorphic pre-treatment could suppress SASP-induced systemic toxicity before senolytic-induced cell clearance, or conversely, senomorphics could be administered post-senolysis to suppress SASP release from surviving or neighboring cells [21,91]. This dual-pronged approach is being actively explored in preclinical models, especially in the context of chemotherapy-induced senescence, metabolic dysfunction, and neuroinflammation.

As the field advances, a deeper mechanistic understanding of SASP regulation, alongside improved in vivo biomarkers of SASP burden, will be essential for rational drug development. Moreover, since the SASP varies depending on tissue context, senescence trigger, and duration, the concept of "context-specific senomorphism" may guide the design of next-generation interventions. In sum, while senolytics represent a bold approach to cellular rejuvenation, senomorphics offer a nuanced and adaptable strategy, one that may hold particular promise for the chronic management of age-related inflammation, frailty, and organ dysfunction without the risks associated with broad cellular depletion.

4. Challenges and Limitations

Despite growing enthusiasm and remarkable progress in the field of senescence-targeting therapies, numerous scientific and clinical challenges remain that hinder the broad implementation of senolytics and senomorphics in aging and disease-modifying interventions. A central obstacle is the inherent heterogeneity of senescent cells across tissues, time, and triggering stimuli. Senescence is not a single, uniform phenotype; instead, it is a context-dependent program influenced by cell type, senescence inducer, micro-environmental signals, and organismal age [92]. As a result, senescent cells exhibit varying dependencies on survival pathways, distinct SASP profiles, and divergent sensitivities to pharmacological agents. This diversity complicates the development of "universal" senolytic or senomorphic drugs and increases the risk of inconsistent efficacy across patient populations or disease models.

Equally limiting is the lack of specific, non-invasive biomarkers that can reliably identify senescent cell burden in vivo (Table 3). While markers such as $p16^{INK4a}$, $p21^{CIP1}$, SA-β-gal activity, and DNA damage foci (γH2AX) are widely used in experimental settings, they are not exclusive to senescence and are often inaccessible in clinical tissues.

Furthermore, no consensus exists on how to quantify systemic senescence load or monitor therapeutic response dynamically. This gap in biomarker development impedes clinical translation by making patient selection, efficacy tracking, and dose optimization difficult in human trials. Without validated biomarkers, it is also challenging to distinguish beneficial, transient senescence (such as that involved in wound repair) from chronic, deleterious senescence driving pathology.

Table 3. Challenges, Implications, and Possible Solutions for Senescence-Targeting Therapies.

Challenge	Description	Implications	Possible Solutions
Heterogeneity of Senescent Cells	Senescent cell features vary by tissue, trigger, and aging context	Limits development of one-size-fits-all senotherapeutics; variable drug response	Use tissue-specific profiling (e.g., single-cell omics); design context-dependent or combinatorial therapies
Lack of Specific Biomarkers	No robust, non-invasive markers to quantify senescent cell burden in vivo	Difficult to identify target patients, track efficacy, or determine optimal dosing	Develop circulating biomarkers, imaging tracers, and senescence-specific transcriptomic signatures
Safety Concerns of Senolytics	Off-target effects on non-senescent cells (e.g., platelets, immune cells)	Heightened toxicity risk, especially in frail elderly patients	Engineer targeted delivery systems (e.g., nanoparticles, prodrugs, ADCs); explore intermittent "hit-and-run" dosing
Limitations of Senomorphics	Broad-acting effects, immunomodulation, lack of clearance	Potential SASP rebound; unclear long-term benefits	Develop pathway-selective senomorphics; combine with senolytics or regenerative therapies
Context-Dependent Role of Senescence	Senescence contributes positively to tissue repair and tumor suppression in some settings	Risk of unintended tissue damage or impaired regeneration if senescent cells are eliminated indiscriminately	Adopt adaptive modulation strategies; tailor timing and duration of intervention to specific physiological contexts
Regulatory and Ethical Uncertainty	Aging is not a recognized medical indication; endpoints and trial designs lack standardization	Slows approval and integration into mainstream care; public skepticism	Define aging-related surrogate endpoints; develop ethical guidelines for preventive gerotherapeutics

The safety profile of senolytic agents represents another critical concern (Table 3). While senolytics aim to selectively eliminate senescent cells, many current compounds exhibit off-target effects, potentially harming non-senescent cells or vital immune populations [93]. For example, BCL-2 family inhibitors like navitoclax, although effective in clearing senescent cells, are associated with dose-limiting thrombocytopenia due to BCL-xL inhibition in platelets [94]. Similarly, broad-acting kinase inhibitors and pro-apoptotic agents can induce systemic toxicity, particularly when used chronically or in combination with other therapeutics [95]. This risk is especially pronounced in aged individuals, who are the primary target population but also the most vulnerable to adverse events due to multimorbidity, frailty, and polypharmacy. As such, improving the selectivity and delivery of senotherapeutics, potentially through tissue-targeted prodrugs, antibody–drug conjugates, or nanoparticle-based systems, remains an urgent priority.

In the case of senomorphics, while they offer a potentially safer approach by modulating rather than eliminating senescent cells, they present their own challenges (Table 3). Most senomorphic compounds exhibit pleiotropic effects, acting on multiple pathways beyond the SASP. This broad mechanism of action increases the risk of unintended consequences, including suppression of beneficial immune responses or interference with cellular stress responses required for tissue integrity. Moreover, long-term administration of senomorphics may create pharmacodynamic tolerance, necessitating escalating doses or combination therapies [96]. The fact that senomorphics do not remove senescent cells also raises the possibility of a rebound in SASP production upon treatment cessation, or continued accumulation of dysfunctional cells, potentially compromising long-term efficacy.

From a translational perspective, one of the most underappreciated barriers is the contextual role of senescence in physiology. Senescence has evolved as a beneficial program in certain settings, including embryogenesis, tissue repair, and tumor suppression. Therefore, indiscriminate suppression or clearance of senescent cells may impair regenerative responses, disrupt tissue homeostasis, or increase cancer risk in certain contexts. For example, in the skin or liver, transient senescence can coordinate repair processes via SASP-mediated recruitment of immune and progenitor cells. Removing these cells prematurely could lead to impaired healing or excessive fibrosis [97]. Understanding the temporal dynamics and reversibility of senescence is essential for ensuring that interventions are not only effective but also aligned with tissue-specific physiological needs.

Regulatory and ethical frameworks for senotherapeutics remain underdeveloped, reflecting the novelty and complexity of targeting aging biology. Unlike traditional therapeutics designed to treat acute or well-defined diseases, senotherapeutics are increasingly being proposed for preventive use to delay or mitigate age-related functional decline in largely asymptomatic individuals. This shift challenges conventional paradigms of clinical trial design, including the selection of endpoints, risk–benefit assessment, and duration of follow-up. The absence of standardized criteria for measuring aging outcomes, such as biological age, functional capacity, or resilience, further complicates regulatory evaluation. Without clear definitions of what constitutes clinical "success" in the context of aging interventions, trials may yield ambiguous results that are difficult to interpret or compare across studies.

To address these issues, a multi-stakeholder approach is urgently needed. Regulatory bodies such as the FDA and EMA should collaborate with geroscience researchers to define age-related surrogate endpoints that are biologically meaningful, clinically actionable, and acceptable for conditional approval. Longitudinal cohorts, biomarker validation studies, and real-world evidence must be leveraged to establish reliable predictors of therapeutic efficacy and long-term safety. Adaptive trial designs, including umbrella and platform trials, may also offer greater flexibility to test multiple agents or dosing strategies in aging populations.

Another critical barrier is the influence of public perception and commercial hype. The growing market for unregulated anti-aging products and exaggerated claims around longevity-enhancing interventions threatens to erode public trust and scientific integrity. To prevent this, stricter oversight of marketing practices, coupled with public education initiatives, is essential. Medical societies and public health institutions should play a central role in communicating the current evidence base, distinguishing between experimental and validated therapies, and ensuring informed decision-making by consumers.

Economic and logistical considerations further complicate the large-scale implementation of senotherapeutics. Aging affects virtually all individuals, implying that successful therapies could generate enormous demand across populations and age groups. This raises concerns about healthcare system capacity, affordability, and equitable access. To address these challenges, health technology assessments (HTAs) should be employed early in drug development to evaluate cost-effectiveness and inform pricing strategies. Moreover, targeting high-risk groups, such as individuals with multi-morbidity, frailty, or accelerated biological aging, may provide the most immediate and measurable benefits, enabling a phased and prioritized rollout of senescence-targeting interventions.

In summary, realizing the clinical potential of senotherapeutics will require more than scientific progress; it will demand regulatory innovation, ethical foresight, public engagement, and strategic health system planning. By addressing these interconnected challenges, the field can move toward the responsible, effective, and equitable integration of anti-aging therapies into mainstream medical care. These span biological uncertainty, technological limitations, clinical safety, and translational logistics. Addressing these barriers will require interdisciplinary collaboration, rigorous biomarker discovery, smarter drug delivery platforms, ethical oversight, and a cautious yet forward-thinking approach to clinical trial design. Only through these efforts can senotherapeutics fulfill their potential as transformative tools for extending health span and mitigating age-related disease burden.

5. Future Directions

Looking ahead, the future of senescence-targeting therapies lies in moving beyond proof-of-concept studies toward clinically actionable, safe, and personalized interventions that can meaningfully extend the human health span. As our understanding of cellular senescence becomes more nuanced, so too must our therapeutic strategies evolve, from blunt cell elimination to context-sensitive modulation, from broad-acting compounds to precision medicine, and from isolated interventions to integrative, systems-level approaches. The next generation of senotherapeutics will likely combine molecular specificity, spatiotemporal control, and functional synergy with other hallmarks of aging to achieve durable rejuvenation without compromising physiological senescence functions [98].

A key frontier in this evolution is the development of next-generation senolytics and senomorphics with enhanced tissue targeting, reduced off-target toxicity, and improved pharmacokinetics (Figure 2). Efforts are underway to engineer senolytic prodrugs that become activated only in senescent cells based on unique metabolic signatures, pH gradients, or enzymatic profiles. Similarly, antibody-drug conjugates (ADCs) that recognize senescence-specific surface markers, such as DPP4, uPAR, or B2M, offer a means of delivering senolytics selectively to affected tissues while sparing healthy cells [99]. Nanoparticle-based delivery systems also hold promise for increasing the precision and local concentration of senotherapeutic agents, especially in hard-to-reach tissues like the brain or bone marrow [100,101]. These approaches not only improve safety profiles but may also allow for combinatorial targeting of different senescent cell populations within the same organism.

At the same time, multi-omics profiling and single-cell technologies are beginning to revolutionize how senescent cells are identified, classified, and understood. Advances in transcriptomics, epigenomics, proteomics, and metabolomics now enable high-resolution mapping of senescence heterogeneity across tissues and disease states (Figure 1). These datasets are facilitating the construction of senescence "atlases" that will guide biomarker discovery, therapeutic targeting, and patient stratification. Moreover, the integration of artificial intelligence and machine learning into omics analysis is accelerating the identification of senescence-specific pathways and predictive biomarkers [102]. These tools will be instrumental in overcoming one of the field's greatest hurdles: the reliable, non-invasive detection and monitoring of senescent cell burden in humans.

In parallel, clinical trial design in geroscience must undergo innovation to accommodate the unique challenges of targeting aging biology. Traditional trial endpoints, such as disease incidence or short-term symptom resolution, may not fully capture the long-term benefits of senotherapeutics. Instead, surrogate endpoints like inflammatory biomarkers, composite frailty indices, epigenetic aging clocks, and digital mobility assessments may offer more sensitive and practical measures of therapeutic efficacy [103]. Adaptive trial designs, decentralized data collection, and real-world evidence frameworks could also help accelerate the regulatory approval process. Importantly, as interventions move from disease-specific indications toward preventive or health-span-enhancing applications, ethical and regulatory frameworks will need to be refined to ensure safety, equity, and accountability.

Beyond pharmacological strategies, combining senotherapeutics with regenerative medicine, immunotherapy, and lifestyle interventions may yield synergistic effects that restore tissue function while removing or suppressing cellular damage. For example, senolytic pre-treatment may improve the engraftment and efficacy of stem cell therapies by clearing the inflammatory milieu, while senomorphics may enhance the benefits of caloric restriction or exercise by reducing systemic SASP-driven inflammation [104]. The convergence of aging research with other cutting-edge domains, such as gene editing, autophagy modulation, and gut microbiome engineering, may unlock entirely new therapeutic paradigms for systemic rejuvenation.

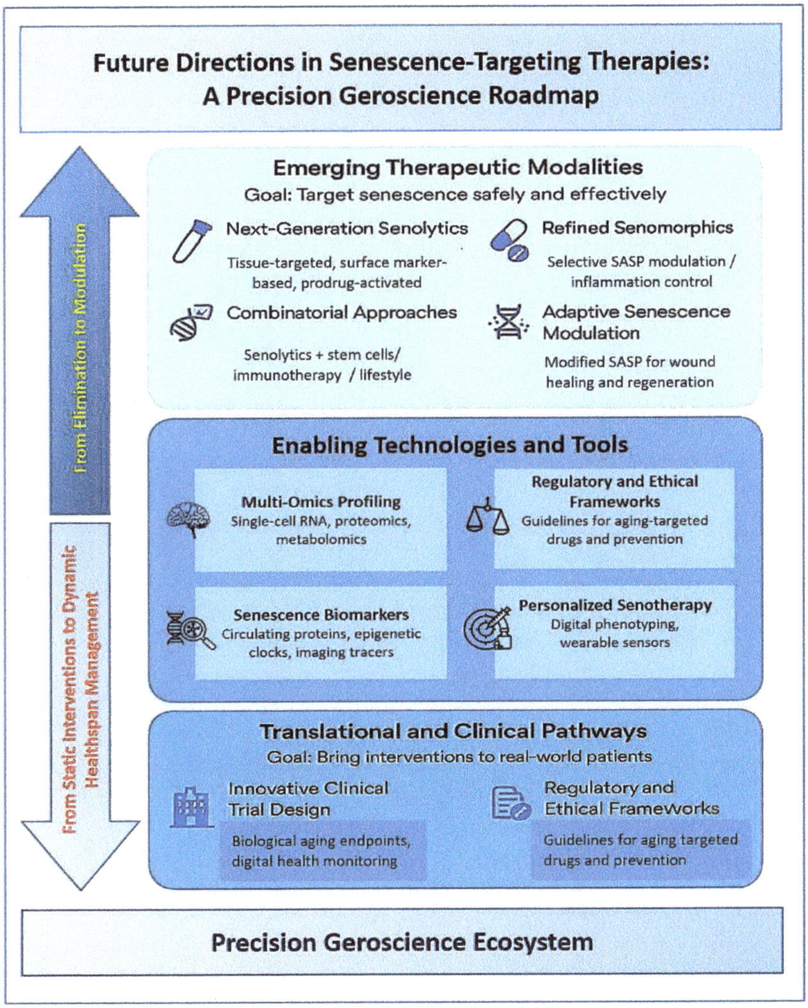

Figure 2. Future Directions in Senescence-Targeting Therapies: A Precision Geroscience Roadmap.

A critical and increasingly recognized frontier in senotherapeutic development involves targeting not only senescent cells themselves but also the microenvironmental context in which they reside, specifically, the aging stem cell niche. Evidence from multiple organ systems demonstrates that senescence within the niche profoundly influences tissue homeostasis, regenerative capacity, and response to therapy. In the neural stem cell niche of the subventricular zone, for instance, senescent astrocytes and microglia disrupt neurogenic signaling and foster a pro-inflammatory milieu that impairs neural regeneration. Similarly, in the hematopoietic stem cell (HSC) niche, age-associated changes in mesenchymal stromal cells, endothelial support cells, and the extracellular matrix reduce HSC self-renewal, skew lineage differentiation, and contribute to immunosenescence.

These insights underscore the need for niche-aware senotherapeutic strategies that address both senescent stem cells and the dysfunctional supportive cells that surround them. Precision senolytic or senomorphic interventions aimed at specific niche-resident populations, such as senescent glia in the CNS or aged stromal cells in the bone marrow, may help restore a pro-regenerative environment. Moreover, the use of localized delivery systems, including niche-targeted nanoparticles, hydrogels, or organ-specific prodrugs, can enhance therapeutic specificity and reduce systemic toxicity. Integrating spatial and biophysical mapping of stem cell niches into emerging senescence atlases, alongside single-cell multi-omics, will further refine our understanding of how senescence unfolds at the tissue level and guide the design of context-specific interventions.

Incorporating the niche perspective into senescence research and therapeutic development represents a vital step toward true precision geroscience. By addressing the organ-specific complexity of senescent cell interactions, future senotherapeutics will be better equipped to rejuvenate regenerative microenvironments and support long-term tissue repair, moving beyond cell elimination to holistic tissue revitalization.

Another promising direction is the concept of "adaptive senescence modulation", which recognizes that not all senescent cells are harmful at all times. Emerging strategies aim to temporally tune senescence responses rather than simply suppress them. For instance, short-term SASP activation may be allowed or even induced to promote regeneration or immune recruitment, followed by controlled silencing to avoid chronic inflammation [83]. This more sophisticated approach to senescence biology will require intelligent biomaterials, feedback-controlled drug delivery systems, and deeper integration of systems biology.

Looking further ahead, there is a growing interest in personalized senotherapy, tailored to an individual's senescence profile, genetic background, and environmental exposures. Just as oncology has moved toward personalized medicine based on tumor subtyping and molecular diagnostics, senescence-targeting interventions may one day be guided by individual aging trajectories mapped through deep longitudinal phenotyping. This shift will depend on the development of robust clinical senescence diagnostics, wearable biosensors for early detection of functional decline, and data platforms that integrate lifestyle, physiological, and molecular aging data.

Ultimately, the future of senescence-targeting interventions lies not in a single molecule or magic bullet, but in a precision geoscience ecosystem, one that harmonizes molecular therapies, diagnostics, digital health, and public policy to compress morbidity, reduces age-related disease burden and enables more people to live longer, healthier, and more functional lives. Realizing this vision will require sustained investment, cross-disciplinary collaboration, and careful ethical stewardship, but the scientific foundation is rapidly solidifying. If successfully implemented, senotherapeutics could represent a defining advance of 21st-century medicine, shifting our healthcare paradigm from reactive disease treatment to proactive health preservation across the lifespan.

6. Conclusions

Targeting cellular senescence has emerged as one of the most compelling strategies for addressing the biological underpinnings of aging and age-related diseases. The development of senolytics and senomorphics offers complementary approaches, either by clearing harmful senescent cells or by modulating their pro-inflammatory and degenerative secretory profiles. Robust preclinical evidence supports the therapeutic potential of these agents in mitigating frailty, improving tissue function, and extending health span in diverse animal models.

However, the path to widespread clinical application is complex. Challenges such as the heterogeneity of senescent cells, the dual role of senescence in physiology and pathology, and the lack of specific in vivo biomarkers underscore the need for cautious and sophisticated intervention strategies. Safety concerns, particularly with senolytic agents, must be rigorously addressed through improved drug design, targeted delivery technologies, and personalized treatment paradigms.

Future directions in the field are likely to include the development of next-generation senotherapeutics with enhanced specificity and minimal toxicity, integration with regenerative medicine and lifestyle interventions, and the implementation of precision senotherapy guided by molecular profiling and AI-driven diagnostics. Additionally, the convergence of geroscience with other domains, such as immunotherapy, epigenetics, and systems biology, will be crucial in creating holistic approaches to aging and longevity. In conclusion, while the translation of senescence-targeting therapies into routine clinical practice remains in its early stages, the conceptual and technological foundation is rapidly strengthening. Senolytics and senomorphics are poised not only to redefine the management of chronic diseases but also to shift the healthcare paradigm from disease treatment to health preservation, heralding a new era in the science of aging.

Author Contributions: T.S.: Writing of original draft, Methodology, Formal analysis, Methodology, Conceptualization. P.B.S.: Writing, Review and editing, Supervision, Project administration. All authors have read and agreed to the published version of the manuscript.

Funding: The study was supported by program-targeted funding from the Ministry of Science and Higher Education of the Republic of Kazakhstan "Prolonging healthy longevity: using new technologies and machine learning to control reversal of aging in old cells" (Grant No. BR24992900).

Data Availability Statement: Not applicable.

Acknowledgments: The authors are thankful to S.D. Asfendiyarov Kazakh National Medical University for the administrative and technical support provided.

Conflicts of Interest: The authors declare no conflicts of interest.

References

1. Xiao, R.P.; Hu, S.A.; Du, X.H.; Wang, Y.W.; Fang, K.; Zhu, Y.B.; Lou, N.B.; Yuan, C.H.; Yang, J. Revolutionizing Senescence Detection: Advancements from Traditional Methods to Cutting-Edge Techniques. *Aging Dis.* **2025**, *16*, 1285–1301. [CrossRef] [PubMed]
2. Li, Q.; Xiao, N.Y.; Zhang, H.; Liang, G.Y.; Lin, Y.; Qian, Z.H.; Yang, X.; Yang, J.K.; Fu, Y.G.; Zhang, C.T.; et al. Systemic aging and aging-related diseases. *FASEB J.* **2025**, *39*, e70430. [CrossRef] [PubMed]
3. Tamatta, R.; Pai, V.R.; Jaiswal, C.; Singh, I.; Singh, A.K. Neuroinflammaging and the Immune Landscape: The Role of Autophagy and Senescence in Aging Brain. *Biogerontology* **2025**, *26*, 52. [CrossRef]
4. García-Domínguez, M. Pathological and Infla mmatory Consequences of Aging. *Biomolecules* **2025**, *15*, 404. [CrossRef] [PubMed]
5. Wang, X.B.; Fukumoto, T.; Noma, K. Therapeutic strategies targeting cellular senescence for cancer and other diseases. *J. Biochem.* **2024**, *175*, 525–537. [CrossRef]
6. Balducci, L.; Falandry, C.; Monfardini, S. Senotherapy, cancer, and aging. *J. Geriatr. Oncol.* **2024**, *15*, 101671. [CrossRef]
7. Wang, B.S.; Han, J.; Elisseeff, J.H.; Demaria, M. The senescence-associated secretory phenotype and its physiological and pathological implications. *Nat. Rev. Mol. Cell Biol.* **2024**, *25*, 958–978. [CrossRef]
8. Noh, S.G.; Kim, H.W.; Kim, S.; Chung, K.W.; Jung, Y.S.; Yoon, J.H.; Yu, B.P.; Lee, J.; Chung, H.Y. Senoinflammation as the underlying mechanism of aging and its modulation by calorie restriction. *Ageing Res. Rev.* **2024**, *101*, 102503. [CrossRef]
9. Imawari, Y.; Nakanishi, M. Senescence and senolysis in cancer: The latest findings. *Cancer Sci.* **2024**, *115*, 2107–2116. [CrossRef]
10. Wang, S.; Huo, T.; Lu, M.; Zhao, Y.; Zhang, J.; He, W.; Chen, H. Recent Advances in Aging and Immunosenescence: Mechanisms and Therapeutic Strategies. *Cells* **2025**, *14*, 499. [CrossRef]
11. Williams, Z.J.; Chow, L.; Dow, S.; Pezzanite, L.M. The potential for senotherapy as a novel approach to extend life quality in veterinary medicine. *Front. Vet. Sci.* **2024**, *11*, 1369153. [CrossRef] [PubMed]
12. Samiminemati, A.; Aprile, D.; Siniscalco, D.; Di Bernardo, G. Methods to Investigate the Secretome of Senescent Cells. *Method Protocol.* **2024**, *7*, 52. [CrossRef]
13. Konstantinou, E.; Longange, E.; Kaya, G. Mechanisms of Senescence and Anti-Senescence Strategies in the Skin. *Biology* **2024**, *13*, 647. [CrossRef]
14. Kondratyeva, L.G.; Matveeva, D.K.; Ezdakova, M.I.; Utkina, M.V.; Ratushnyy, A.Y. The influence of the geroprotective cytokine on the transcriptome of young and senescent mesenchymal stem cells. *BMC Res. Notes* **2025**, *18*, 195. [CrossRef] [PubMed]
15. Wyles, S.P.; Yu, G.T.; Gold, M.; Behfar, A. Topical Platelet Exosomes Reduce Senescence Signaling in Human Skin: An Exploratory Prospective Trial. *Dermatol. Surg.* **2024**, *50*, S160–S165. [CrossRef] [PubMed]

16. Nishizawa, H.; Matsumoto, M.; Yamanaka, M.; Irikura, R.; Nakajima, K.; Tada, K.; Nakayama, Y.; Konishi, M.; Itoh, N.; Funayama, R.; et al. BACH1 inhibits senescence, obesity, and short lifespan by ferroptotic FGF21 secretion. *Cell Rep.* **2024**, *43*, 114403. [CrossRef]
17. Poblocka, M.; Bassey, A.L.; Smith, V.M.; Falcicchio, M.; Manso, A.S.; Althubiti, M.; Sheng, X.; Kyle, A.; Barber, R.; Frigerio, M.; et al. Targeted clearance of senescent cells using an antibody-drug conjugate against a specific membrane marker. *Sci. Rep.* **2021**, *11*, 20358. [CrossRef]
18. Zumerle, S.; Sarill, M.; Saponaro, M.; Colucci, M.; Contu, L.; Lazzarini, E.; Sartori, R.; Pezzini, C.; Rinaldi, A.; Scanu, A.; et al. Targeting senescence induced by age or chemotherapy with a polyphenol-rich natural extract improves longevity and healthspan in mice. *Nat. Aging* **2024**, *4*, 1231–1248. [CrossRef]
19. Dorronsoro, A.; Santiago, F.E.; Grassi, D.; Zhang, T.; Lai, R.C.; McGowan, S.J.; Angelini, L.; Lavasani, M.; Corbo, L.; Lu, A.; et al. Mesenchymal stem cell-derived extracellular vesicles reduce senescence and extend health span in mouse models of aging. *Aging Cell* **2021**, *20*, e13337. [CrossRef]
20. Sun, W.T.; Gao, Y.; Wu, Y.B.; Wu, W.; Wang, C.F.; Chen, J.X.; Luan, C.J.; Hua, M.; Liu, W.L.; Gong, W.J.; et al. Targeted apoptosis of senescent cells by valproic acid alleviates therapy-induced cellular senescence and lung aging. *Phytomedicine* **2024**, *135*, 156131. [CrossRef]
21. Mansfield, L.; Ramponi, V.; Gupta, K.; Stevenson, T.; Mathew, A.B.; Barinda, A.J.; Herbstein, F.; Morsli, S. Emerging insights in senescence: Pathways from preclinical models to therapeutic innovations. *NPJ Aging* **2024**, *10*, 53. [CrossRef] [PubMed]
22. Liu, B.; Peng, Z.; Zhang, H.; Zhang, N.; Liu, Z.; Xia, Z.; Huang, S.; Luo, P.; Cheng, Q. Regulation of cellular senescence in tumor progression and therapeutic targeting: Mechanisms and pathways. *Mol. Cancer* **2025**, *24*, 106. [CrossRef]
23. Li, C.; Yuan, Y.; Jia, Y.; Zhou, Q.; Wang, Q.; Jiang, X. Cellular senescence: From homeostasis to pathological implications and therapeutic strategies. *Front. Immunol.* **2025**, *16*, 1534263. [CrossRef]
24. Melo dos Santos, L.S.; Trombetta-Lima, M.; Eggen, B.J.L.; Demaria, M. Cellular senescence in brain aging and neurodegeneration. *Ageing Res. Rev.* **2024**, *93*, 102141. [CrossRef] [PubMed]
25. Zeng, Y.; Buonfiglio, F.; Li, J.Y.; Pfeiffer, N.; Gericke, A. Mechanisms Underlying Vascular Inflammaging: Current Insights and Potential Treatment Approaches. *Aging Dis.* **2025**. *ahead of print*.
26. Stojanovic, S.D.; Thum, T.; Bauersachs, J. Anti-senescence therapies: A new concept to address cardiovascular disease. *Cardiovasc. Res.* **2025**, *121*, 730–747. [CrossRef]
27. Qin, Y.; Liu, H.X.; Wu, H.L. Cellular Senescence in Health, Disease, and Lens Aging. *Pharmaceuticals* **2025**, *18*, 244. [CrossRef] [PubMed]
28. Burdusel, D.; Doeppner, T.R.; Surugiu, R.; Hermann, D.M.; Olaru, D.G.; Popa-Wagner, A. The Intersection of Epigenetics and Senolytics in Mechanisms of Aging and Therapeutic Approaches. *Biomolecules* **2025**, *15*, 18. [CrossRef] [PubMed]
29. Shah, J.; Al-Hashimi, A.; Benedetto, M.; Ruchaya, P.J. From bench to bedside: The critical need for standardized senescence detection. *Arch. Cardiovasc. Dis.* **2025**, *118*, 205–211. [CrossRef]
30. Baier, M.P.; Ranjit, R.; Owen, D.B.; Wilson, J.L.; Stiles, M.A.; Masingale, A.M.; Thomas, Z.; Bredegaard, A.; Sherry, D.M.; Logan, S. Cellular Senescence Is a Central Driver of Cognitive Disparities in Aging. *Aging Cell* **2025**. *ahead of print*. [CrossRef]
31. Gadecka, A.; Nowak, N.; Bulanda, E.; Janiszewska, D.; Dudkowska, M.; Sikora, E.; Bielak-Zmijewska, A. The senolytic cocktail, dasatinib and quercetin, impacts the chromatin structure of both young and senescent vascular smooth muscle cells. *Geroscience* **2025**. *ahead of print*. [CrossRef]
32. Tang, J.R.; Li, J.Y.; Hou, Z.Y.; He, R.; Li, B.Z.; Gong, J.J.; Xie, Y.H.; Meng, W.R.; Liu, Y.K.; Ouchi, T.; et al. Dasatinib and Quercetin Mitigate Age-Related Alveolar Bone Inflammaging and Neutrophil Infiltration. *Oral. Dis.* **2025**. *ahead of print*. [CrossRef] [PubMed]
33. Heinrichs, D.P.; Maldonado, V.V.; Ardana, I.K.K.; Porter, R.M.; Samsonraj, R.M. Assessing the Effects of Dasatinib on Mesenchymal Stem/Stromal Cells. *Cell Mol. Bioeng.* **2024**, *17*, 609–618. [CrossRef]
34. Maurer, S.; Kirsch, V.; Ruths, L.; Brenner, R.E.; Riegger, J. Senolytic therapy combining Dasatinib and Quercetin restores the chondrogenic phenotype of human osteoarthritic chondrocytes by the release of pro-anabolic mediators. *Aging Cell* **2025**, *24*, e14361. [CrossRef]
35. Greenberg, E.F.; Voorbach, M.J.; Smith, A.; Reuter, D.R.; Zhuang, Y.C.; Wang, J.Q.; Wooten, D.W.; Asque, E.; Hu, M.; Hoft, C.; et al. Navitoclax safety, tolerability, and effect on biomarkers of senescence and neurodegeneration in aged nonhuman primates. *Heliyon* **2024**, *10*, e36483. [CrossRef]
36. Born, E.; Lipskaia, L.; Breau, M.; Houssaini, A.; Beaulieu, D.; Marcos, E.; Pierre, R.; Do Cruzeiro, M.; Lefevre, M.; Derumeaux, G.; et al. Eliminating Senescent Cells Can Promote Pulmonary Hypertension Development and Progression. *Circulation* **2023**, *147*, 650–666. [CrossRef]
37. Hassan, J.W.; Bhatwadekar, A.D. Senolytics in the treatment of diabetic retinopathy. *Front. Pharmacol.* **2022**, *13*, 896907. [CrossRef]
38. Chin, A.F.; Han, J.; Clement, C.C.; Choi, Y.; Zhang, H.; Browne, M.; Jeon, O.H.; Elisseeff, J.H. Senolytic treatment reduces oxidative protein stress in an aging male murine model of post-traumatic osteoarthritis. *Aging Cell* **2023**, *22*, 13979. [CrossRef]
39. Aleksandrova, Y.; Neganova, M. Antioxidant Senotherapy by Natural Compounds: A Beneficial Partner in Cancer Treatment. *Antioxidants* **2025**, *14*, 199. [CrossRef]

40. Guo, Y.L.; Fenwick, P.; Viola, P.; Al-Sahaf, M.; Barnes, P.J.; Donnelly, L.E. Navitoclax inhibits senescence in human alveolar type 2 cells from COPD patients. *Eur. Respir. J.* **2024**, *64*, PA897. [CrossRef]
41. Agha-Mir-Salim, D.; Bhayadia, R.; Heckl, D.; Klusmann, J.H. Evaluation of Navitoclax (ABT-263) as a promising Therapy for Pediatric Acute Myeloid Leukemia. *Klin. Padiatr.* **2024**, *236*, 200. [CrossRef]
42. Kolodkin-Gal, D.; Roitman, L.; Ovadya, Y.; Azazmeh, N.; Assouline, B.; Schlesinger, Y.; Kalifa, R.; Horwitz, S.; Khalatnik, Y.; Hochner-Ger, A.; et al. Senolytic elimination of Cox2-expressing senescent cells inhibits the growth of premalignant pancreatic lesions. *Gut* **2022**, *71*, 345–355. [CrossRef]
43. Rad, A.N.; Grillari, J. Current senolytics: Mode of action, efficacy and limitations, and their future. *Mech. Ageing Dev.* **2024**, *217*, 111888. [CrossRef]
44. Liu, Y.; Hou, Q.H.; Wang, R.; Liu, Y.; Cheng, Z.S. FOXO4-D-Retro-Inverso targets extracellular matrix production in fibroblasts and ameliorates bleomycin-induced pulmonary fibrosis in mice. *Arch. Pharmacol.* **2023**, *396*, 2393–2403. [CrossRef]
45. Xu, X.F.; Lou, Z.L.; Li, J.S.; Liang, F.X.; Yu, Y.F.; Wu, M. Inhibition of Hsp90 Alleviates Necroptosis and Inflammation in Lung Epithelial Cells During Pulmonary Ischemia-Reperfusion Injury. *Clin. Exp. Pharmacol. Physiol.* **2025**, *52*, e70037. [CrossRef]
46. Qin, M.T.; Ren, J.; Chen, X.D.; Zhou, W.; Zhang, S.Y.; Zhang, W.L.; Shi, M.X.; Zhang, M.Z.; Liu, H.S.; Ma, Y.F.; et al. CD19-targeted HSP90 inhibitor nanoparticle combined with TKIs reduces tumor burden and enhances T-cell immunity in murine B-cell malignancies. *Theranostics* **2025**, *15*, 3589–3609. [CrossRef]
47. Yang, Z.Y.; Zhang, J.X.; Yuan, Q.Y.; Wang, X.Y.; Zeng, W.D.; Mi, Y.L.; Zhang, C.Q. Flavonoid Fisetin Alleviates Ovarian Aging of Laying Chickens by Enhancing Antioxidant Capacity and Glucose Metabolic Homeostasis. *Antioxidants* **2024**, *13*, 1432. [CrossRef] [PubMed]
48. Wang, X.G.; Zhang, X.Y.; Liao, Q.Y.; Rui, X.L.; Wang, R. Study on the Mechanism of Raspberry (Rubi fructus) in Treating Type 2 Diabetes Based on UPLC-Q-Exactive Orbitrap MS, Network Pharmacology, and Experimental Validation. *Phytochem. Anal.* **2025**, *36*, 744–758. [CrossRef]
49. Das, G.; Kameswaran, S.; Ramesh, B.; Bangeppagari, M.; Nath, R.; Das Talukdar, A.; Shin, H.S.; Patra, J.K. Anti-Aging Effect of Traditional Plant-Based Food: An Overview. *Foods* **2024**, *13*, 3785. [CrossRef]
50. Lamichhane, G.; Liu, J.; Lee, S.J.; Lee, D.Y.; Zhang, G.L.; Kim, Y. Curcumin Mitigates the High-Fat High-Sugar Diet-Induced Impairment of Spatial Memory, Hepatic Metabolism, and the Alteration of the Gut Microbiome in Alzheimer's Disease-Induced (3xTg-AD) Mice. *Nutrients* **2024**, *16*, 240. [CrossRef] [PubMed]
51. Riessland, M.; Orr, M.E. Translating the Biology of Aging into New Therapeutics for Alzheimer's Disease: Senolytics. *J. Prev. Alzheimer's Dis.* **2023**, *10*, 633–646. [CrossRef] [PubMed]
52. Poisa-Beiro, L.; Landry, J.J.M.; Yan, B.; Kardorff, M.; Eckstein, V.; Villacorta, L.; Krammer, P.H.; Zaugg, J.; Gavin, A.-C.; Benes, V.; et al. A Senescent Cluster in Aged Human Hematopoietic Stem Cell Compartment as Target for Senotherapy. *Int. J. Mol. Sci.* **2025**, *26*, 787. [CrossRef] [PubMed]
53. Sarad, K.; Jankowska, U.; Skupien-Rabian, B.; Babler, A.; Kramann, R.; Dulak, J.; Jaźwa-Kusior, A. Senescence of endothelial cells promotes phenotypic changes in adventitial fibroblasts: Possible implications for vascular aging. *Mol. Cell Biochem.* **2024**, *480*, 1027–1043. [CrossRef] [PubMed]
54. Chang, M.Y.; Dong, Y.; Cruickshank-Taylor, A.B.; Gnawali, G.; Bi, F.C.; Wang, W. Senolytic Prodrugs: A Promising Approach to Enhancing Senescence-Targeting Intervention. *Chembiochem* **2024**, *25*, e202400355. [CrossRef] [PubMed]
55. Martín-Vicente, P.; López-Martínez, C.; Rioseras, B.; Albaiceta, G.M. Activation of senescence in critically ill patients: Mechanisms, consequences and therapeutic opportunities. *Ann. Intensive Care* **2024**, *14*, 2. [CrossRef]
56. Imb, M.; Véghelyi, Z.; Maurer, M.; Kühnel, H. Exploring Senolytic and Senomorphic Properties of Medicinal Plants for Anti-Aging Therapies. *Int. J. Mol. Sci.* **2024**, *25*, 10419. [CrossRef]
57. Wan, T.; Fielder, E.; Alimohammadiha, G.; Ishaq, A.; Low, E.; Weigand, B.M.; Kelly, G.; Parker, C.; Griffin, B.; Jurk, D.; et al. Short senolytic or senostatic interventions rescue progression of radiation-induced frailty and premature ageing in mice. *eLife* **2022**, *11*, 75492. [CrossRef]
58. Mikawa, T.; Yoshida, K.; Kondoh, H. Senotherapy preserves resilience in aging. *Geriatr. Gerontol. Int.* **2024**, *24*, 845–849. [CrossRef]
59. McHugh, D.; Durán, I.; Gil, J. Senescence as a therapeutic target in cancer and age-related diseases. *Nat. Rev. Drug Discov.* **2025**, *24*, 57–71. [CrossRef]
60. Ya, J.; Bayraktutan, U. Senolytics and Senomorphics Targeting p38MAPK/NF-κB Pathway Protect Endothelial Cells from Oxidative Stress-Mediated Premature Senescence. *Cells* **2024**, *13*, 1292. [CrossRef]
61. Kita, A.; Yamamoto, S.; Saito, Y.; Chikenji, T.S. Cellular senescence and wound healing in aged and diabetic skin. *Front. Physiol.* **2024**, *15*, 1344116. [CrossRef]
62. Gallage, S.; Irvine, E.E.; Avila, J.E.B.; Reen, V.; Pedroni, S.M.A.; Duran, I.; Ranvir, V.; Khadayate, S.; Pombo, J.; Brookes, S.; et al. Ribosomal S6 kinase 1 regulates inflammaging via the senescence secretome. *Nat. Aging* **2024**, *4*, 1544–1561. [CrossRef] [PubMed]
63. Lin, H.; Salech, F.; Lim, A.; Vogrin, S.; Duque, G. The effect of rapamycin and its analogues on age-related musculoskeletal diseases: A systematic review. *Aging Clin. Exp. Res.* **2022**, *34*, 2317–2333. [CrossRef]

64. Selvarani, R.; Mohammed, S.; Richardson, A. Effect of rapamycin on aging and age-related diseases-past and future. *Geroscience* **2021**, *43*, 1135–1158. [CrossRef]
65. Politano, D.; Tonduti, D.; Battini, R.; Fazzi, E.; Orcesi, S. Exploring emerging JAK inhibitors in the treatment of Aicardi-Goutières syndrome. *Expert. Opin. Emerg. Drugs* **2025**, *30*, 21–39. [CrossRef]
66. Hu, Z.H.; Lu, L.; Feng, J.D.; Song, H.B.; Zhang, S.Y.; Yang, L.; Liu, Y.H.; Wang, T. Low-Dose Baricitinib Plus Narrow-Band Ultraviolet B for the Treatment of Progressive Non-Segmental Vitiligo: A Prospective, Controlled, Open-Label Study. *Pigment. Cell Melanoma Res.* **2025**, *38*, 13209. [CrossRef]
67. Utama, A.; Wijesinghe, R.; Thng, S. Janus kinase inhibitors and the changing landscape of vitiligo management: A scoping review. *Int. J. Dermatol.* **2024**, *63*, 1020–1035. [CrossRef] [PubMed]
68. Kim, S.; Ruminski, P.; Singh, M.; Staser, K.; Ashami, K.; Ritchey, J.; Lim, S.; Dipersio, J.F.; Choi, J. Novel JAK Inhibitors to Reduce Graft-Versus-Host Disease after Allogeneic Hematopoietic Cell Transplantation in a Preclinical Mouse Model. *Molecules* **2024**, *29*, 1801. [CrossRef] [PubMed]
69. Solignac, M.; Cabrera, N.; Fouillet-Desjonqueres, M.; Duquesne, A.; Laurent, A.; Foray, A.P.; Viel, S.; Zekre, F.; Belot, A. JAK inhibitors in refractory juvenile rheumatic diseases: Efficacy, tolerance and type-I interferon profiling, a single center retrospective study. *J. Autoimmun.* **2024**, *147*, 103248. [CrossRef]
70. Mansilla-Polo, M.; Morgado-Carrasco, D. Biologics Versus JAK Inhibitors. Part I: Cancer Risk. A Narrative Review. *Dermatology Ther.* **2024**, *14*, 1389–1442. [CrossRef]
71. Chrienova, Z.; Rysanek, D.; Oleksak, P.; Stary, D.; Bajda, M.; Reinis, M.; Mikyskova, R.; Novotny, O.; Andrys, R.; Skarka, A.; et al. Discovery of small molecule mechanistic target of rapamycin inhibitors as anti-aging and anti-cancer therapeutics. *Front. Aging Neurosci.* **2022**, *14*, 1048260. [CrossRef] [PubMed]
72. Frede, N.; Lorenzetti, R.; Hüppe, J.M.; Janowska, I.; Troilo, A.; Schleyer, M.T.; Venhoff, A.C.; Voll, R.E.; Thiel, J.; Venhoff, N.; et al. JAK inhibitors differentially modulate B cell activation, maturation and function: A comparative analysis of five JAK inhibitors in an B cell differentiation model and in patients with rheumatoid arthritis. *Front. Immunol.* **2023**, *14*, 1087986. [CrossRef] [PubMed]
73. Yang, F.W.; Zhao, L.J.; Xu, Q.L.; Zhao, J.H. The journey of p38 MAP kinase inhibitors: From bench to bedside in treating inflammatory diseases. *Eur. J. Med. Chem.* **2024**, *280*, 116950. [CrossRef]
74. Leal, A.S.; Liby, K.T. The BRD4 Inhibitor I-BET-762 Reduces HO-1 Expression in Macrophages and the Pancreas of Mice. *Int. J. Mol. Sci.* **2024**, *25*, 9985. [CrossRef]
75. Cheng, J.J.; Wu, L.; Chen, X.W.; Li, S.; Sun, R.J.; Huang, Y.W.; Wang, P.; Ouyang, J.W.; Pei, P.P.; Yang, H.C.; et al. Polo-like kinase 2 promotes microglial activation via regulation of the HSP90a/IKKR a /IKK R pathway. *Cell Rep.* **2024**, *43*, 114827. [CrossRef]
76. Thrikawala, S.U.; Anderson, M.H.; Rosowski, E.E. Glucocorticoids Suppress NF-kB-Mediated Neutrophil Control of Hyphal Growth. *J. Immunol.* **2024**, *213*, 2400021. [CrossRef] [PubMed]
77. Sui, Q.H.; Yang, H.Q.; Hu, Z.Y.; Jin, X.; Chen, Z.C.; Jiang, W.; Sun, F.H. The Research Progress of Metformin Regulation of Metabolic Reprogramming in Malignant Tumors. *Pharm. Res.* **2024**, *41*, 2143–2159. [CrossRef]
78. Tshivhase, A.M.; Matsha, T.; Raghubeer, S. Resveratrol attenuates high glucose-induced inflammation and improves glucose metabolism in HepG2 cells. *Sci. Rep.* **2024**, *14*, 1106. [CrossRef]
79. Makgoba, T.B.; Kapp, E.; Egieyeh, S.; Joubert, J. HDAC3 inhibitors: A patent review of their broad-spectrum applications as therapeutic agents. *Expert Opin. Ther. Pat.* **2024**, *34*, 273–295. [CrossRef]
80. Han, Z.Q.; Wang, L.J.; Xu, S.Y.; Zhang, H.R.; Cheng, J.; Pan, S.F. Microvesicle-Shuttled microRNA-130b Activates the Hepatic Inflammation by Inhibiting Glucocorticoid-Receptor-Mediated Immunosuppression in High-Fat Diet-Induced Obese Mice. *Vet. Sci.* **2024**, *11*, 565. [CrossRef]
81. Meng, Q.Q.; Bai, M.; Guo, M.L.; Li, Z.X.; Liu, W.W.; Fan, X.J.; Sun, R.; Yang, X.R.; Yuan, D.F.; Shi, Y.L.; et al. Inhibition of Serum- and Glucocorticoid-Regulated Protein Kinase-1 Aggravates Imiquimod-Induced Psoriatic Dermatitis and Enhances Proinflammatory Cytokine Expression through the NF-kB Pathway. *J. Investig. Dermatol.* **2023**, *143*, 954–964. [CrossRef] [PubMed]
82. Bansal, A.; Mostafa, M.M.; Kooi, C.; Sasse, S.K.; Michi, A.N.; Shah, S.V.; Leigh, R.; Gerber, A.N.; Newton, R. Interplay between nuclear factor-KB, p38 MAPK, and glucocorticoid receptor signaling synergistically induces functional TLR2 in lung epithelial cells. *J. Biol. Chem.* **2022**, *298*, 101747. [CrossRef] [PubMed]
83. Alqahtani, S.; Alqahtani, T.; Venkatesan, K.; Sivadasan, D.; Ahmed, R.; Sirag, N.; Elfadil, H.; Abdullah Mohamed, H.; TA, H.; Elsayed Ahmed, R.; et al. SASP Modulation for Cellular Rejuvenation and Tissue Homeostasis: Therapeutic Strategies and Molecular Insights. *Cells* **2025**, *14*, 608. [CrossRef]
84. Tang, Q.Q.; Xiao, D.Y.; Veviorskiy, A.; Xin, Y.; Lok, S.W.Y.; Pulous, F.E.; Zhang, P.R.; Zhu, Y.F.; Ma, Y.M.; Hu, X.; et al. AI-Driven Robotics Laboratory Identifies Pharmacological TNIK Inhibition as a Potent Senomorphic Agent. *Aging Dis.* **2026**. *ahead of print*. [CrossRef]
85. Malaquin, N.; Rodier, F. In *Cellular Senescence and Aging*; Dynamic and scalable assessment of the senescence-associated secretory phenotype (SASP). Academic Press, Elsevier: London, UK, 2024; pp. 181–195. [CrossRef]

86. Giroud, J.; Bouriez, I.; Paulus, H.; Pourtier, A.; Debacq-Chainiaux, F.; Pluquet, O. Exploring the Communication of the SASP: Dynamic, Interactive, and Adaptive Effects on the Microenvironment. *Int. J. Mol. Sci.* **2023**, *24*, 10788. [CrossRef] [PubMed]
87. Huang, W.; Hickson, L.J.; Eirin, A.; Kirkland, J.L.; Lerman, L.O. Cellular senescence: The good, the bad and the unknown. *Nat. Rev. Nephrol.* **2022**, *18*, 611–627. [CrossRef]
88. Giannoula, Y.; Kroemer, G.; Pietrocola, F. Cellular senescence and the host immune system in aging and age-related disorders. *Biomed. J.* **2023**, *46*, 100581. [CrossRef]
89. Wang, L.; Hong, W.; Zhu, H.; He, Q.; Yang, B.; Wang, J.; Weng, Q. Macrophage senescence in health and diseases. *Acta Pharm. Sin. B* **2024**, *14*, 1508–1524. [CrossRef] [PubMed]
90. Sun, Y.; Li, Q.; Kirkland, J.L. Targeting senescent cells for a healthier longevity: The roadmap for an era of global aging. *Life Med.* **2022**, *1*, 103–119. [CrossRef]
91. Xia, W.; Chen, H.; Yang, H.; Zhu, L.; Xie, C.; Hou, M. Depletion of SASP senescent cardiomyocytes with senolytic drugs confers therapeutic effects in doxorubicin-related cardiotoxicity. *FEBS J.* **2024**, *291*, 4029–4042. [CrossRef]
92. Kumari, R.; Jat, P. Mechanisms of Cellular Senescence: Cell Cycle Arrest and Senescence Associated Secretory Phenotype. *Front. Cell Dev. Biol.* **2021**, *9*, 645593. [CrossRef]
93. Revskij, D.; Woitas, A.; Koelle, B.; Umstätter, C.; Zechner, D.; Khan, F.M.; Fuellen, G.; Jaster, R. Effects of triggers of senescence and senolysis in murine pancreatic cancer cells. *Hepatobiliary Pancreat. Dis.* **2024**, *23*, 628–637. [CrossRef]
94. McCorkle, J.R.; Ahn, R.; Cao, C.D.; Hill, K.S.; Dietrich, C.S.; Kolesar, J.M. Antineoplastic Drug Synergy of Artesunate with Navitoclax in Models of High-Grade Serous Ovarian Cancer. *Cancers* **2024**, *16*, 1321. [CrossRef] [PubMed]
95. Narayanan, J.; Tamilanban, T.; Kumar, P.S.; Guru, A.; Muthupandian, S.; Kathiravan, M.K.; Arockiaraj, J. Role and mechanistic actions of protein kinase inhibitors as an effective drug target for cancer and COVID. *Arch. Microbiol.* **2023**, *205*, 238. [CrossRef]
96. Chaib, S.; Tchkonia, T.; Kirkland, J.L. Cellular senescence and senolytics: The path to the clinic. *Nat. Med.* **2022**, *28*, 1556–1568. [CrossRef] [PubMed]
97. Ring, N.A.R.; Valdivieso, K.; Grillari, J.; Redl, H.; Ogrodnik, M. The role of senescence in cellular plasticity: Lessons from regeneration and development and implications for age-related diseases. *Dev. Cell* **2022**, *57*, 1083–1101. [CrossRef] [PubMed]
98. Fu, T.E.; Zhou, Z. Senescent cells as a target for anti-aging interventions: From senolytics to immune therapies. *J. Transl. Intern. Med.* **2025**, *13*, 33–47. [CrossRef]
99. Zhang, L.; Pitcher, L.E.; Yousefzadeh, M.J.; Niedernhofer, L.J.; Robbins, P.D.; Zhu, Y. Cellular senescence: A key therapeutic target in aging and diseases. *J. Clin. Investig.* **2022**, *132*, e158450. [CrossRef]
100. Bi, J.; Zeng, J.; Liu, X.; Mo, C.; Yao, M.; Zhang, J.; Yuan, P.; Jia, B.; Xu, S. Drug delivery for age-related bone diseases: From therapeutic targets to common and emerging therapeutic strategies. *Saudi Pharm. J.* **2024**, *32*, 102209. [CrossRef]
101. Wilar, G.; Suhandi, C.; Wathoni, N.; Fukunaga, K.; Kawahata, I. Nanoparticle-Based Drug Delivery Systems Enhance Treatment of Cognitive Defects. *Int. J. Nanomed.* **2024**, *19*, 11357–11378. [CrossRef]
102. Lin, M.; Guo, J.; Gu, Z.; Tang, W.; Tao, H.; You, S.; Jia, D.; Sun, Y.; Jia, P. Machine learning and multi-omics integration: Advancing cardiovascular translational research and clinical practice. *J. Transl. Med.* **2025**, *23*, 388. [CrossRef]
103. Silva, N.; Rajado, A.T.; Esteves, F.; Brito, D.; Apolónio, J.; Roberto, V.P.; Binnie, A.; Araújo, I.; Nóbrega, C.; Bragança, J.; et al. Measuring healthy ageing: Current and future tools. *Biogerontology* **2023**, *24*, 845–866. [CrossRef] [PubMed]
104. Ji, S.; Xiong, M.; Chen, H.; Liu, Y.; Zhou, L.; Hong, Y.; Wang, M.; Wang, C.; Fu, X.; Sun, X. Cellular rejuvenation: Molecular mechanisms and potential therapeutic interventions for diseases. *Signal Transduct. Target. Ther.* **2023**, *8*, 116. [CrossRef] [PubMed]

biomolecules

MDPI

Review

Pathological and Inflammatory Consequences of Aging

Mario García-Domínguez [1,2,3]

[1] Program of Immunology and Immunotherapy, CIMA-Universidad de Navarra, 31008 Pamplona, Spain; mgdom@unav.es
[2] Department of Immunology and Immunotherapy, Clínica Universidad de Navarra, 31008 Pamplona, Spain
[3] Centro de Investigación Biomédica en Red de Cáncer (CIBERONC), 28029 Madrid, Spain

Abstract: Aging is a complex, progressive, and irreversible biological process that entails numerous structural and functional changes in the organism. These changes affect all bodily systems, reducing their ability to respond and adapt to the environment. Chronic inflammation is one of the key factors driving the development of age-related diseases, ultimately causing a substantial decline in the functional abilities of older individuals. This persistent inflammatory state (commonly known as "inflammaging") is characterized by elevated levels of pro-inflammatory cytokines, an increase in oxidative stress, and a perturbation of immune homeostasis. Several factors, including cellular senescence, contribute to this inflammatory milieu, thereby amplifying conditions such as cardiovascular disease, neurodegeneration, and metabolic disorders. Exploring the mechanisms of chronic inflammation in aging is essential for developing targeted interventions aimed at promoting healthy aging. This review explains the strong connection between aging and chronic inflammation, highlighting potential therapeutic approaches like pharmacological treatments, dietary strategies, and lifestyle changes.

Keywords: inflammation; inflammaging; aging; proteostasis; telomere shortening; cellular senescence; immunosenescence; adipaging; parthanatos

1. Introduction

The process of aging constitutes a multifaceted and intricate phenomenon characterized by a gradual deterioration of several biological functions [1,2], thereby rendering humans more susceptible to a range of age-associated diseases [3]. This mechanism involves many molecular, cellular, and systemic alterations that progressively impair overall functionality [4–6]. At the molecular level, aging is linked to genomic instability [7], resulting in the accumulation of DNA damage [8]. Consequently, the DNA damage leads to telomere shortening [9] and epigenetic alterations that regulate gene expression [10]. On the other hand, aging is linked to strong reduction in proteostasis, which impairs cells' ability to maintain a functional proteome [11]. Additionally, mitochondrial dysfunction occurs, leading to a decrease in energy production and an increased generation of reactive oxygen species (ROS) that damage cellular structures [12].

These changes result in cellular senescence, a state in which cells cease to divide but remain metabolically active [13]. Senescent cells exhibit a senescence-associated secretory phenotype (SASP), characterized by the secretion of pro-inflammatory cytokines (such as IL-1α, IL-1β, IL-6, IL-8, and IL-18), growth factors (e.g., VEGF, TGF-β, GM-CSF, IGFBP-2, and GDF-15), and proteases [14–16]. The accumulation of senescent cells within tissues is now widely recognized as a key driver of aging and the development of many conditions like cardiovascular diseases, neurodegeneration, and cancer [17–19]. The sustained secretion of the mentioned SASP factors occupies an essential position in the formation and continuation of a chronic inflammatory milieu; this environment contributes to both peripheral and central inflammatory mechanisms [20].

At the peripheral level, the release of SASP mediators significantly amplifies systemic inflammation, triggering a cascade of physiopathological events that disrupt normal tissue homeostasis [21]. However, this persistent inflammatory state causes structural and functional alterations in tissues, driving pathological remodeling in organs such as the heart, liver, and kidneys. Pro-inflammatory mediators promote fibrosis and ischemic injury by inducing maladaptive repair processes while simultaneously impairing the regenerative processes necessary for maintaining tissue integrity. As a result, this situation creates a cycle of dysfunction that accelerates tissue damage and hinders proper recovery [22,23]. This biological process, known as inflammaging, is characterized by chronic low-grade inflammation associated with aging, sustained by the SASP mediators [24,25]. This condition not only perpetuates tissue damage but also amplifies the degenerative changes associated with aging, establishing a vicious cycle of inflammation and dysfunction that impedes tissue repair and accelerates the progression of age-related diseases.

At the central nervous system (CNS) level, SASP mediators can exert significant effects on the brain. One of the major consequences is the disruption of the blood–brain barrier (BBB), a critical defense mechanism that protects the brain from harmful substances present in the bloodstream [26]. The prolonged inflammation caused by SASP mediators can compromise the integrity of the BBB; thus, it facilitates the infiltration of peripheral immune cells, toxins, and pro-inflammatory cytokines into the CNS [27]. This dysregulation within the CNS not only initiates neuroinflammation but also strongly activates microglial cells. In their activated state, these cells release additional pro-inflammatory cytokines, further exacerbating neurodegenerative processes linked to aging such as Parkinson's and Alzheimer's diseases [28,29]. Microglial activation induces several effects, like synaptic dysfunction, neuronal cell death, and disruptions in neuronal signaling [30,31]. This aberrant inflammatory response may accelerate neuronal loss, impair cognitive function, and evoke the accumulation of toxic protein aggregates, such as β-amyloid plaques in Alzheimer's disease and α-synuclein in Parkinson's disease [32,33].

This review will provide a comprehensive analysis of the pathophysiological consequences of aging, emphasizing the molecular, cellular, and systemic alterations that drive age-related deterioration. Furthermore, this review will investigate the inflammatory processes linked to aging, both at the peripheral and central levels. Particular emphasis will be placed on the underlying mechanisms that drive inflammation, including immune cell activation, oxidative stress, and the accumulation of senescent cells. By synthesizing current research findings, this review aims to provide a deeper understanding of the interplay between aging and inflammation, shedding light on potential therapeutic interventions to mitigate the adverse effects of these processes.

2. Aging: Physiopathological Consequences

Aging is a process that leads to functional physiological changes, including increased susceptibility to many chronic conditions [3]. The complex physiopathological processes linked to aging influence multiple body systems, such as the cardiovascular, nervous, immune, and musculoskeletal systems, thereby elevating the risk of neurodegeneration, cardiovascular diseases, and metabolic disorders, among others [34]. Gaining a deep understanding of the physiological consequences of aging is essential for developing effective therapeutic approaches that not only support healthy aging but also mitigate the impact of age-related diseases. As the population ages, it becomes increasingly important to identify and address the underlying biological mechanisms that contribute to the decline in organ function, immune system efficiency, and cognitive abilities. By prioritizing these processes, researchers can design interventions that promote longevity, enhance quality of life, and reduce the burden of chronic diseases, ultimately helping older individuals maintain their independence and well-being as they age [35].

2.1. Nervous System

The human body undergoes numerous alterations within its CNS throughout the aging process, leading to changes in cognitive functions, memory capabilities, and overall brain health status [36]. A key factor in this process is the increase in oxidative stress. On a molecular level, oxidative stress is critical, as aging neurons accumulate ROS due to mitochondrial dysfunction [37]. The brain experiences elevated levels of malondialdehyde (MDA) along with lipid peroxidation products such as 4-hydroxy-2-nonenal (4-HNE), which induce DNA damage and promote lipid peroxidation and protein misfolding factors involved in neurodegenerative processes [38,39].

Telomere shortening is a fundamental marker of aging that amplifies neuronal vulnerability by restricting the regenerative potential of neural stem cells and reducing synaptic plasticity while also inhibiting neurogenesis in the hippocampus, a crucial brain region for learning and memory [40,41]. Conversely, the balance of proteostasis, involving protein synthesis, folding, and degradation, is disrupted during aging as the functional efficiency of both the ubiquitin–proteasome system and autophagy declines [42]. This deficiency facilitates the accumulation of protein aggregates such as β-amyloid plaques seen in Alzheimer's disease (Figure 1) and α-synuclein aggregates found in Parkinson's disease [32,33].

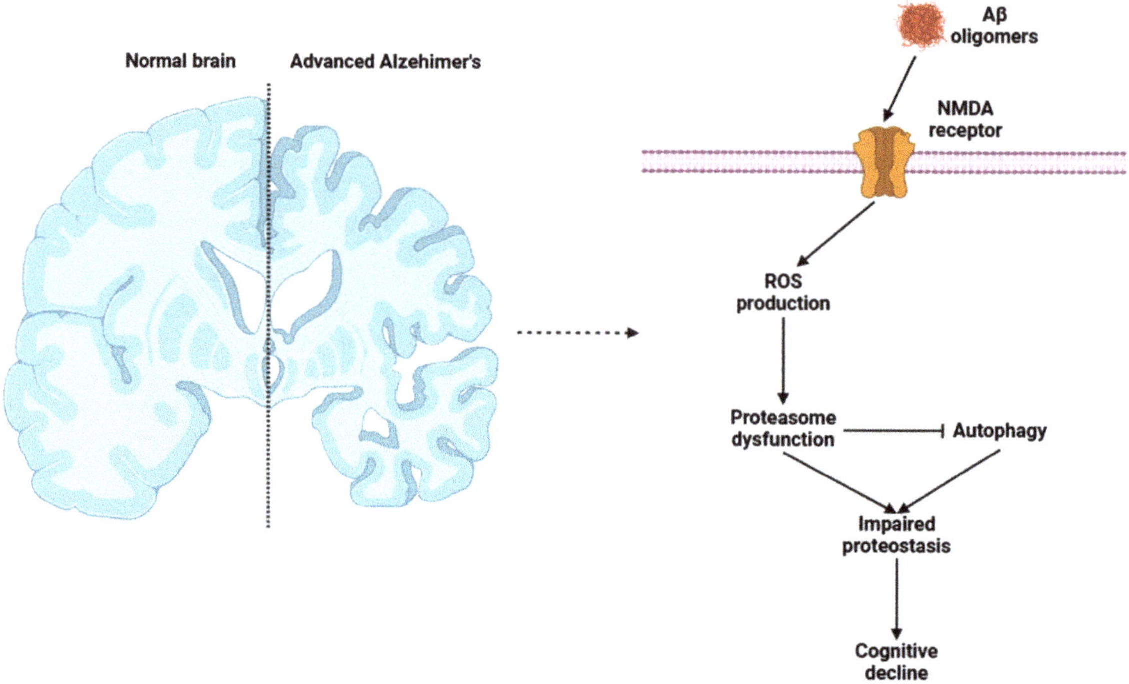

Figure 1. Cellular mechanisms that occur in the CNS neurons of individuals with Alzheimer's disease. Abbreviations: NMDA (N-methyl-D-aspartate) and ROS (reactive oxygen species).

Furthermore, reduced levels of neurotransmitters such as dopamine, acetylcholine, and serotonin can disrupt signal transmission, resulting in several problems with motor coordination, mood regulation, and cognitive function [43]. The myelin sheath is likewise affected by dysfunctions in oligodendrocytes, resulting in delayed neural conduction and increased susceptibility to demyelinating diseases [44]. Neuroinflammation, primarily driven by activated microglia and astrocytes, accelerates neural aging through the release of various pro-inflammatory cytokines such as IL-6, IL-1β, and TNF-α [45]. These cytokines can impair synaptic function and contribute to neuronal degeneration [46]. Furthermore, aging influences several signaling pathways that are fundamental for normal neuronal function, including the PI3K-AKT and MAPK pathways [47]. Additionally, the age-related decline in BBB integrity aggravates neurodegenerative conditions by facilitating the infiltration of peripheral immune cells and toxins into the central nervous system, triggering chronic inflammation and neuronal injury [27].

On the other hand, the buildup of advanced glycation end-products (AGEs) also negatively impacts neuronal function by causing oxidative stress and persistent inflammation [48]. Finally, the meninges, the protective layers surrounding the brain and spinal cord, undergo age-related changes in gene expression that may modify the brain's microenvironment and contribute to age-related neurological decline [49].

2.2. *Cardiovascular System*

The aging process profoundly affects the cardiovascular system through a complex interplay of molecular and cellular mechanisms, resulting in structural and functional alterations that compromise both cardiac and vascular integrity [50]. At the molecular level, collagen accumulation and impaired cardiomyocyte function play a critical role in the development of left ventricular hypertrophy and myocardial fibrosis [51]. Additionally, mitochondrial dysfunction, which involves decreased ATP production and the increased generation of ROS, disrupts cellular energy balance and worsens oxidative stress [52]. The blood vessels also experience significant transformations, such as reduced nitric oxide availability due to lower activity of endothelial nitric oxide synthase (eNOS) and increased nitric oxide depletion from ROS [53]. Chronic low-grade inflammation, driven by the SASP phenotype, along with impaired angiogenesis resulting from diminished vascular endothelial growth factor (VEGF) signaling, further exacerbates cardiovascular dysfunction [54].

Telomere shortening in vascular cells accelerates cellular senescence and increases the risk of apoptosis [55]. Epigenetic alterations, including changes in DNA methylation, histone modifications, and dysregulated microRNA expression, modulate gene activity linked to oxidative stress, fibrosis, and inflammation [56]. The accumulation of AGEs promotes the crosslinking of extracellular matrix proteins, leading to increased arterial stiffness and diastolic dysfunction [57]. Additionally, impaired protein maintenance and autophagy lead to the accumulation of misfolded proteins and damaged organelles, accelerating cellular dysfunction and disease progression [58].

2.3. *Gastrointestinal System*

The aging process has a significant effect on the gastrointestinal (GI) system, leading to functional decline and a higher risk of diseases through various molecular mechanisms [59]. One major factor is the degeneration of neuromuscular function, which includes the loss of interstitial cells of Cajal and a reduction in acetylcholine signaling, resulting in impaired peristalsis and contributing to constipation [60,61]. Esophageal issues, particularly the weakening of the lower esophageal sphincter, are influenced by changes in nitric oxide signaling, which raises the risk of gastroesophageal reflux disease (GERD) [62]. Furthermore, gastric acid secretion decreases due to atrophic gastritis, which is frequently linked to inflammation caused by *Helicobacter pylori* and an increase in pro-inflammatory cytokines like IL-1β, IL-2, IL-4, IL-6, IL-8, IL-10, IFN-γ, and TNF-α [63]. The reduction in acid secretion can interfere with nutrient absorption [64].

In the pancreas, mitochondrial dysfunction and endoplasmic reticulum stress lead to a significant reduction in enzyme secretion from the pancreatic islets, resulting in the malabsorption of macronutrients [65,66]. The liver also suffers from decreased blood flow and reduced activity of cytochrome P450 enzymes, especially CYP3A4, which affects drug metabolism and raises the risk of medication toxicity [67]. On the other hand, the colon undergoes structural weakening due to strong alterations in extracellular matrix remodeling, driven by matrix metalloproteinases, thereby increasing the risk of diverticulosis and inflammatory complications [68]. The risk of colorectal cancer also rises due to accumulated DNA damage, telomere shortening, and many mutations in tumor suppressor genes such as *TP53*, worsened by the chronic activation of NF-κB and the overproduction of prostaglandin E2 [69,70]. Moreover, polypharmacy can disrupt gut health, as NSAIDs inhibit COX-1 and raise the risk of ulcers, while antibiotics can lead to dysbiosis, increasing susceptibility to *Clostridioides difficile* infections [71,72]. Dehydration, resulting from reduced sensitivity of hypothalamic osmoreceptors, exacerbates colonic transit, leading to more severe constipation and increased systemic inflammation [73].

Ultimately, the gut–brain axis plays a critical role in the aging process, impacting cognitive function, mental health, and overall well-being [74]. The gut microbiome, a complex community of trillions of microorganisms residing in the digestive tract, promotes bidirectional communication with the brain via the gut–brain axis, an interconnected network that includes the vagus nerve, immune system, and neuroactive compounds such as neurotransmitters and short-chain fatty acids (SCFAs) [75]. With age, the gut microbiome undergoes profound changes, often exhibiting decreased diversity due to factors such as an imbalanced diet, medication use (particularly antibiotics), and decreased physical activity [76]. Alterations in the gut microbiota, characterized by a strong reduction in *Bifidobacteria* and *Lactobacillus* populations, activate Toll-like receptor (TLR) pathways and compromise gut barrier integrity by downregulating tight junction proteins such as occludin and claudin-1 [77–79]. This situation contributes to systemic inflammation, immune dysregulation, and cognitive decline [80]. Additionally, the gut microbiota regulates the production of some neurotransmitters, including serotonin and dopamine, which participate in mood regulation and cognitive function [81]. Finally, aging is associated with a reduction in gut barrier integrity, resulting in increased intestinal permeability, which enables harmful substances to enter the bloodstream and trigger immune responses that could accelerate neurodegeneration [82].

2.4. Respiratory System

As individuals age, the respiratory system undergoes several physiological changes that progressively reduce pulmonary function, increasing susceptibility to respiratory illnesses and impairing oxygenation efficiency [83,84]. The primary age-related alteration is the gradual loss of lung elasticity, driven by structural modifications in elastin and collagen fibers within lung tissue [85], due to the deposition of collagen and the degradation of elastin fibers. This change is caused by an imbalance between matrix metalloproteinases (MMPs), enzymes responsible for degrading extracellular matrix (ECM) components, and their tissue inhibitors (TIMPs) [86], which disrupt the structural integrity of the lungs, leading to decreased lung compliance and a reduction in alveolar surface area [87].

In the alveolar region, aging is associated with a reduction in the production of surfactants (such as dipalmitoylphosphatidylcholine—DPPC) and several surfactant proteins [88]. This leads to an altered composition of the surfactant, resulting in a less efficient reduction in surface tension in the alveolar sacs. Consequently, this alteration in surfactant composition not only compromises alveolar integrity but also hinders effective gas exchange by reducing the surface area [89].

The chest wall becomes increasingly rigid due to costal cartilage calcification and structural changes in the intercostal muscles, limiting thoracic expansion and decreasing the capacity for deep inhalation [90]. The diaphragm may also experience functional decline due to age-related sarcopenia, further weakening inspiratory and expiratory forces [91]. Moreover, aging negatively impacts pulmonary immune defense mechanisms, including a significant reduction in mucociliary clearance, thereby increasing vulnerability to respiratory infections such as pneumonia and chronic obstructive pulmonary disease (COPD) [92]. Finally, the sensitivity of central and peripheral chemoreceptors responsible for detecting fluctuations in blood oxygen and carbon dioxide levels diminishes with age, leading to a blunted ventilatory response to hypoxia and hypercapnia [93].

2.5. Urogenital System

In the urinary tract, age-related changes manifest as a decrease in renal mass, particularly in the renal cortex, leading to a reduction in the number of functioning nephrons [94,95]. This decline in the nephron population contributes to a gradual decrease in the glomerular filtration rate and renal blood flow, impairing the kidneys' ability to concentrate urine, regulate fluid balance, and maintain electrolyte homeostasis [96]. As a result, older adults may experience diminished renal clearance of waste products, making them more susceptible to dehydration, drug accumulation, and electrolyte imbalances, which can further exacerbate comorbid conditions such as hypertension and diabetes [97,98]. Additionally, the dysregulation of the renin–angiotensin–aldosterone system (RAAS) and endothelial dysfunction, marked by reduced nitric oxide production, further compromise renal function in older individuals [99,100].

The bladder also undergoes significant structural and functional changes with aging. A decrease in detrusor muscle mass, along with increased collagen deposition in the bladder wall, leads to reduced bladder compliance and impaired contractility [101]. These alterations, combined with numerous age-related neurological changes affecting autonomic bladder control, contribute to lower bladder capacity and an increased prevalence of urinary symptoms like frequency, urgency, nocturia, and incomplete emptying [102]. Additionally, age-related weakening of the urethral sphincter and pelvic floor muscles in men and women increases the risk of urinary incontinence, which can impact the overall quality of life and psychological well-being [103,104].

In men, the prostate gland commonly enlarges with age due to benign prostatic hyperplasia (BPH), which may produce urinary obstruction, hesitancy, a weak stream, and post-void dribbling [105]. In BPH, the overexpression of HSP70 subfamily members contributes to cell survival, proliferation, and epithelial–mesenchymal transition, promoting prostate enlargement [106]. Oxidative stress plays a fundamental role, with a considerable accumulation of ROS resulting in mitochondrial dysfunction and, subsequently, cellular damage [107]. Additionally, the activation of the CXCL12/CXCR4 axis moderates prostate myofibroblast phenoconversion through non-canonical EGFR/MEK/ERK signaling, contributing to tissue fibrosis [108]. If BPH is left untreated, severe cases lead to chronic urinary retention and an increased risk of urinary tract infections, bladder stones, and kidney damage [109–111]. In contrast, the urogenital system in older women undergoes numerous changes, particularly post menopause, due to reducing estrogen levels [112]. This hormonal shift significantly contributes to the atrophy of the vaginal and urethral epithelia, leading to dyspareunia, vaginal dryness, increased susceptibility to bacterial and fungal infections, and decreased urethral resistance, thereby increasing the risk of stress and incontinence [113]. The loss of estrogen also accelerates connective tissue degeneration, increasing the risk of pelvic organ prolapse [114].

The impact of aging extends to the reproductive system, where a progressive decline in fertility is observed in both sexes [115]. In men, testicular function diminishes, resulting in reduced testosterone production, a lower sperm count, and a decrease in sperm motility, which can contribute to subfertility [116]. Although older men retain reproductive potential throughout life, these changes may lead to decreased sexual function, including reduced libido and an increased prevalence of erectile dysfunction, often influenced by vascular, neurological, and several metabolic factors [117,118]. Women experience a more abrupt transition with menopause, marked by the cessation of ovarian function and reproductive capability [119]. The depletion of ovarian follicles leads to a substantial decline in estrogen and progesterone levels, which may reduce antioxidant capacity, thereby increasing oxidative damage. This process promotes a pro-inflammatory state characterized by elevated cytokine production, including IL-6 and TNF-α, with systemic effects on bone density and cardiovascular function [119–121].

2.6. Endocrine System

The aging process profoundly affects the endocrine system, resulting in significant physiological and biochemical changes that disrupt homeostasis, metabolism, and overall health [122]. At a molecular level, aging results in the dysregulation of hormonal signaling pathways, often manifested by diminished hormone production, reduced receptor sensitivity, and impaired feedback systems [123]. A critical factor in this decline is the gradual deterioration of hypothalamic function, which leads to a decrease in the secretion of regulatory hormones such as gonadotropin-releasing hormone (GnRH), growth hormone-releasing hormone (GHRH), and corticotropin-releasing hormone (CRH) [124]. This reduction leads to lower levels of peripheral hormones, including growth hormone (GH), insulin-like growth factor 1 (IGF-1), and sex steroid hormones (androgens, estrogens, and progestogens) [125].

Several factors, such as cellular senescence, oxidative stress, and inflammaging, further worsen endocrine dysfunction by impairing the responsiveness of endocrine glands [126]. Furthermore, aging induces certain post-translational modifications (PTMs), leading to alterations in proteins that affect cellular signaling, enzymatic activity, and hormone receptor functionality, further contributing to endocrine decline [127]. These PTMs include glycation, a non-enzymatic process in which reducing sugars, such as glucose and fructose, covalently bind to proteins, lipids, or nucleic acids [128]. This process results in the formation of advanced glycation end-products (AGEs), a diverse group of molecular complexes that progressively accumulate with age [129]. AGEs play a key role in cellular dysfunction, particularly in the endocrine system, as they can modify hormone receptors, disrupt their normal functioning, and alter the structure of circulating hormones [130].

The aging of pancreatic β-cells is associated with mitochondrial impairment, endoplasmic reticulum stress, and epigenetic changes, which collectively diminish insulin secretion and heighten the risk of developing type 2 diabetes [131]. In the adrenal glands, age-related modifications include disrupted cortisol secretion due to altered sensitivity to adrenocorticotropic hormone (ACTH), exacerbating stress responses and metabolic irregularities [132]. Similarly, the reduction in dehydroepiandrosterone (DHEA) secretion negatively impacts immune function and tissue repair [133]. On the other hand, changes in thyroid function, particularly the reduced bioconversion of thyroxine (T4) to triiodothyronine (T3), lead to metabolic slowdowns and greater vulnerability to age-related illnesses [134].

2.7. Musculoskeletal System

The aging process evokes a cascade of structural and functional changes that significantly reduce mobility, strength, and overall quality of life [135]. These modifications are driven by a complex interplay of mechanisms that together contribute to the progressive decline in musculoskeletal health. Among the most prominent of these factors is oxidative stress. Oxidative damage affects mitochondria, decreasing ATP production and impairing muscle function while reducing cellular energy. Moreover, the decline in repair mechanisms with age worsens the buildup of damaged molecules, organelles, and tissues [136,137].

In skeletal muscle, one of the hallmark consequences of aging is sarcopenia, characterized by the progressive loss of muscle mass and function [138]. Sarcopenia is driven by several factors, including a decrease in the number and size of muscle fibers, particularly fast-twitch type II fibers, which are responsible for rapid, high-intensity muscle contractions. This decline in muscle fibers results in reduced strength and endurance, making daily activities more difficult [139]. The ability of muscle tissue to repair itself also diminishes with age, due in part to a decline in satellite cell function [140]. Satellite cells play a critical role in muscle regeneration and repair; however, with advancing age, their ability to proliferate and differentiate into new muscle fibers decreases substantially [141]. Aging also leads to an imbalance between muscle protein synthesis and protein degradation. The reduced biosynthesis of muscle proteins and the increased breakdown of existing muscle proteins contribute to muscle atrophy [142]. Moreover, age-related hormonal changes, including a reduction in GH and testosterone levels, further impair muscle anabolism and regeneration, intensifying the overall loss of muscle mass and function [143].

Simultaneously, aging has significant consequences for bone health. In bone tissue, aging leads to an imbalance between bone formation and resorption, favoring bone loss and increasing the risk of fractures [144]. Osteoblasts, the cells responsible for bone formation, decrease in number and activity with age, while osteoclasts, the cells responsible for breaking down bone, either remain unchanged or increase in activity [145]. Oxidative stress is a key factor in this process, where increased levels of ROS cause cellular damage and induce apoptosis in osteocytes [146]. Simultaneously, there is a reduction in the osteogenic capacity of bone marrow stromal cells, indicated by a reduced expression of Runx2, an essential regulator of osteoblast differentiation [147]. The alteration in bone metabolism is also influenced by age-related changes in systemic hormones, such as estrogen, testosterone, and growth hormone, as well as local growth factors that regulate bone turnover [148,149]. Another contributor to age-related bone fragility is the accumulation of AGEs in collagen fibers within the bone matrix, leading to the increased rigidity and decreased elasticity of collagen. This process causes the bone matrix to become more brittle and more prone to fractures, further weakening bone strength [150].

The ECM also undergoes several changes with age in both muscle and bone [151,152]. In muscle tissue, the ECM becomes increasingly fibrotic, which reduces its elasticity and impairs its ability to support muscle regeneration and contraction [153]. Muscle fibrosis limits the capacity of muscle tissue to contract efficiently, contributing to reduced strength and function [154]. Additionally, changes in the composition and function of MMPs contribute to the remodeling of muscle tissue and the degradation of some structural proteins [155]. Similarly, in bone tissue, aging leads to diverse alterations in collagen crosslinking and biomineralization patterns, influencing the mechanical properties of bone [156]. These changes in the bone ECM are driven by shifts in the expression and activity of MMPs, further contributing to the increased fragility of bones in older individuals [157].

In muscle and bone tissues, epigenetic changes can disrupt the production of essential proteins necessary for tissue regeneration and maintenance, accelerating the decline in tissue function [158]. The accumulation of senescent cells in muscle and bone tissues releases several pro-inflammatory cytokines and MMPs, which aggravate tissue degeneration and impair the repair capacity of the musculoskeletal system [159].

Finally, as individuals age, the connection between the brain and muscles becomes progressively crucial for maintaining overall physical health and functional independence [160]. The brain regulates muscle movement through a network of neurons that transmit electrical signals from the CNS to activate muscle fibers and coordinate voluntary movements. However, with aging, this communication system may become less efficient [161]. The motor cortex, which is responsible for planning and executing movement, may experience a reduction in both the quantity and effectiveness of neural connections, leading to slower reaction times, impaired coordination, and a decreased ability to perform complex motor tasks [162]. The decline in neural and muscular functions elevates the risk of falls, diminishes mobility, and adversely impacts quality of life [163].

3. Chronic Inflammation in Aging

The relationship between aging and peripheral inflammation represents a complex and multifactorial process, with many molecular mechanisms contributing to a prolonged state of chronic low-grade inflammation, also called inflammaging [24,25,164–166]. In contrast to acute inflammation, which is a transient response to infection or injury [167], inflammaging is a persistent, low-grade inflammatory state that develops as a result of the combined influence of internal and external factors accumulated throughout life. This process is characterized by sustained immune pathway activation, the increased production of pro-inflammatory cytokines, and the dysregulation of immune homeostasis, all of which contribute to the progressive functional decline associated with aging [168].

Aging affects multiple peripheral organs, including the liver, adipose tissue, skeletal muscles, and gastrointestinal tract, all of which play a crucial role in modulating systemic inflammation [17–19]. The progressive dysfunction of these organs with age is primarily caused by molecular and cellular alterations, including oxidative stress, genomic instability, epigenetic changes, mitochondrial impairment, and cellular senescence [1,2,169]. All of these create an inflammatory microenvironment that evokes tissue damage, ultimately contributing to the onset and progression of many age-related diseases, including cardiovascular disorders, neurodegenerative conditions, and cancer [170–172]. At the molecular level, inflammaging involves a complex network of inflammatory mediators, including cytokines, acute-phase proteins, and DAMPs, which activate various intracellular signaling pathways [7–12,173].

A defining feature of inflammaging is the SASP phenotype. As aging occurs, senescent cells accumulate in several tissues, promoting a pro-inflammatory environment that supports immune system activation and drives tissue remodeling [13–16,174]. An additional factor in inflammaging is gut microbiota dysbiosis, which has become increasingly recognized as a significant regulator of systemic inflammation in aging individuals [175]. Age-related alterations in gut microbiota composition can result in increased intestinal permeability, facilitating the translocation of bacterial endotoxins, such as lipopolysaccharide (LPS), into the circulation [176]. This process, called metabolic endotoxemia, triggers sustained immune cell activation, further enhancing systemic inflammation [177].

Nevertheless, chronic inflammation in the CNS of older adults is primarily mediated by the accumulation of SASP mediators, which can cross the BBB and activate glial cells, thereby contributing to the detrimental effects of the inflammatory process. These factors not only exacerbate the neuroinflammatory response but also create a cycle of sustained immune activation that further impairs brain health, making it a significant factor in age-related neurological disorders [26,27].

3.1. Peripheral Inflammation

Chronic inflammation is considered one of the hallmarks of aging, and a key player in this process is the accumulation of senescent cells [20]. Cellular senescence is a state in which cells cease proliferation, frequently in response to many stressors, including DNA damage, telomere shortening, and oxidative stress [7–12]. These cells show high metabolic activity; they lose their ability to replicate, which makes them resistant to apoptotic signals [13]. Senescent cells (Figure 2) adopt an active state, undergoing some changes in gene expression rather than remaining inactive, thereby promoting the senescence-associated secretory phenotype (SASP). Senescent cells secrete several pro-inflammatory cytokines, including IL-6 and TNF-α, both of which play essential roles in inflammation and immune responses [178,179]. In addition to IL-6 and TNF-α, SASP factors encompass other molecules, such as IL-1β and IL-8, chemokines (e.g., CCL2 and CCL4), MMPs (e.g., MMP-3 and MMP-13), and growth factors (e.g., TGF-β and IGF-1), amplifying local inflammation and contributing to tissue remodeling and degradation [178,180–182]. As people age, the activity of the ubiquitin–proteasome system (UPS) decreases, leading to the strong accumulation of polyubiquitinated proteins. This decline is associated with structural changes in the proteasome, the reduced expression of proteasome subunits, and oxidative damage to proteasomal components [183]. In parallel, both autophagy and chaperone-mediated autophagy (CMA) show age-related dysfunctions, leading to a decreased formation of autophagic vacuoles and delayed fusion of autophagosomes with lysosomes, worsening the inflammatory process [184]. Furthermore, the accumulation of ROS stimulates the DNA damage response (DDR) pathway, leading to the stabilization of p53 and the upregulation of p16 and p21 [185]. The activation of p16 and p21 leads to a permanent and stable cell cycle arrest, which is a critical characteristic of senescent cells [186]. This growth arrest is mediated through the inhibition of cyclin-dependent kinases, particularly CDK4/6, leading to the inhibition of retinoblastoma protein (Rb) phosphorylation, a vital regulator of the G1/S transition in the cell cycle [187]. Furthermore, the activation of TLRs and the NLRP3 inflammasome by DAMPs released from senescent cells further amplifies the inflammatory response [188].

A key characteristic of SASP mediators is their capacity to evoke autocrine and paracrine signaling, where the signals released by senescent cells affect healthy cells [185,189]. For instance, the pro-inflammatory cytokines secreted by senescent cells can promote the activation of immune cells, such as macrophages and T lymphocytes, which in turn release more pro-inflammatory molecules, exacerbating the inflammatory environment [190,191]. The presence of SASP mediators in some tissues, such as adipose tissue, skeletal muscle, liver, and the vasculature, has been linked to several major age-related diseases [192]. In atherosclerosis, senescent cells located in the blood vessel walls contribute to the chronic inflammation that underlies plaque formation [14]. Furthermore, in osteoarthritis, senescent cells situated in the cartilage secrete factors that degrade the ECM, leading to joint dysfunction and pain [193]. Although senescence serves as a protective function by preventing the proliferation of damaged cells, its presence has detrimental long-term effects. The aging immune system becomes less efficient at clearing senescent cells, which leads to the exacerbation of inflammation [194]. In this context, aging significantly influences the clearance of damaged cells. The reduced effectiveness of immunosurveillance leads to a decreased ability of the body to identify and eliminate damaged cells [195].

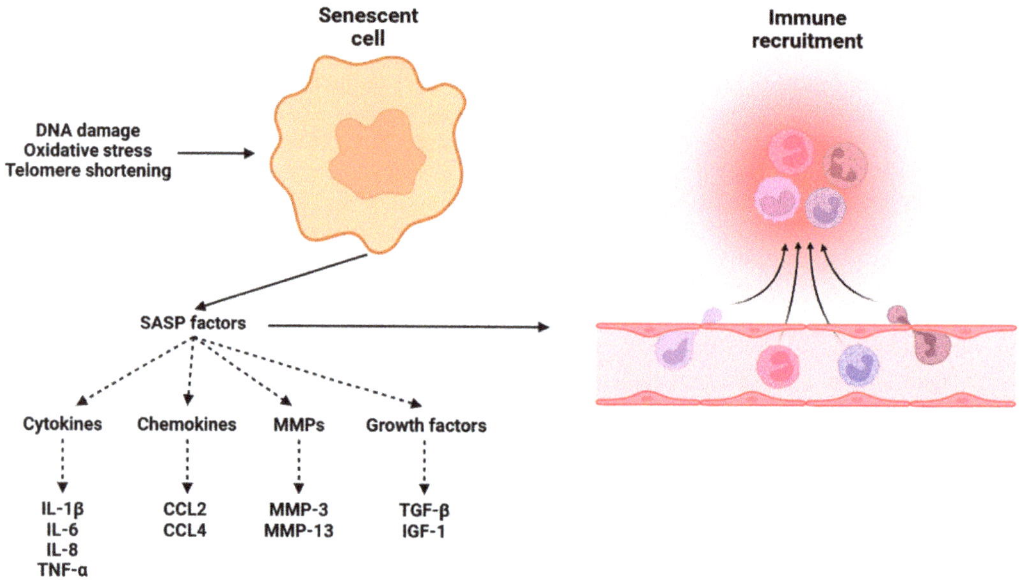

Figure 2. The process through which senescent cells secrete SASP mediators in response to damage caused by oxidative stress, telomere shortening, and DNA impairment. Abbreviations: DNA (deoxyribonucleic acid), IL-1β (Interleukin 1 beta), IL-6 (Interleukin 6), IL-8 (Interleukin 8), TNF-α (tumor necrosis factor alpha), CCL2 (C-C motif chemokine ligand 2), CCL4 (C-C motif chemokine ligand 4), MMP-3 (matrix metalloproteinase 3), MMP-13 (matrix metalloproteinase 13), TGF-β (transforming growth factor beta), and IGF-1 (insulin-like growth factor 1).

A critical molecular pathway that contributes significantly to aging-related inflammation is the involvement of the NF-κB transcription factor, which plays a crucial role in the immune system's response to stress, injury, and infection [196]. While this response is protective in younger individuals, its persistent activation in the elderly leads to an overzealous immune reaction that contributes to chronic low-grade inflammation [197]. NF-κB, activated by several stimuli (such as pro-inflammatory cytokines and ROS), triggers a cascade of downstream signaling events that enhance the production of pro-inflammatory mediators and adhesion molecules in neutrophils (such as E-selectin and VCAM-1) [198]. Conversely, one of the most consequential adverse effects of prolonged activation is the dysregulation and functional impairment of regulatory $CD4^+$ $CD25^+$ T cells (Tregs), which are essential for maintaining immune balance by controlling excessive immune responses [199]. With advancing age, Tregs experience a decline in function, reducing their ability to regulate immune activation. This dysfunction in Tregs leads to the breakdown of the finely tuned regulation of immune responses, promoting an imbalance in CD4+ T-cell activity and exacerbating tissue damage [199,200]. Additionally, $CD4^+$ $CD25^+$ Tregs are susceptible to mitochondrial dysfunction, which is characterized by decreased mitochondrial protein levels and impaired oxidative phosphorylation, along with increased mitochondrial oxidative stress [201]. Tregs undergo alterations in mitochondrial dynamics, including a reduction in mitochondrial mass, a compromised membrane potential, and the accumulation of damaged mitochondria with abnormal morphology, resulting in instability in FoxP3 expression and a reduced capacity to sustain immune homeostasis [202]. Furthermore, mitochondrial dysfunction in aging Tregs is further amplified by impairments in mitophagy (the process responsible for eliminating damaged mitochondria), leading to a self-perpetuating cycle of oxidative damage and cellular senescence [201].

Nevertheless, the aging immune system undergoes additional functional alterations, commonly referred to as immunosenescence. This process encompasses the reduction in the efficiency and capacity of immune responses [203]. In the innate immune system, macrophage and neutrophil activity decreases due to reduced signaling from GM-CSF and impaired TLRs function in monocytes, macrophages, and dendritic cells (DCs). Therefore, these immune cells become less efficient at detecting and responding to pathogens, resulting in a delayed immune response [204]. Likewise, the adaptive immune system suffers a decline in T- and B-cell functionality, compromising their ability to recognize specific pathogens and generate many specific immune responses [205]. This impairment manifests as a reduced ability to produce antibodies after vaccination or infection, making older individuals more susceptible to many infections, including influenza and pneumonia [206].

On the other hand, the molecular mechanisms driving age-related inflammation and vascular dysfunction involve a network of interconnected pathways, each contributing to the multifaceted phenotype of aging. At the epigenetic level, DNA methylation undergoes significant changes with age, marked by global hypomethylation and site-specific hypermethylation [207]. These alterations lead to the activation of numerous pro-inflammatory genes, such as *IL-6*, *TNF-α*, and *COX-2*, while simultaneously suppressing anti-inflammatory factors like IL-10 and TGF-β [208,209]. Histone modifications are vital in chromatin remodeling during aging, with a strong reduction in H3K9 trimethylation and an increase in H4K16 acetylation being particularly notable [210–212]. Moreover, the expression of non-coding RNAs, especially microRNAs, is altered with age. For instance, the upregulation of miR-21 promotes vascular inflammation by targeting PTEN and activating the AKT pathway, while the downregulation of miR-126 impairs endothelial function by reducing VEGF signaling [213,214].

The interaction between inflammation and obesity in aging creates a network of molecular processes that significantly impacts health and longevity. Obesity, a metabolic disorder in the elderly [215], induces adipocyte hypertrophy and hyperplasia, leading to a significant increase in the secretion of pro-inflammatory adipokines such as leptin and resistin [216,217]. These adipokines initiate a cascade of inflammatory events, activating mainly mast cells and $CD4^+$ T cells to release numerous pro-inflammatory cytokines, including IL-6, IL-12, IL-17, IL-18, and TNF-α [218]. Simultaneously, the expansion of adipose tissue linked to obesity results in localized hypoxia, activating the NF-κB pathway and further enhancing the production of pro-inflammatory mediators [219]. Furthermore, excess adipose tissue acts as an endocrine organ, secreting bioactive molecules that disturb the metabolic balance, including altered lipid metabolism and increased circulating free fatty acids, exacerbating oxidative stress and mitochondrial dysfunction in immune cells [220]. The cumulative effect of these processes establishes a self-perpetuating cycle of inflammation and metabolic dysregulation, known as adipaging, which not only accelerates the aging process but also elevates the risk of age-related diseases such as type 2 diabetes [221].

3.2. Central Inflammation

The detrimental influence of SASP mediators on the CNS arises through intricate molecular mechanisms that undermine the structural and functional integrity of the BBB. The BBB is a highly specialized and selective permeability barrier composed of endothelial cells interconnected by tight junctions, primarily consisting of transmembrane proteins such as claudins, occludin, and zonula occludens (ZO) proteins, supported by pericytes and astrocytes [222]. Under physiological conditions, this barrier plays an essential role in maintaining CNS homeostasis by regulating the exchange of ions, nutrients, and signaling molecules while restricting the entry of circulating immune cells and neurotoxic agents [223]. However, chronic exposure of the BBB to SASP factors impairs its selective permeability, resulting in widespread neurovascular dysfunction [224]. As the BBB integrity declines with age, peripheral inflammatory mediators and immune cells are able to infiltrate the CNS [225].

Among the key SASP factors implicated in BBB breakdown are a range of pro-inflammatory cytokines, including IL-1β, IL-6, and TNF-α, in addition to MMPs (such as MMP-2 and MMP-9), which directly degrade tight junction proteins [226]. This disruption facilitates the increased permeability of the BBB, allowing the infiltration of peripheral immune cells, ROS, and additional inflammatory mediators into the CNS parenchyma [227]. The extravasation of these harmful agents exacerbates neuroinflammation by triggering the activation of resident microglial cells, which play a crucial role in the innate immune response within the CNS [228]. Microglial activation is mainly mediated via TLR signaling cascades and the NF-κB transcriptional pathway, leading to a feedforward loop of pro-inflammatory cytokine release that reinforces neuronal damage [229]. The mentioned cytokines activate microglia through TLR4, resulting in the nuclear translocation of NF-κB and subsequent amplification of inflammatory responses [230]. Simultaneously, the activation of the NLRP3 inflammasome in microglia enhances the release of mature IL-1β, perpetuating neuroinflammation [231].

A consequence of sustained microglial activation (Figure 3) is the excessive production of reactive oxygen and nitrogen species (ROS and RNS), mainly mediated by inducible nitric oxide synthase (iNOS) and NADPH oxidase [232,233]. This overproduction creates a highly pro-inflammatory and cytotoxic environment within the CNS, disrupting cellular homeostasis. The accumulation of ROS exacerbates mitochondrial dysfunction, leading to impaired ATP synthesis and energy deficits in neurons. The energy deficiency compromises several neuronal processes, including synaptic transmission, plasticity, and cell viability [234]. Oxidative stress disrupts several redox-sensitive signaling pathways, triggering molecular damage like DNA oxidation, lipid peroxidation, and protein misfolding [235].

In addition to mitochondrial dysfunction, the interplay between neuroinflammation and oxidative stress exacerbates excitotoxicity. Overactivated microglia evoke excessive glutamate release while simultaneously impairing its reuptake by astrocytes, resulting in glutamate accumulation within the synaptic cleft [236]. This causes the overactivation of NMDA and AMPA receptors, leading to an aberrant influx of calcium into neurons [237]. The increased intracellular calcium activates multiple neurotoxic pathways, including the stimulation of calpains and caspases (proteolytic enzymes that degrade essential cellular components and drive apoptotic and necrotic cell death) [238,239].

At the molecular level, elevated calcium influx disrupts calcium-dependent signaling cascades, exacerbating mitochondrial stress and stimulating the additional release of ROS [240]. This feedback loop perpetuates oxidative damage and cellular dysfunction, creating a self-sustaining cycle of neurodegeneration. Moreover, neuronal death amplifies microglial activation, exacerbating inflammation and oxidative stress, which in turn accelerates disease progression [241].

With respect to astroglia, this cell type plays a key role in the inflammatory processes within the CNS. Astrocytes are crucial in driving chronic neuroinflammation, thereby disturbing homeostasis and increasing the vulnerability of neurons to degeneration and cell death [242]. Astrocytes, the most abundant glial cells in the CNS, contribute to neuroinflammation via the activation of fundamental signaling pathways. Two major transcription factors, NF-κB and AHR, are central players in this regulatory network. NF-κB is a rapidly inducible factor that governs the expression of many pro-inflammatory genes, including cytokines (such as *IL-6* and *TNF-α*), chemokines (such as *CCL2* and *CXCL10*), and adhesion molecules (e.g., *ICAM-1*) [243]. In contrast, AHR has been identified as a potential negative regulator of NF-κB-mediated inflammation, modulating the intensity of inflammatory responses through strong interactions with cytochrome P450 enzymes [244]. The dysregulation of these pathways in astrocytes contributes to a persistent pro-inflammatory state, exacerbating neurodegeneration [245]. Moreover, astrocytes respond to these cytokines, further amplifying inflammatory signaling through JAK/STAT pathways [246].

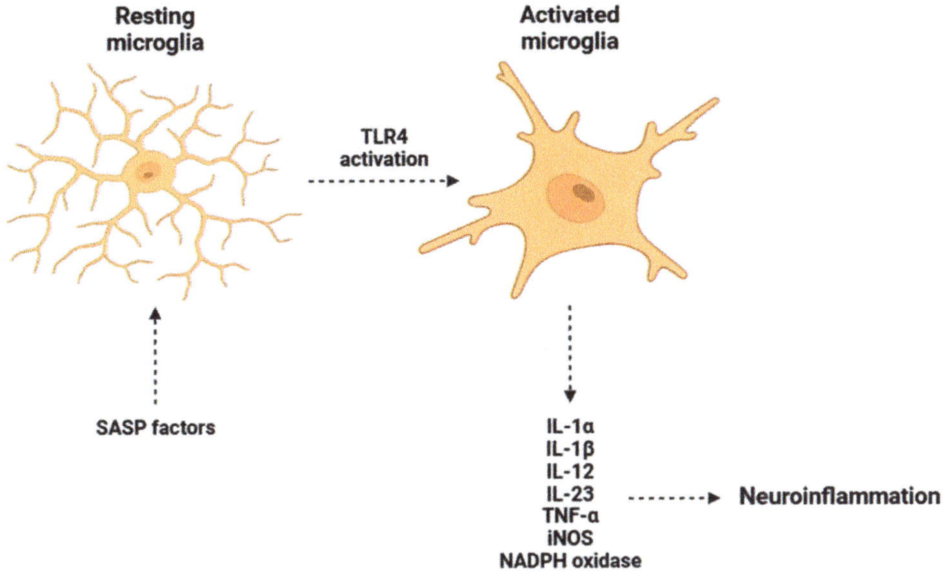

Figure 3. Microglial activation induced by SASP mediators in resting-state microglia. The activation of TLR4 receptors by SASP factors stimulates microglial activation and the release of various pro-inflammatory cytokines. Abbreviations: IL-1α (Interleukin 1 alpha), IL-1β (Interleukin 1 beta), IL-12 (Interleukin 12), IL-23 (Interleukin 23), TNF-α (tumor necrosis factor alpha), iNOS (inducible nitric oxide synthase), and NADPH (nicotinamide adenine dinucleotide phosphate—reduced form).

Oxidative stress is another critical component of age-related neuroinflammation. The accumulation of ROS due to mitochondrial dysfunction leads to the oxidative damage of cellular components, such as lipids, proteins, and DNA [12]. Aged microglia and astrocytes exhibit impaired antioxidant responses, further exacerbating ROS accumulation [247,248]. This oxidative stress not only directly damages neurons but also activates NF-κβ, thereby sustaining the inflammatory cycle. Dysfunctional mitochondria in aging CNS cells exhibit reduced oxidative phosphorylation efficiency, increased mitochondrial DNA mutations, and enhanced cytochrome c release, triggering apoptotic pathways [249].

Aging-related inflammation also affects neurotransmitter systems, disrupting synaptic plasticity and neuronal communication. Pro-inflammatory cytokines modulate neurotransmitter release and receptor expression, leading to excitotoxicity and impaired synaptic function. For example, TNF-α has been shown to disrupt glutamate homeostasis by decreasing the expression of various glutamate transporters (GLT-1 and GLAST) in astrocytes, leading to elevated extracellular glutamate levels and facilitating excitotoxic neuronal damage through excessive NMDA receptor stimulation [250]. Persistent neuroinflammatory and oxidative conditions alter synaptic plasticity by modulating the expression and functionality of key synaptic proteins, including postsynaptic density protein 95 (PSD-95), synaptophysin, and glutamatergic receptor subunits [251]. The breakdown of these proteins impairs long-term potentiation (LTP) and long-term depression (LTD), critical mechanisms underlying learning and memory [252,253].

Moreover, chronic neuroinflammation affects dendritic spine morphology, contributing to the synaptic loss and cognitive decline observed in numerous neurodegenerative diseases [254]. Beyond synaptic dysfunction, the sustained presence of SASP mediators promotes τ hyperphosphorylation, a hallmark of Alzheimer's disease, through the dysregulation of kinases such as glycogen synthase kinase 3β (GSK-3β) and cyclin-dependent kinase 5 (CDK5). Hyperphosphorylated τ aggregates into insoluble neurofibrillary tangles (NFTs), breaking axonal transport and neuronal communication [255,256]. In Parkinson's disease, SASP-mediated inflammation enhances α-synuclein misfolding and aggregation via impaired autophagy–lysosomal degradation and proteasomal dysfunction, leading to the accumulation of oligomers that intensify dopaminergic neurodegeneration [257,258]. The persistence of misfolded protein aggregates, coupled with chronic inflammatory and oxidative stress conditions, drives neuronal apoptosis through both caspase-dependent and caspase-independent pathways [259]. Mitochondrial outer membrane permeabilization (MOMP), mediated by the pro-apoptotic Bcl-2 family proteins BAX and BAK, initiates apoptosis by releasing cytochrome c and activating caspase-3 and caspase-9 [260]. Additionally, the excessive activation of PARP-1 in response to DNA damage contributes to parthanatos, a form of programmed cell death distinct from classical apoptosis, further exacerbating neuronal loss [261]. In contrast to classical apoptosis, which is marked by well-defined morphological and biochemical hallmarks such as cell shrinkage, chromatin condensation, and membrane blebbing [262], parthanatos is characterized by a unique series of cellular events, including the overproduction of PAR chains, mitochondrial dysfunction, and the release of apoptosis-inducing factor (AIF) from mitochondria, which subsequently translocates to the nucleus to trigger chromatin condensation and DNA fragmentation [263]. Parthanatos causes irreversible neuronal damage and worsens neuronal loss, particularly in neurodegenerative diseases and acute neuronal injuries, where it interacts with other cell death pathways to accelerate tissue degeneration [264].

4. Current Anti-Inflammatory Strategies for Aging

Adopting a healthy lifestyle has long been recognized as one of the most effective strategies for maintaining overall well-being and reducing the impact of aging. A growing body of research supports the idea that some lifestyle factors (e.g., proper nutrition, moderate physical activity, and mental stability) contribute to the mitigation of the aging process. A well-balanced and sufficient nutrient intake has been shown to positively impact aging, with numerous studies suggesting that nutritional factors play an integral role in the modulation of age-related molecular mechanisms. Long-term adherence to a diet rich in polyphenols, naturally occurring compounds in fruits and vegetables, has been particularly associated with significant improvements in intestinal permeability [265,266]. This effect may help maintain the integrity of the intestinal barrier, thereby reducing systemic inflammation [267]. Moreover, polyphenolic compounds have been demonstrated to activate many antioxidant and anti-inflammatory pathways, thus mitigating oxidative stress and promoting cellular resilience [268].

On the other hand, research has revealed that the intake of specific probiotic strains, particularly *Lactobacillus pentosus* var. *plantarum* C29, can reduce SASP factors [269]. The modulation of the gut microbiota through probiotic supplementation results in a reduction in pro-inflammatory cytokine levels and has the potential to decrease the expression of age-associated biomarkers, including tumor suppressor proteins like p16 and p53 [270]. These markers are crucial for cellular senescence and apoptosis, and their regulation has been linked to slowing age-associated cellular dysfunction [271].

Similarly, the intake of polyunsaturated fatty acids (PUFAs), particularly ω-3 fatty acids, has been shown to reduce the levels of pro-inflammatory cytokines [272]. It is thought that reducing inflammation helps preserve metabolic and immune balance, thereby supporting healthier aging [273]. The anti-inflammatory effects of PUFAs extend beyond their role in lipid metabolism and influence gene expression by activating some nuclear receptors such as peroxisome proliferator-activated receptors (PPARs), which regulate lipid homeostasis and inflammatory signaling pathways [274].

Furthermore, the frequent intake of essential vitamins such as vitamin C and vitamin E has been demonstrated to enhance immune cell functions in older adults. Vitamin C, a significant antioxidant, is vital for the proper function of neutrophils, which are critical to immune defense [275]. Specifically, vitamin C facilitates neutrophil chemotaxis and phagocytosis, thereby augmenting the body's ability to fight infections [276]. Vitamin E, known for its potent antioxidant properties, also plays a fundamental role in stabilizing cellular membranes and protecting immune cells from oxidative damage [277].

Beyond these vitamins, mineral supplementation, particularly zinc, has been identified as a critical factor in supporting immune homeostasis during aging [278]. Zinc is involved in the regulation of a variety of immune processes, including the differentiation and function of T cells [279]. Studies indicate that zinc supplementation can increase the naive T-cell population, which is essential for the body's ability to respond to new pathogens [280]. Furthermore, zinc contributes to balancing the activity of $CD4^+$ T cells, enhancing the immune response and reducing the negative impact of age-related immune dysfunction [281].

Exercise is widely recognized as a highly effective and accessible strategy for mitigating the aging process, functioning through complex biological mechanisms that counteract DNA damage, oxidative stress, and the overall decline in cellular function that often accompanies aging [282]. Recent studies have underscored its role in delaying the aging process at the molecular level, especially through its capacity to regulate telomere length and the activity of crucial enzymes such as telomerase [283]. Research examining middle-aged marathon runners and triathletes has shown that these individuals exhibit significantly higher telomerase activity along with longer telomeres in their circulating white blood cells compared to their sedentary counterparts. This finding suggests that endurance exercise not only helps maintain telomere integrity but may also promote the rejuvenation of immune cells, enhancing overall immune function and mitigating age-related declines in immune response [284]. This evidence indicates that endurance exercise helps preserve telomere integrity and rejuvenates immune cells, improving immune function and delaying age-related immune decline. The underlying mechanisms likely involve reducing oxidative stress, modulating inflammation, and activating genes that promote cell survival and repair [285,286]. In contrast, resistance training offers benefits for combatting aging at the cellular level, with studies showing significant improvements in the health of adipose tissue in older and obese patients [287]. It was discovered that resistance exercise led to a reduction in the number of cells expressing p16, a protein marker commonly associated with cellular senescence [288]. By reducing the number of p16-expressing cells in thigh adipose tissue, resistance training appears to foster a more youthful and functional tissue environment, potentially by promoting the turnover of damaged cells and alleviating inflammation in adipose tissue [289].

Emerging research has increasingly highlighted the pro-inflammatory cytokine network as a crucial target for anti-aging interventions, focusing on the potential to alleviate chronic inflammation, which is one of the key drivers of age-related diseases and cellular dysfunction [290]. In light of this, various anti-inflammatory compounds have garnered significant attention for their ability to counteract inflammation, promote cellular health, and potentially delay the onset of age-related decline. Among these, metformin, a widely used medication for managing type 2 diabetes, has emerged as a promising candidate for promoting healthy aging [291]. Metformin has been shown to influence the IKK/NF-κB pathway, a primary regulator of inflammation, as well as the GPX7/NRF2 axis, which is essential for cellular defense against oxidative stress and the regulation of antioxidant responses [292,293]. Recent research has revealed that metformin interacts with PEN2, an identified target involved in age-related inflammation [294]. By modulating all of these pathways, metformin reduces inflammation and promotes healthier aging, potentially extending lifespan [291]. Similarly, aspirin reduces oxidative damage, maintains tissue function, promotes cellular regeneration, and eliminates SASP mediators, thereby presenting a potential therapeutic strategy for age-related tissue degeneration [295].

Finally, among the broad spectrum of cellular anti-aging therapies, senolytics (agents designed to selectively eliminate senescent cells) emerge as one of the most promising yet highly debated strategies in gerontology [296]. Since their discovery in 2015 [297], several compounds have moved from initial discovery to clinical trials, resulting in considerable enthusiasm within the scientific community. Senolytic drugs can be classified into phytochemicals and proprietary compounds, differing in their origin, composition, regulatory oversight, and mechanisms of action. Phytochemicals are naturally occurring compounds derived from plants, often associated with potential health benefits and therapeutic properties. They are found in several foods, herbal medicines, and dietary supplements, with their biological effects typically attributed to the synergistic interactions of multiple bioactive constituents [297–299]. In contrast, proprietary drugs are synthetic compounds developed by pharmaceutical companies. These drugs undergo thorough testing, including clinical trials, to assess their safety, efficacy, and standardized dosing before gaining regulatory approval [299,300].

Among the most studied senolytic agents are dasatinib and quercetin. Dasatinib has shown effectiveness in eliminating senescent human adipocyte progenitor cells, which reside in adipose tissue and contribute to age-related metabolic dysfunction [301,302]. Quercetin has proven potent in clearing senescent endothelial cells, which line the blood vessels, as well as bone marrow stem cells in murine models [303]. These two drugs constitute a combination therapy that has shown significant efficacy in targeting various types of senescent cells, thereby reducing inflammation, enhancing tissue function, and alleviating the burden of age-related diseases in animal models [301,304–306]. Fenofibrate, primarily used for cholesterol management, has emerged as a promising therapeutic candidate due to its ability to clear senescent cells by upregulating PPARα [307]. Ultimately, taurine and spermidine have emerged as promising senolytic agents in the context of aging, demonstrating potential in promoting cellular homeostasis and mitigating age-related decline. Taurine, a conditionally essential amino acid, has been shown to enhance mitochondrial function, reduce oxidative stress, and modulate inflammatory responses, thereby contributing to the clearance of senescent cells [308]. Similarly, spermidine, a polyamine present in youthful cells, has been strongly implicated in autophagy induction, chromatin stabilization, and the suppression of pro-inflammatory pathways, all of which are fundamental for maintaining cellular integrity and delaying age-associated pathologies [309].

5. Conclusions

In conclusion, this review presents a thorough analysis of the complex biological processes that underpin aging and its connection to various pathological conditions. Aging encompasses several intricate physiological changes that significantly impact the body's ability to maintain homeostasis. This loss of homeostasis plays a role in the development of chronic diseases commonly associated with aging, such as cardiovascular disease and neurodegenerative disorders, by disrupting cellular activities, impairing the body's ability to repair tissues, and increasing its vulnerability to inflammation and oxidative stress.

One of the key themes explored in this article is the role of inflammation in the aging process. Chronic low-grade inflammation, also known as inflammaging, is highlighted as a hallmark of aging that drives the progression of these age-related diseases. Unlike the acute, short-lived inflammation that arises from injury or infection, inflammaging is a sustained, low-grade inflammation that can inflict lasting damage on the body. This chronic inflammatory state accelerates the decline of cellular and tissue function by inducing tissue damage and impairing the body's capacity for self-repair.

Furthermore, this article explores the connection between aging and immune system dysfunction, which plays a key role in the initiation and progression of diseases in older adults. Immunosenescence, the progressive deterioration of immune function, impairs the body's ability to protect against pathogens and injuries. As the immune system weakens, older individuals become more susceptible to infections, autoimmune disorders, and cancer. These processes not only induce inflammation but also result in the accumulation of dysfunctional cells, thereby exacerbating the pathological effects of aging.

This research highlights the importance of identifying therapeutic strategies to mitigate the effects of aging and chronic inflammation. Targeting specific inflammatory mediators and promoting immune rejuvenation could play a fundamental role in extending healthspan (the period of life spent in good health). Furthermore, delaying cellular senescence, the process by which cells lose their ability to divide and function, could help maintain tissue integrity and enhance overall health in aging populations.

Finally, this review calls for a more cohesive healthcare approach for the aging population, one that addresses not only the symptoms of age-related diseases but also the root causes of aging itself. By focusing on the fundamental mechanisms of aging and chronic inflammation and implementing personalized interventions, it is possible to encourage healthier aging, enhance quality of life, and increase the resilience of older adults.

Funding: This research received no external funding.

Institutional Review Board Statement: Not applicable.

Informed Consent Statement: Not applicable.

Data Availability Statement: Not applicable.

Conflicts of Interest: The author declares no conflicts of interest.

Abbreviations

The following abbreviations are used in this manuscript:

4-HNE	4-hydroxy-2-nonenal
ACTH	Adrenocorticotropic hormone
AGE	Advanced glycation end-product
AHR	Aryl hydrocarbon receptor
AIF	Apoptosis-inducing factor
AKT/PKB	Protein kinase B
AMPA	Alpha-amino-3-hydroxy-5-methyl-4-isoxazolepropionic acid
ATP	Adenosine triphosphate
BAK	Bcl-2 homologous antagonist/killer
BAX	Bcl-2-associated X protein
BBB	Blood–brain barrier
Bcl-2	B-cell lymphoma 2
BPH	Benign prostatic hyperplasia

CCL2	C-C motif chemokine ligand 2
CCL4	C-C motif chemokine ligand 4
CD25	Cluster of differentiation 25
CD4	Cluster of differentiation 4
CDK4/6	Cyclin-dependent kinases 4 and 6
CDK5	Cyclin-dependent kinase 5
CMA	Chaperone-mediated autophagy
CNS	Central nervous system
COPD	Chronic obstructive pulmonary disease
COX-1	Cyclooxygenase 1
COX-2	Cyclooxygenase 2
CRH	Corticotropin-releasing hormone
CXCL10	C-X-C motif chemokine ligand 10
CXCL12	C-X-C motif chemokine ligand 12
CXCR4	C-X-C chemokine receptor type 4
CYP3A4	Cytochrome P450 3A4
DAMPs	Damage-associated molecular patterns
DC	Dendritic cell
DDR	DNA damage response
DHEA	Dehydroepiandrosterone
DNA	Deoxyribonucleic acid
DPPC	Dipalmitoylphosphatidylcholine
ECM	Extracellular matrix
EGFR	Epidermal growth factor receptor
eNOS	Endothelial nitric oxide synthase
ERK	Extracellular signal-regulated kinase 1/2
FoxP3	Forkhead box P3
GDF-15	Growth differentiation factor 15
GERD	Gastroesophageal reflux disease
GH	Growth hormone
GHRH	Growth hormone-releasing hormone
GI	Gastrointestinal system
GLAST	Glutamate aspartate transporter
GLT-1	Glutamate transporter 1
GM-CSF	Granulocyte–macrophage colony-stimulating factor
GnRH	Gonadotropin-releasing hormone
GPX7	Glutathione peroxidase 7
GSK-3β	Glycogen synthase kinase 3 beta
HSP70	70 kDa heat shock protein
ICAM-1	Intercellular adhesion molecule 1
IFN-γ	Interferon gamma
IGF-1	Insulin-like growth factor 1
IGFBP-2	Insulin-like growth factor binding protein 2
IKK	IκB Kinase
IL-10	Interleukin 10
IL-12	Interleukin 12
IL-17	Interleukin 17
IL-18	Interleukin 18
IL-1α	Interleukin 1 alpha
IL-1β	Interleukin 1 beta
IL-2	Interleukin 2
IL-4	Interleukin 4
IL-6	Interleukin 6

IL-8	Interleukin 8
iNOS	Inducible nitric oxide synthase
LPS	Lipopolysaccharide
LTD	Long-term depression
LTP	Long-term potentiation
MAPK	Mitogen-activated protein kinase
MDA	Malondialdehyde
MEK	Mitogen-activated protein kinase kinase
MMP	Matrix metalloproteinase
MMP-13	Matrix metalloproteinase 13
MMP-2	Matrix metalloproteinase 2
MMP-3	Matrix metalloproteinase 3
MMP-9	Matrix metalloproteinase 9
MOMP	Mitochondrial outer membrane permeabilization
NADPH	Nicotinamide adenine dinucleotide phosphate (reduced form)
NFT	Neurofibrillary tangle
NF-κB	Nuclear factor kappa-light-chain-enhancer of activated B cells
NLRP3	NLR family pyrin domain containing 3
NMDA	N-methyl-D-aspartate
NRF2	Nuclear factor erythroid 2-related factor 2
NSAID	Non-steroidal anti-inflammatory drug
p16	Cyclin-dependent kinase inhibitor 2A
p21	Cyclin-dependent kinase inhibitor 1
P450	Cytochrome P450
p53	Tumor protein P53
PAR	Poly (ADP-ribose)
PARP-1	Poly (ADP-ribose) polymerase 1
PEN2	Presenilin enhancer 2
PI3K	Phosphoinositide 3-kinase
PPAR	Peroxisome proliferator-activated receptor
PPARα	Peroxisome proliferator-activated receptor alpha
PSD-95	Postsynaptic density protein 95
PTEN	Phosphatase and tensin homolog
PTM	Post-translational modification
PUFA	Polyunsaturated fatty acid
RAAS	Renin–angiotensin–aldosterone system
Rb	Retinoblastoma
RNA	Ribonucleic acid
RNS	Reactive nitrogen species
ROS	Reactive oxygen species
Runx2	Runt-related transcription factor 2
SASP	Senescence-associated secretory phenotype
SCFA	Short-chain fatty acid
T3	Triiodothyronine
T4	Thyroxine
TGF-β	Transforming growth factor beta
TIMP	Tissue inhibitor of metalloproteinase
TLR	Toll-like receptor
TLR4	Toll-like receptor 4
TNFR	Tumor necrosis factor receptor
TNF-α	Tumor necrosis factor alpha
TP53	Tumor protein p53
Treg	Regulatory T cell

UPS	Ubiquitin–proteasome system
VCAM-1	Vascular cell adhesion molecule 1
VEGF	Vascular endothelial growth factor
ZO	Zonula occludens

References

1. Guo, J.; Huang, X.; Dou, L.; Yan, M.; Shen, T.; Tang, W.; Li, J. Aging and aging-related diseases: From molecular mechanisms to interventions and treatments. *Signal Transduct. Target. Ther.* **2022**, *7*, 391. [CrossRef]
2. López-Otín, C.; Blasco, M.A.; Partridge, L.; Serrano, M.; Kroemer, G. Hallmarks of aging: An expanding universe. *Cell* **2023**, *186*, 243–278. [CrossRef] [PubMed]
3. Jaul, E.; Barron, J. Age-Related Diseases and Clinical and Public Health Implications for the 85 Years Old and Over Population. *Front. Public Health* **2017**, *5*, 335. [CrossRef] [PubMed]
4. Li, Y.; Tian, X.; Luo, J.; Bao, T.; Wang, S.; Wu, X. Molecular mechanisms of aging and anti-aging strategies. *Cell Commun. Signal.* **2024**, *22*, 285. [CrossRef]
5. da Silva, P.F.L.; Schumacher, B. Principles of the Molecular and Cellular Mechanisms of Aging. *J. Investig. Dermatol.* **2021**, *141*, 951–960. [CrossRef] [PubMed]
6. Preston, J.; Biddell, B. The physiology of ageing and how these changes affect older people. *Medicine* **2020**, *49*, 1–5. [CrossRef]
7. Vijg, J.; Suh, Y. Genome instability and aging. *Annu. Rev. Physiol.* **2013**, *75*, 645–668. [CrossRef]
8. Schumacher, B.; Pothof, J.; Vijg, J.; Hoeijmakers, J.H.J. The central role of DNA damage in the ageing process. *Nature* **2021**, *592*, 695–703. [CrossRef]
9. Rossiello, F.; Jurk, D.; Passos, J.F.; d'Adda di Fagagna, F. Telomere dysfunction in ageing and age-related diseases. *Nat. Cell Biol.* **2022**, *24*, 135–147. [CrossRef]
10. Dhar, P.; Moodithaya, S.S.; Patil, P. Epigenetic alterations-The silent indicator for early aging and age-associated health-risks. *Aging Med.* **2022**, *5*, 287–293. [CrossRef]
11. Hipp, M.S.; Kasturi, P.; Hartl, F.U. The proteostasis network and its decline in ageing. *Nat. Rev. Mol. Cell Biol.* **2019**, *20*, 421–435. [CrossRef] [PubMed]
12. Srivastava, S. The Mitochondrial Basis of Aging and Age-Related Disorders. *Genes* **2017**, *8*, 398. [CrossRef]
13. Zhang, L.; Pitcher, L.E.; Yousefzadeh, M.J.; Niedernhofer, L.J.; Robbins, P.D.; Zhu, Y. Cellular senescence: A key therapeutic target in aging and diseases. *J. Clin. Investig.* **2022**, *132*, e158450. [CrossRef] [PubMed]
14. Sun, Y.; Wang, X.; Liu, T.; Zhu, X.; Pan, X. The multifaceted role of the SASP in atherosclerosis: From mechanisms to therapeutic opportunities. *Cell Biosci.* **2022**, *12*, 74. [CrossRef] [PubMed]
15. Evans, D.S.; Young, D.; Tanaka, T.; Basisty, N.; Bandinelli, S.; Ferrucci, L.; Campisi, J.; Schilling, B. Proteomic Analysis of the Senescence-Associated Secretory Phenotype: GDF-15, IGFBP-2, and Cystatin-C Are Associated with Multiple Aging Traits. *J. Gerontol. A Biol. Sci. Med. Sci.* **2024**, *79*, glad265. [CrossRef] [PubMed]
16. Basisty, N.; Kale, A.; Jeon, O.H.; Kuehnemann, C.; Payne, T.; Rao, C.; Holtz, A.; Shah, S.; Sharma, V.; Ferrucci, L.; et al. A proteomic atlas of senescence-associated secretomes for aging biomarker development. *PLoS Biol.* **2020**, *18*, e3000599. [CrossRef]
17. Hu, C.; Zhang, X.; Teng, T.; Ma, Z.G.; Tang, Q.Z. Cellular Senescence in Cardiovascular Diseases: A Systematic Review. *Aging Dis.* **2022**, *13*, 103–128. [CrossRef]
18. Martínez-Cué, C.; Rueda, N. Cellular Senescence in Neurodegenerative Diseases. *Front. Cell. Neurosci.* **2020**, *14*, 16. [CrossRef]
19. Schmitt, C.A.; Wang, B.; Demaria, M. Senescence and cancer—role and therapeutic opportunities. *Nat. Rev. Clin. Oncol.* **2022**, *19*, 619–636. [CrossRef]
20. Escobar, A.P.; Bonansco, C.; Cruz, G.; Dagnino-Subiabre, A.; Fuenzalida, M.; Negrón, I.; Sotomayor-Zárate, R.; Martínez-Pinto, J.; Jorquera, G. Central and Peripheral Inflammation: A Common Factor Causing Addictive and Neurological Disorders and Aging-Related Pathologies. *Int. J. Mol. Sci.* **2023**, *24*, 10083. [CrossRef]
21. Li, X.; Li, C.; Zhang, W.; Wang, Y.; Qian, P.; Huang, H. Inflammation and aging: Signaling pathways and intervention therapies. *Signal Transduct. Target Ther.* **2023**, *8*, 239. [CrossRef] [PubMed]
22. Vannella, K.M.; Wynn, T.A. Mechanisms of Organ Injury and Repair by Macrophages. *Annu. Rev. Physiol.* **2017**, *79*, 593–617. [CrossRef] [PubMed]
23. Scalise, R.F.M.; De Sarro, R.; Caracciolo, A.; Lauro, R.; Squadrito, F.; Carerj, S.; Bitto, A.; Micari, A.; Bella, G.D.; Costa, F.; et al. Fibrosis after Myocardial Infarction: An Overview on Cellular Processes, Molecular Pathways, Clinical Evaluation and Prognostic Value. *Med. Sci.* **2021**, *9*, 16. [CrossRef] [PubMed]
24. Franceschi, C.; Garagnani, P.; Parini, P.; Giuliani, C.; Santoro, A. Inflammaging: A new immune-metabolic viewpoint for age-related diseases. *Nat. Rev. Endocrinol.* **2018**, *14*, 576–590. [CrossRef]
25. Dugan, B.; Conway, J.; Duggal, N.A. Inflammaging as a target for healthy ageing. *Age Ageing* **2023**, *52*, afac328. [CrossRef]

26. Wu, D.; Chen, Q.; Chen, X.; Han, F.; Chen, Z.; Wang, Y. The blood-brain barrier: Structure, regulation, and drug delivery. *Signal. Transduct. Target. Ther.* **2023**, *8*, 217. [CrossRef]
27. Silva, B.A.; Farías, M.I.; Miglietta, E.A.; Leal, M.C.; Ávalos, J.C.; Pitossi, F.J.; Ferrari, C.C. Understanding the role of the blood brain barrier and peripheral inflammation on behavior and pathology on ongoing confined cortical lesions. *Mult. Scler. Relat. Disord.* **2022**, *57*, 103346. [CrossRef]
28. Miao, J.; Ma, H.; Yang, Y.; Liao, Y.; Lin, C.; Zheng, J.; Yu, M.; Lan, J. Microglia in Alzheimer's disease: Pathogenesis, mechanisms, and therapeutic potentials. *Front. Aging Neurosci.* **2023**, *15*, 1201982. [CrossRef]
29. Ho, M.S. Microglia in Parkinson's Disease. *Adv. Exp. Med. Biol.* **2019**, *1175*, 335–353.
30. Gao, C.; Jiang, J.; Tan, Y.; Chen, S. Microglia in neurodegenerative diseases: Mechanism and potential therapeutic targets. *Signal Transduct. Target Ther.* **2023**, *8*, 359. [CrossRef]
31. Xu, Y.; Gao, W.; Sun, Y.; Wu, M. New insight on microglia activation in neurodegenerative diseases and therapeutics. *Front. Neurosci.* **2023**, *17*, 1308345. [CrossRef]
32. Zhang, Y.; Chen, H.; Li, R.; Sterling, K.; Song, W. Amyloid β-based therapy for Alzheimer's disease: Challenges, successes and future. *Signal Transduct. Target. Ther.* **2023**, *8*, 248. [CrossRef]
33. Calabresi, P.; Mechelli, A.; Natale, G.; Volpicelli-Daley, L.; Di Lazzaro, G.; Ghiglieri, V. Alpha-synuclein in Parkinson's disease and other synucleinopathies: From overt neurodegeneration back to early synaptic dysfunction. *Cell Death Dis* **2023**, *14*, 176. [CrossRef] [PubMed]
34. Tian, Y.E.; Cropley, V.; Maier, A.B.; Lautenschlager, N.T.; Breakspear, M.; Zalesky, A. Heterogeneous aging across multiple organ systems and prediction of chronic disease and mortality. *Nat. Med.* **2023**, *29*, 1221–1231. [CrossRef] [PubMed]
35. Khalil, R.; Diab-Assaf, M.; Lemaitre, J.M. Emerging Therapeutic Approaches to Target the Dark Side of Senescent Cells: New Hopes to Treat Aging as a Disease and to Delay Age-Related Pathologies. *Cells* **2023**, *12*, 915. [CrossRef] [PubMed]
36. Kesidou, E.; Theotokis, P.; Damianidou, O.; Boziki, M.; Konstantinidou, N.; Taloumtzis, C.; Sintila, S.A.; Grigoriadis, P.; Evangelopoulos, M.E.; Bakirtzis, C.; et al. CNS Ageing in Health and Neurodegenerative Disorders. *J. Clin. Med.* **2023**, *12*, 2255. [CrossRef]
37. Salim, S. Oxidative Stress and the Central Nervous System. *J. Pharmacol. Exp. Ther.* **2017**, *360*, 201–205. [CrossRef]
38. Jové, M.; Mota-Martorell, N.; Pradas, I.; Martín-Gari, M.; Ayala, V.; Pamplona, R. The Advanced Lipoxidation End-Product Malondialdehyde-Lysine in Aging and Longevity. *Antioxidants* **2020**, *9*, 1132. [CrossRef]
39. Li, Y.; Zhao, T.; Li, J.; Xia, M.; Li, Y.; Wang, X.; Liu, C.; Zheng, T.; Chen, R.; Kan, D.; et al. Oxidative Stress and 4-hydroxy-2-nonenal (4-HNE): Implications in the Pathogenesis and Treatment of Aging-related Diseases. *J. Immunol. Res.* **2022**, *2022*, 2233906. [CrossRef]
40. Liu, M.Y.; Nemes, A.; Zhou, Q.G. The Emerging Roles for Telomerase in the Central Nervous System. *Front. Mol. Neurosci.* **2018**, *11*, 160. [CrossRef]
41. Palmos, A.B.; Duarte, R.R.R.; Smeeth, D.M.; Hedges, E.C.; Nixon, D.F.; Thuret, S.; Powell, T.R. Telomere length and human hippocampal neurogenesis. *Neuropsychopharmacology* **2020**, *45*, 2239–2247. [CrossRef] [PubMed]
42. Hetz, C. Adapting the proteostasis capacity to sustain brain healthspan. *Cell* **2021**, *184*, 1545–1560. [CrossRef] [PubMed]
43. Lee, J.; Kim, H.J. Normal Aging Induces Changes in the Brain and Neurodegeneration Progress: Review of the Structural, Biochemical, Metabolic, Cellular, and Molecular Changes. *Front. Aging Neurosci.* **2022**, *14*, 931536. [CrossRef] [PubMed]
44. Chapman, T.W.; Hill, R.A. Myelin plasticity in adulthood and aging. *Neurosci. Lett.* **2020**, *715*, 134645. [CrossRef]
45. Latham, A.S.; Moreno, J.A.; Geer, C.E. Biological agents and the aging brain: Glial inflammation and neurotoxic signaling. *Front. Aging* **2023**, *4*, 1244149. [CrossRef]
46. Zhang, W.; Xiao, D.; Mao, Q.; Xia, H. Role of neuroinflammation in neurodegeneration development. *Signal. Transduct. Target. Ther.* **2023**, *8*, 267. [CrossRef]
47. Liu, Y.; Liu, Q.; Zhang, Z.; Yang, Y.; Zhou, Y.; Yan, H.; Wang, X.; Li, X.; Zhao, J.; Hu, J.; et al. The regulatory role of PI3K in ageing-related diseases. *Ageing Res. Rev.* **2023**, *88*, 101963. [CrossRef]
48. Byun, K.; Yoo, Y.; Son, M.; Lee, J.; Jeong, G.B.; Park, Y.M.; Salekdeh, G.H.; Lee, B. Advanced glycation end-products produced systemically and by macrophages: A common contributor to inflammation and degenerative diseases. *Pharmacol. Ther.* **2017**, *177*, 44–55. [CrossRef]
49. Neutzner, M.; Kohler, C.; Frank, S.; Killer, H.E.; Neutzner, A. Impact of aging on meningeal gene expression. *Fluids Barriers CNS* **2023**, *20*, 12. [CrossRef]
50. Singam, N.S.V.; Fine, C.; Fleg, J.L. Cardiac changes associated with vascular aging. *Clin. Cardiol.* **2020**, *43*, 92–98. [CrossRef]
51. Anwar, I.; Wang, X.; Pratt, R.E.; Dzau, V.J.; Hodgkinson, C.P. The impact of aging on cardiac repair and regeneration. *J. Biol. Chem.* **2024**, *300*, 107682. [CrossRef] [PubMed]
52. Sagar, S.; Gustafsson, A.B. Cardiovascular aging: The mitochondrial influence. *J. Cardiovasc. Aging* **2023**, *3*, 33. [CrossRef] [PubMed]

53. González, L.D.M.; Romero-Orjuela, S.P.; Rabeya, F.J.; Del Castillo, V.; Echeverri, D. Age and vascular aging: An unexplored frontier. *Front. Cardiovasc. Med.* **2023**, *10*, 1278795. [CrossRef] [PubMed]
54. Han, Y.; Kim, S.Y. Endothelial senescence in vascular diseases: Current understanding and future opportunities in senotherapeutics. *Exp. Mol. Med.* **2023**, *55*, 1–12. [CrossRef]
55. De Meyer, T.; Nawrot, T.; Bekaert, S.; De Buyzere, M.L.; Rietzschel, E.R.; Andrés, V. Telomere Length as Cardiovascular Aging Biomarker: JACC Review Topic of the Week. *J. Am. Coll. Cardiol.* **2018**, *72*, 805–813. [CrossRef]
56. Herman, A.B.; Occean, J.R.; Sen, P. Epigenetic dysregulation in cardiovascular aging and disease. *J. Cardiovasc. Aging* **2021**, *1*, 10. [CrossRef]
57. Kim, M.; Kim, H.L.; Lim, W.H.; Seo, J.B.; Kim, S.H.; Kim, M.A.; Zo, J.H. Association between arterial stiffness and left ventricular diastolic function: A large population-based cross-sectional study. *Front. Cardiovasc. Med.* **2022**, *9*, 1001248. [CrossRef]
58. Miyamoto, S. Autophagy and cardiac aging. *Cell Death Differ.* **2019**, *26*, 653–664. [CrossRef]
59. Soenen, S.; Rayner, C.K.; Jones, K.L.; Horowitz, M. The ageing gastrointestinal tract. *Curr. Opin. Clin. Nutr. Metab. Care* **2016**, *19*, 12–18. [CrossRef]
60. Sun, T.; Li, D.; Hu, S.; Huang, L.; Sun, H.; Yang, S.; Wu, B.; Ji, F.; Zhou, D. Aging-dependent decrease in the numbers of enteric neurons, interstitial cells of Cajal and expression of connexin43 in various regions of gastrointestinal tract. *Aging* **2018**, *10*, 3851–3865. [CrossRef]
61. Nguyen, T.T.; Baumann, P.; Tüscher, O.; Schick, S.; Endres, K. The Aging Enteric Nervous System. *Int. J. Mol. Sci.* **2023**, *24*, 9471. [CrossRef] [PubMed]
62. Kurin, M.; Fass, R. Management of Gastroesophageal Reflux Disease in the Elderly Patient. *Drugs Aging* **2019**, *36*, 1073–1081. [CrossRef] [PubMed]
63. Yu, B.; Xiang, L.; Peppelenbosch, M.P.; Fuhler, G.M. Overlapping cytokines in H. pylori infection and gastric cancer: A tandem meta-analysis. *Front. Immunol.* **2023**, *14*, 1125658. [CrossRef] [PubMed]
64. Skokowski, J.; Vashist, Y.; Girnyi, S.; Cwalinski, T.; Mocarski, P.; Antropoli, C.; Brillantino, A.; Boccardi, V.; Goyal, A.; Ciarleglio, F.A.; et al. The Aging Stomach: Clinical Implications of H. pylori Infection in Older Adults-Challenges and Strategies for Improved Management. *Int. J. Mol. Sci.* **2024**, *25*, 12826. [CrossRef]
65. Amorim, J.A.; Coppotelli, G.; Rolo, A.P.; Palmeira, C.M.; Ross, J.M.; Sinclair, D.A. Mitochondrial and metabolic dysfunction in ageing and age-related diseases. *Nat. Rev. Endocrinol.* **2022**, *18*, 243–258. [CrossRef]
66. Mbara, K.C.; Fotsing, M.C.D.; Ndinteh, D.T.; Mbeb, C.N.; Nwagwu, C.S.; Khan, R.; Mokhetho, K.C.; Baijnath, H.; Nlooto, M.; Mokhele, S.; et al. Endoplasmic reticulum stress in pancreatic β-cell dysfunction: The potential therapeutic role of dietary flavonoids. *Curr. Res. Pharmacol. Drug Discov.* **2024**, *6*, 100184. [CrossRef]
67. Konstandi, M.; Johnson, E.O. Age-related modifications in CYP-dependent drug metabolism: Role of stress. *Front. Endocrinol.* **2023**, *14*, 1143835. [CrossRef]
68. Wang, Q.; Qi, Y.; Shen, W.; Xu, J.; Wang, L.; Chen, S.; Hou, T.; Si, J. The Aged Intestine: Performance and Rejuvenation. *Aging Dis.* **2021**, *12*, 1693–1712. [CrossRef]
69. Yue, T.; Chen, S.; Zhu, J.; Guo, S.; Huang, Z.; Wang, P.; Zuo, S.; Liu, Y. The aging-related risk signature in colorectal cancer. *Aging* **2021**, *13*, 7330–7349. [CrossRef]
70. Pretzsch, E.; Nieß, H.; Bösch, F.; Westphalen, C.B.; Jacob, S.; Neumann, J.; Werner, J.; Heinemann, V.; Angele, M.K. Age and metastasis—How age influences metastatic spread in cancer. Colorectal cancer as a model. *Cancer Epidemiol.* **2022**, *77*, 102112. [CrossRef]
71. Sohail, R.; Mathew, M.; Patel, K.K.; Reddy, S.A.; Haider, Z.; Naria, M.; Habib, A.; Abdin, Z.U.; Razzaq Chaudhry, W.; Akbar, A. Effects of Non-steroidal Anti-inflammatory Drugs (NSAIDs) and Gastroprotective NSAIDs on the Gastrointestinal Tract: A Narrative Review. *Cureus* **2023**, *15*, e37080. [CrossRef] [PubMed]
72. Jump, R.L. *Clostridium difficile* infection in older adults. *Aging Health* **2013**, *9*, 403–414. [CrossRef] [PubMed]
73. Li, S.; Xiao, X.; Zhang, X. Hydration Status in Older Adults: Current Knowledge and Future Challenges. *Nutrients* **2023**, *15*, 2609. [CrossRef]
74. Rutsch, A.; Kantsjö, J.B.; Ronchi, F. The Gut-Brain Axis: How Microbiota and Host Inflammasome Influence Brain Physiology and Pathology. *Front. Immunol.* **2020**, *11*, 604179. [CrossRef] [PubMed]
75. Afzaal, M.; Saeed, F.; Shah, Y.A.; Hussain, M.; Rabail, R.; Socol, C.T.; Hassoun, A.; Pateiro, M.; Lorenzo, J.M.; Rusu, A.V.; et al. Human gut microbiota in health and disease: Unveiling the relationship. *Front. Microbiol.* **2022**, *13*, 999001. [CrossRef]
76. Wu, Y.L.; Xu, J.; Rong, X.Y.; Wang, F.; Wang, H.J.; Zhao, C. Gut microbiota alterations and health status in aging adults: From correlation to causation. *Aging Med.* **2021**, *4*, 206–213. [CrossRef]
77. Kossowska, M.; Olejniczak, S.; Karbowiak, M.; Mosiej, W.; Zielińska, D.; Brzezicka, A. The Interplay between Gut Microbiota and Cognitive Functioning in the Healthy Aging Population: A Systematic Review. *Nutrients* **2024**, *16*, 852. [CrossRef]
78. Zhang, Y.; Wei, S.; Zhang, H.; Jo, Y.; Kang, J.S.; Ha, K.T.; Joo, J.; Lee, H.J.; Ryu, D. Gut microbiota-generated metabolites: Missing puzzles to hosts' health, diseases, and aging. *BMB Rep.* **2024**, *57*, 207–215. [CrossRef]

79. Menezes, A.A.; Shah, Z.A. A Review of the Consequences of Gut Microbiota in Neurodegenerative Disorders and Aging. *Brain Sci.* **2024**, *14*, 1224. [CrossRef]
80. Ghosh, T.S.; Shanahan, F.; O'Toole, P.W. The gut microbiome as a modulator of healthy ageing. *Nat. Rev. Gastroenterol. Hepatol.* **2022**, *19*, 565–584. [CrossRef]
81. Chen, Y.; Xu, J.; Chen, Y. Regulation of Neurotransmitters by the Gut Microbiota and Effects on Cognition in Neurological Disorders. *Nutrients* **2021**, *13*, 2099. [CrossRef] [PubMed]
82. Salazar, A.M.; Aparicio, R.; Clark, R.I.; Rera, M.; Walker, D.W. Intestinal barrier dysfunction: An evolutionarily conserved hallmark of aging. *Dis. Model Mech.* **2023**, *16*, dmm049969. [CrossRef] [PubMed]
83. Lalley, P.M. The aging respiratory system--pulmonary structure, function and neural control. *Respir. Physiol. Neurobiol.* **2013**, *187*, 199–210. [CrossRef] [PubMed]
84. Wang, Y.; Huang, X.; Luo, G.; Xu, Y.; Deng, X.; Lin, Y.; Wang, Z.; Zhou, S.; Wang, S.; Chen, H.; et al. The aging lung: Microenvironment, mechanisms, and diseases. *Front. Immunol.* **2024**, *15*, 1383503. [CrossRef]
85. Ulldemolins, A.; Narciso, M.; Sanz-Fraile, H.; Otero, J.; Farré, R.; Gavara, N.; Almendros, I. Effects of aging on the biomechanical properties of the lung extracellular matrix: Dependence on tissular stretch. *Front. Cell. Dev. Biol.* **2024**, *12*, 1381470. [CrossRef]
86. Vitenberga, Z.; Pilmane, M. Age-related lung tissue remodeling due to the local distribution of MMP-2, TIMP-2, TGF-β and Hsp70. *Biotech. Histochem.* **2018**, *93*, 239–248. [CrossRef]
87. Torrelles, J.B.; Restrepo, B.I.; Bai, Y.; Ross, C.; Schlesinger, L.S.; Turner, J. The Impact of Aging on the Lung Alveolar Environment, Predetermining Susceptibility to Respiratory Infections. *Front. Aging.* **2022**, *3*, 818700. [CrossRef]
88. Ratjen, F.; Rehn, B.; Costabel, U.; Bruch, J. Age-dependency of surfactant phospholipids and surfactant protein A in bronchoalveolar lavage fluid of children without bronchopulmonary disease. *Eur. Respir. J.* **1996**, *9*, 328–333. [CrossRef]
89. Hentschel, R.; Bohlin, K.; van Kaam, A.; Fuchs, H.; Danhaive, O. Surfactant replacement therapy: From biological basis to current clinical practice. *Pediatr. Res.* **2020**, *88*, 176–183. [CrossRef]
90. Adachi, D.; Yamada, M.; Nishiguchi, S.; Fukutani, N.; Hotta, T.; Tashiro, Y.; Morino, S.; Shirooka, H.; Nozaki, Y.; Hirata, H.; et al. Age-related decline in chest wall mobility: A cross-sectional study among community-dwelling elderly women. *J. Am. Osteopath. Assoc.* **2015**, *115*, 384–389. [CrossRef]
91. Yamada, T.; Minami, T.; Shinohara, T.; Ouchi, S.; Mabuchi, S.; Yoshino, S.; Emoto, K.; Nakagawa, K.; Yoshimi, K.; Saito, M.; et al. The Impact of Ageing on Diaphragm Function and Maximal Inspiratory Pressure: A Cross-Sectional Ultrasound Study. *Diagnostics* **2025**, *15*, 163. [CrossRef] [PubMed]
92. Shaikh, S.B.; Goracci, C.; Tjitropranoto, A.; Rahman, I. Impact of aging on immune function in the pathogenesis of pulmonary diseases: Potential for therapeutic targets. *Expert Rev. Respir. Med.* **2023**, *17*, 351–364. [CrossRef] [PubMed]
93. Paleczny, B.; Niewiński, P.; Rydlewska, A.; Piepoli, M.F.; Borodulin-Nadzieja, L.; Jankowska, E.A.; Ponikowska, B. Age-related reflex responses from peripheral and central chemoreceptors in healthy men. *Clin. Auton. Res.* **2014**, *24*, 285–296. [CrossRef] [PubMed]
94. Vetterlein, M.W.; Jindal, T.; Becker, A.; Regier, M.; Kluth, L.A.; Tilki, D.; Chun, F.K. Small renal masses in the elderly: Contemporary treatment approaches and comparative oncological outcomes of nonsurgical and surgical strategies. *Investig. Clin. Urol.* **2016**, *57*, 231–239. [CrossRef]
95. Alam, R.; Yerrapragada, A.; Wlajnitz, T.; Watts, E.; Pallauf, M.; Enikeev, D.; Chang, P.; Wagner, A.A.; McKiernan, J.M.; Pierorazio, P.M.; et al. Evaluation of Growth Rates for Small Renal Masses in Elderly Patients Undergoing Active Surveillance. *Eur. Urol. Open Sci.* **2023**, *50*, 78–84. [CrossRef]
96. Wang, Y.; Zhang, W.; Qian, T.; Sun, H.; Xu, Q.; Hou, X.; Hu, W.; Zhang, G.; Drummond, G.R.; Sobey, C.G.; et al. Reduced renal function may explain the higher prevalence of hyperuricemia in older people. *Sci. Rep.* **2021**, *11*, 1302. [CrossRef]
97. Ray, N.; Reddy, P.H. Structural and physiological changes of the kidney with age and its impact on chronic conditions and COVID-19. *Ageing Res. Rev.* **2023**, *88*, 101932. [CrossRef]
98. Dybiec, J.; Szlagor, M.; Młynarska, E.; Rysz, J.; Franczyk, B. Structural and Functional Changes in Aging Kidneys. *Int. J. Mol. Sci.* **2022**, *23*, 15435. [CrossRef]
99. Monteonofrio, L.; Florio, M.C.; AlGhatrif, M.; Lakatta, E.G.; Capogrossi, M.C. Aging- and gender-related modulation of RAAS: Potential implications in COVID-19 disease. *Vasc. Biol.* **2020**, *3*, R1–R14. [CrossRef]
100. Li, M.; Wang, D.; Liu, Z.; Huang, Y.; Zhang, Q.; Pan, C.; Lin, Y.; Sun, L.; Zheng, Y. Assessing the effects of aging on the renal endothelial cell landscape using single-cell RNA sequencing. *Front. Genet.* **2023**, *14*, 1175716. [CrossRef]
101. Nishii, H. A Review of Aging and the Lower Urinary Tract: The Future of Urology. *Int. Neurourol. J.* **2021**, *25*, 273–284. [CrossRef]
102. Lee, J.; Yoo, S.; Cho, M.C.; Jeong, H.; Choo, M.S.; Son, H. Significance of a decrease in the proportion of detrusor muscle to bladder wall for non-invasive diagnosis of detrusor underactivity in men with lower urinary tract symptoms. *Sci. Rep.* **2022**, *12*, 5237. [CrossRef] [PubMed]
103. Hardy, C.C.; Korstanje, R. Aging and urinary control: Alterations in the brain-bladder axis. *Aging Cell* **2023**, *22*, e13990. [CrossRef] [PubMed]

104. Batmani, S.; Jalali, R.; Mohammadi, M.; Bokaee, S. Prevalence and factors related to urinary incontinence in older adults women worldwide: A comprehensive systematic review and meta-analysis of observational studies. *BMC Geriatr.* **2021**, *21*, 212. [CrossRef] [PubMed]
105. Sayegh, N.; Gross, K. Benign prostatic hyperplasia: A global challenge of the ageing population. *Lancet Healthy Longev.* **2022**, *3*, e725–e726. [CrossRef]
106. Fu, X.; Liu, H.; Liu, J.; DiSanto, M.E.; Zhang, X. The Role of Heat Shock Protein 70 Subfamily in the Hyperplastic Prostate: From Molecular Mechanisms to Therapeutic Opportunities. *Cells* **2022**, *11*, 2052. [CrossRef]
107. Olascoaga, S.; Castañeda-Sánchez, J.I.; Königsberg, M.; Gutierrez, H.; López-Diazguerrero, N.E. Oxidative stress-induced gene expression changes in prostate epithelial cells in vitro reveal a robust signature of normal prostatic senescence and aging. *Biogerontology* **2024**, *25*, 1145–1169. [CrossRef]
108. Rodríguez-Nieves, J.A.; Patalano, S.C.; Almanza, D.; Gharaee-Kermani, M.; Macoska, J.A. CXCL12/CXCR4 Axis Activation Mediates Prostate Myofibroblast Phenoconversion through Non-Canonical EGFR/MEK/ERK Signaling. *PLoS ONE* **2016**, *11*, e0159490. [CrossRef]
109. Rodriguez-Mañas, L. Urinary tract infections in the elderly: A review of disease characteristics and current treatment options. *Drugs Context.* **2020**, *9*, 2020-4-13. [CrossRef]
110. Maresca, G.; Mc Clinton, S.; Swami, S.; El-Mokadem, I.; Donaldson, J.F. Do men with bladder stones benefit from treatment of benign prostatic obstruction? *BJU Int.* **2022**, *130*, 619–627. [CrossRef]
111. Liu, P.; Quinn, R.R.; Lam, N.N.; Elliott, M.J.; Xu, Y.; James, M.T.; Manns, B.; Ravani, P. Accounting for Age in the Definition of Chronic Kidney Disease. *JAMA Intern. Med.* **2021**, *181*, 1359–1366. [CrossRef] [PubMed]
112. Calleja-Agius, J.; Brincat, M.P. The urogenital system and the menopause. *Climacteric* **2015**, *18* (Suppl. S1), 18–22. [CrossRef] [PubMed]
113. Liu, Y.; Li, C. Hormone Therapy and Biological Aging in Postmenopausal Women. *JAMA Netw. Open* **2024**, *7*, e2430839. [CrossRef] [PubMed]
114. Critchlow, A.J.; Hiam, D.; Williams, R.; Scott, D.; Lamon, S. The role of estrogen in female skeletal muscle aging: A systematic review. *Maturitas* **2023**, *178*, 107844. [CrossRef]
115. Wei, S.; Tang, W.; Chen, D.; Xiong, J.; Xue, L.; Dai, Y.; Guo, Y.; Wu, C.; Dai, J.; Wu, M.; et al. Multiomics insights into the female reproductive aging. *Ageing Res. Rev.* **2024**, *95*, 102245. [CrossRef]
116. Dong, S.; Chen, C.; Zhang, J.; Gao, Y.; Zeng, X.; Zhang, X. Testicular aging, male fertility and beyond. *Front. Endocrinol.* **2022**, *13*, 1012119. [CrossRef]
117. Kaufman, J.M.; Lapauw, B.; Mahmoud, A.; T'Sjoen, G.; Huhtaniemi, I.T. Aging and the Male Reproductive System. *Endocr. Rev.* **2019**, *40*, 906–972. [CrossRef]
118. Jimbo, M.; Kunisaki, J.; Ghaed, M.; Yu, V.; Flores, H.A.; Hotaling, J.M. Fertility in the aging male: A systematic review. *Fertil. Steril.* **2022**, *118*, 1022–1034. [CrossRef]
119. Davis, S.R.; Pinkerton, J.; Santoro, N.; Simoncini, T. Menopause-Biology, consequences, supportive care, and therapeutic options. *Cell* **2023**, *186*, 4038–4058. [CrossRef]
120. Andrews, R.; Lacey, A.; Bache, K.; Kidd, E.J. The role of menopausal symptoms on future health and longevity: A systematic scoping review of longitudinal evidence. *Maturitas* **2024**, *190*, 108130. [CrossRef]
121. Bi, J.; Zhang, C.; Lu, C.; Mo, C.; Zeng, J.; Yao, M.; Jia, B.; Liu, Z.; Yuan, P.; Xu, S. Age-related bone diseases: Role of inflammaging. *J. Autoimmun.* **2024**, *143*, 103169. [CrossRef] [PubMed]
122. Xing, Y.; Xuan, F.; Wang, K.; Zhang, H. Aging under endocrine hormone regulation. *Front. Endocrinol.* **2023**, *14*, 1223529. [CrossRef] [PubMed]
123. van den Beld, A.W.; Kaufman, J.M.; Zillikens, M.C.; Lamberts, S.W.J.; Egan, J.M.; van der Lely, A.J. The physiology of endocrine systems with ageing. *Lancet Diabetes Endocrinol.* **2018**, *6*, 647–658. [CrossRef] [PubMed]
124. Hajdarovic, K.H.; Yu, D.; Webb, A.E. Understanding the aging hypothalamus, one cell at a time. *Trends Neurosci.* **2022**, *45*, 942–954. [CrossRef]
125. Pataky, M.W.; Young, W.F.; Nair, K.S. Hormonal and Metabolic Changes of Aging and the Influence of Lifestyle Modifications. *Mayo Clin. Proc.* **2021**, *96*, 788–814. [CrossRef]
126. Cappola, A.R.; Auchus, R.J.; El-Hajj Fuleihan, G.; Handelsman, D.J.; Kalyani, R.R.; McClung, M.; Stuenkel, C.A.; Thorner, M.O.; Verbalis, J.G. Hormones and Aging: An Endocrine Society Scientific Statement. *J. Clin. Endocrinol Metab.* **2023**, *108*, 1835–1874. [CrossRef]
127. Santos, A.L.; Lindner, A.B. Protein Posttranslational Modifications: Roles in Aging and Age-Related Disease. *Oxid. Med. Cell. Longev.* **2017**, *2017*, 5716409. [CrossRef]
128. Kim, C.S.; Park, S.; Kim, J. The role of glycation in the pathogenesis of aging and its prevention through herbal products and physical exercise. *J. Exerc. Nutr. Biochem.* **2017**, *21*, 55–61. [CrossRef]

129. Chaudhuri, J.; Bains, Y.; Guha, S.; Kahn, A.; Hall, D.; Bose, N.; Gugliucci, A.; Kapahi, P. The Role of Advanced Glycation End Products in Aging and Metabolic Diseases: Bridging Association and Causality. *Cell. Metab.* **2018**, *28*, 337–352. [CrossRef]
130. Gill, V.; Kumar, V.; Singh, K.; Kumar, A.; Kim, J.J. Advanced Glycation End Products (AGEs) May Be a Striking Link Between Modern Diet and Health. *Biomolecules* **2019**, *9*, 888. [CrossRef]
131. Tudurí, E.; Soriano, S.; Almagro, L.; Montanya, E.; Alonso-Magdalena, P.; Nadal, Á.; Quesada, I. The pancreatic β-cell in ageing: Implications in age-related diabetes. *Ageing Res. Rev.* **2022**, *80*, 101674. [CrossRef] [PubMed]
132. Warde, K.M.; Smith, L.J.; Basham, K.J. Age-related Changes in the Adrenal Cortex: Insights and Implications. *J. Endocr. Soc.* **2023**, *7*, bvad097. [CrossRef] [PubMed]
133. Powrie, Y.S.L.; Smith, C. Central intracrine DHEA synthesis in ageing-related neuroinflammation and neurodegeneration: Therapeutic potential? *J. Neuroinflamm.* **2018**, *15*, 289. [CrossRef] [PubMed]
134. van Heemst, D. The ageing thyroid: Implications for longevity and patient care. *Nat. Rev. Endocrinol.* **2024**, *20*, 5–15. [CrossRef]
135. Cai, Y.; Han, Z.; Cheng, H.; Li, H.; Wang, K.; Chen, J.; Liu, Z.X.; Xie, Y.; Lin, Y.; Zhou, S.; et al. The impact of ageing mechanisms on musculoskeletal system diseases in the elderly. *Front. Immunol.* **2024**, *15*, 1405621. [CrossRef]
136. Xu, H.; Brown, J.L.; Bhaskaran, S.; Van Remmen, H. Reactive oxygen species in the pathogenesis of sarcopenia. *Free Radic. Biol. Med.* **2025**, *227*, 446–458. [CrossRef]
137. Marques-Carvalho, A.; Kim, H.N.; Almeida, M. The role of reactive oxygen species in bone cell physiology and pathophysiology. *Bone Rep.* **2023**, *19*, 101664. [CrossRef]
138. Dao, T.; Green, A.E.; Kim, Y.A.; Bae, S.J.; Ha, K.T.; Gariani, K.; Lee, M.R.; Menzies, K.J.; Ryu, D. Sarcopenia and Muscle Aging: A Brief Overview. *Endocrinol. Metab.* **2020**, *35*, 716–732. [CrossRef]
139. Cho, M.R.; Lee, S.; Song, S.K. A Review of Sarcopenia Pathophysiology, Diagnosis, Treatment and Future Direction. *J. Korean Med. Sci.* **2022**, *37*, e146. [CrossRef]
140. Muñoz-Cánoves, P.; Neves, J.; Sousa-Victor, P. Understanding muscle regenerative decline with aging: New approaches to bring back youthfulness to aged stem cells. *FEBS J.* **2020**, *287*, 406–416. [CrossRef]
141. Sousa-Victor, P.; García-Prat, L.; Muñoz-Cánoves, P. Control of satellite cell function in muscle regeneration and its disruption in ageing. *Nat. Rev. Mol. Cell. Biol.* **2022**, *23*, 204–226. [CrossRef] [PubMed]
142. Bodine, S.C.; Sinha, I.; Sweeney, H.L. Mechanisms of Skeletal Muscle Atrophy and Molecular Circuitry of Stem Cell Fate in Skeletal Muscle Regeneration and Aging. *J. Gerontol. A Biol. Sci. Med. Sci.* **2023**, *78*, 14–18. [CrossRef] [PubMed]
143. Gupta, P.; Kumar, S. Sarcopenia and Endocrine Ageing: Are They Related? *Cureus* **2022**, *14*, e28787. [CrossRef] [PubMed]
144. Ravazzano, L.; Colaianni, G.; Tarakanova, A.; Xiao, Y.B.; Grano, M.; Libonati, F. Multiscale and multidisciplinary analysis of aging processes in bone. *NPJ Aging* **2024**, *10*, 28. [CrossRef]
145. Becerikli, M.; Jaurich, H.; Schira, J.; Schulte, M.; Döbele, C.; Wallner, C.; Abraham, S.; Wagner, J.M.; Dadras, M.; Kneser, U.; et al. Age-dependent alterations in osteoblast and osteoclast activity in human cancellous bone. *J. Cell. Mol. Med.* **2017**, *21*, 2773–2781. [CrossRef]
146. Werner, S.L.; Sharma, R.; Woodruff, K.; Horn, D.; Harris, S.E.; Gorin, Y.; Lee, D.Y.; Hua, R.; Gu, S.; Fajardo, R.J.; et al. CSF-1 in Osteocytes Inhibits Nox4-mediated Oxidative Stress and Promotes Normal Bone Homeostasis. *JBMR Plus* **2019**, *4*, e10080. [CrossRef]
147. Cheng, Y.H.; Liu, S.F.; Dong, J.C.; Bian, Q. Transcriptomic alterations underline aging of osteogenic bone marrow stromal cells. *World J. Stem Cells* **2021**, *13*, 128–138. [CrossRef]
148. Cannarella, R.; Barbagallo, F.; Condorelli, R.A.; Aversa, A.; La Vignera, S.; Calogero, A.E. Osteoporosis from an Endocrine Perspective: The Role of Hormonal Changes in the Elderly. *J. Clin. Med.* **2019**, *8*, 1564. [CrossRef]
149. Mills, E.G.; Yang, L.; Nielsen, M.F.; Kassem, M.; Dhillo, W.S.; Comninos, A.N. The Relationship Between Bone and Reproductive Hormones Beyond Estrogens and Androgens. *Endocr. Rev.* **2021**, *42*, 691–719. [CrossRef]
150. Vashishth, D.; Dhaliwal, R.; Rubin, M. AGEs (Advanced Glycation End-products) in bone come of age. *Bone* **2025**, *190*, 117301. [CrossRef]
151. Rahmati, M.; Nalesso, G.; Mobasheri, A.; Mozafari, M. Aging and osteoarthritis: Central role of the extracellular matrix. *Ageing Res. Rev.* **2017**, *40*, 20–30. [CrossRef]
152. Sinha, U.; Malis, V.; Chen, J.S.; Csapo, R.; Kinugasa, R.; Narici, M.V.; Sinha, S. Role of the Extracellular Matrix in Loss of Muscle Force With Age and Unloading Using Magnetic Resonance Imaging, Biochemical Analysis, and Computational Models. *Front. Physiol.* **2020**, *11*, 626. [CrossRef] [PubMed]
153. Petrocelli, J.J.; McKenzie, A.I.; de Hart, N.M.M.P.; Reidy, P.T.; Mahmassani, Z.S.; Keeble, A.R.; Kaput, K.L.; Wahl, M.P.; Rondina, M.T.; Marcus, R.L.; et al. Disuse-induced muscle fibrosis, cellular senescence, and senescence-associated secretory phenotype in older adults are alleviated during re-ambulation with metformin pre-treatment. *Aging Cell* **2023**, *22*, e13936. [CrossRef] [PubMed]
154. Tieland, M.; Trouwborst, I.; Clark, B.C. Skeletal muscle performance and ageing. *J. Cachexia Sarcopenia Muscle* **2018**, *9*, 3–19. [CrossRef] [PubMed]

155. Chen, W.J.; Lin, I.H.; Lee, C.W.; Chen, Y.F. Aged Skeletal Muscle Retains the Ability to Remodel Extracellular Matrix for Degradation of Collagen Deposition after Muscle Injury. *Int. J. Mol. Sci.* **2021**, *22*, 2123. [CrossRef]
156. Kanazawa, Y.; Takahashi, T.; Inoue, T.; Nagano, M.; Koinuma, S.; Eiyo, H.; Tamura, Y.; Miyachi, R.; Iida, N.; Miyahara, K.; et al. Effects of Aging on Intramuscular Collagen-Related Factors After Injury to Mouse Tibialis Anterior Muscle. *Int. J. Mol. Sci.* **2025**, *26*, 801. [CrossRef]
157. Bektan Kanat, B.; Ulugerger Avci, G.; Bayramlar, O.F.; Unal, D.; Sonmez, O.; Bolayirli, I.M.; Doventas, A.; Erdincler, D.S.; Yavuzer, H. Matrix Metalloproteinases/Tissue Inhibitors of Metalloproteinases Ratio: A Biomarker of Bone Resorption in Geriatric Osteoporosis? *Ann. Geriatr. Med. Res.* **2023**, *27*, 151–157. [CrossRef]
158. Fuggle, N.R.; Laskou, F.; Harvey, N.C.; Dennison, E.M. A review of epigenetics and its association with ageing of muscle and bone. *Maturitas* **2022**, *165*, 12–17. [CrossRef]
159. Gerosa, L.; Malvandi, A.M.; Malavolta, M.; Provinciali, M.; Lombardi, G. Exploring cellular senescence in the musculoskeletal system: Any insights for biomarkers discovery? *Ageing Res. Rev.* **2023**, *88*, 101943. [CrossRef]
160. Li, Z.; Wu, X.; Yan, Z.; Cui, Y.; Liu, Y.; Cui, S.; Wang, Y.; Liu, T. Unveiling the muscle-brain axis: A bidirectional mendelian randomization study investigating the causal relationship between sarcopenia-related traits and brain aging. *Arch. Gerontol. Geriatr.* **2024**, *123*, 105412. [CrossRef]
161. Kumar, A.; Vaca-Dempere, M.; Mortimer, T.; Deryagin, O.; Smith, J.G.; Petrus, P.; Koronowski, K.B.; Greco, C.M.; Segalés, J.; Andrés, E.; et al. Brain-muscle communication prevents muscle aging by maintaining daily physiology. *Science* **2024**, *384*, 563–572. [CrossRef] [PubMed]
162. Bhattacharjee, S.; Kashyap, R.; Abualait, T.; Annabel Chen, S.H.; Yoo, W.K.; Bashir, S. The Role of Primary Motor Cortex: More Than Movement Execution. *J. Mot. Behav.* **2021**, *53*, 258–274. [CrossRef] [PubMed]
163. Han, X.; Ashraf, M.; Tipparaju, S.M.; Xuan, W. Muscle-Brain crosstalk in cognitive impairment. *Front. Aging Neurosci.* **2023**, *15*, 1221653. [CrossRef] [PubMed]
164. Saleh, Z.; Mirzazadeh, S.; Mirzaei, F.; Heidarnejad, K.; Meri, S.; Kalantar, K. Alterations in metabolic pathways: A bridge between aging and weaker innate immune response. *Front. Aging* **2024**, *5*, 1358330. [CrossRef]
165. Bachmann, M.C.; Bellalta, S.; Basoalto, R.; Gómez-Valenzuela, F.; Jalil, Y.; Lépez, M.; Matamoros, A.; von Bernhardi, R. The Challenge by Multiple Environmental and Biological Factors Induce Inflammation in Aging: Their Role in the Promotion of Chronic Disease. *Front. Immunol.* **2020**, *11*, 570083. [CrossRef]
166. Chung, H.Y.; Kim, D.H.; Lee, E.K.; Chung, K.W.; Chung, S.; Lee, B.; Seo, A.Y.; Chung, J.H.; Jung, Y.S.; Im, E.; et al. Redefining Chronic Inflammation in Aging and Age-Related Diseases: Proposal of the Senoinflammation Concept. *Aging Dis.* **2019**, *10*, 367–382. [CrossRef]
167. Soliman, A.M.; Barreda, D.R. Acute Inflammation in Tissue Healing. *Int. J. Mol. Sci.* **2022**, *24*, 641. [CrossRef]
168. Santoro, A.; Bientinesi, E.; Monti, D. Immunosenescence and inflammaging in the aging process: Age-related diseases or longevity? *Ageing Res. Rev.* **2021**, *71*, 101422. [CrossRef]
169. Di Micco, R.; Krizhanovsky, V.; Baker, D.; d'Adda di Fagagna, F. Cellular senescence in ageing: From mechanisms to therapeutic opportunities. *Nat. Rev. Mol. Cell. Biol.* **2021**, *22*, 75–95. [CrossRef]
170. Abdellatif, M.; Rainer, P.P.; Sedej, S.; Kroemer, G. Hallmarks of cardiovascular ageing. *Nat. Rev. Cardiol.* **2023**, *20*, 754–777. [CrossRef]
171. Hou, Y.; Dan, X.; Babbar, M.; Wei, Y.; Hasselbalch, S.G.; Croteau, D.L.; Bohr, V.A. Ageing as a risk factor for neurodegenerative disease. *Nat. Rev. Neurol.* **2019**, *15*, 565–581. [CrossRef] [PubMed]
172. López-Otín, C.; Pietrocola, F.; Roiz-Valle, D.; Galluzzi, L.; Kroemer, G. Meta-hallmarks of aging and cancer. *Cell Metab.* **2023**, *35*, 12–35. [CrossRef] [PubMed]
173. Chung, H.Y.; Sung, B.; Jung, K.J.; Zou, Y.; Yu, B.P. The molecular inflammatory process in aging. *Antioxid. Redox Signal.* **2006**, *8*, 572–581. [CrossRef] [PubMed]
174. Fulop, T.; Larbi, A.; Pawelec, G.; Khalil, A.; Cohen, A.A.; Hirokawa, K.; Witkowski, J.M.; Franceschi, C. Immunology of Aging: The Birth of Inflammaging. *Clin. Rev. Allergy Immunol.* **2023**, *64*, 109–122. [CrossRef]
175. Molinero, N.; Antón-Fernández, A.; Hernández, F.; Ávila, J.; Bartolomé, B.; Moreno-Arribas, M.V. Gut Microbiota, an Additional Hallmark of Human Aging and Neurodegeneration. *Neuroscience* **2023**, *518*, 141–161. [CrossRef]
176. Bosco, N.; Noti, M. The aging gut microbiome and its impact on host immunity. *Genes Immun.* **2021**, *22*, 289–303. [CrossRef]
177. Chmielarz, M.; Sobieszczańska, B.; Środa-Pomianek, K. Metabolic Endotoxemia: From the Gut to Neurodegeneration. *Int. J. Mol. Sci.* **2024**, *25*, 7006. [CrossRef]
178. Tylutka, A.; Walas, Ł.; Zembron-Lacny, A. Level of IL-6, TNF, and IL-1β and age-related diseases: A systematic review and meta-analysis. *Front. Immunol.* **2024**, *15*, 1330386. [CrossRef]
179. Davizon-Castillo, P.; McMahon, B.; Aguila, S.; Bark, D.; Ashworth, K.; Allawzi, A.; Campbell, R.A.; Montenont, E.; Nemkov, T.; D'Alessandro, A.; et al. TNF-α-driven inflammation and mitochondrial dysfunction define the platelet hyperreactivity of aging. *Blood* **2019**, *134*, 727–740. [CrossRef]

180. Rea, I.M.; Gibson, D.S.; McGilligan, V.; McNerlan, S.E.; Alexander, H.D.; Ross, O.A. Age and Age-Related Diseases: Role of Inflammation Triggers and Cytokines. *Front. Immunol.* **2018**, *9*, 586. [CrossRef]
181. Tominaga, K.; Suzuki, H.I. TGF-β Signaling in Cellular Senescence and Aging-Related Pathology. *Int. J. Mol. Sci.* **2019**, *20*, 5002. [CrossRef] [PubMed]
182. Freitas-Rodríguez, S.; Folgueras, A.R.; López-Otín, C. The role of matrix metalloproteinases in aging: Tissue remodeling and beyond. *Biochim. Biophys. Acta Mol. Cell Res.* **2017**, *1864*, 2015–2025. [CrossRef] [PubMed]
183. Neves, D.; Duarte-Pereira, S.; Matos, S.; Silva, R.M. Proteostasis networks in aging: Novel insights from text-mining approaches. *Biogerontology* **2023**, *24*, 555–562. [CrossRef] [PubMed]
184. Ruano, D. Proteostasis Dysfunction in Aged Mammalian Cells. The Stressful Role of Inflammation. *Front. Mol. Biosci.* **2021**, *8*, 658742. [CrossRef]
185. Kumari, R.; Jat, P. Mechanisms of Cellular Senescence: Cell Cycle Arrest and Senescence Associated Secretory Phenotype. *Front. Cell. Dev. Biol.* **2021**, *9*, 645593. [CrossRef]
186. Huang, W.; Hickson, L.J.; Eirin, A.; Kirkland, J.L.; Lerman, L.O. Cellular senescence: The good, the bad and the unknown. *Nat. Rev. Nephrol.* **2022**, *18*, 611–627. [CrossRef]
187. Bonelli, M.; La Monica, S.; Fumarola, C.; Alfieri, R. Multiple effects of CDK4/6 inhibition in cancer: From cell cycle arrest to immunomodulation. *Biochem. Pharmacol.* **2019**, *170*, 113676. [CrossRef]
188. Liang, R.; Qi, X.; Cai, Q.; Niu, L.; Huang, X.; Zhang, D.; Ling, J.; Wu, Y.; Chen, Y.; Yang, P.; et al. The role of NLRP3 inflammasome in aging and age-related diseases. *Immun. Ageing* **2024**, *21*, 14. [CrossRef]
189. Ohtani, N. The roles and mechanisms of senescence-associated secretory phenotype (SASP): Can it be controlled by senolysis? *Inflamm. Regen.* **2022**, *42*, 11. [CrossRef]
190. Elder, S.S.; Emmerson, E. Senescent cells and macrophages: Key players for regeneration? *Open Biol.* **2020**, *10*, 200309. [CrossRef]
191. Marin, I.; Serrano, M.; Pietrocola, F. Recent insights into the crosstalk between senescent cells and CD8 T lymphocytes. *NPJ Aging* **2023**, *9*, 8. [CrossRef] [PubMed]
192. Wang, B.; Han, J.; Elisseeff, J.H.; Demaria, M. The senescence-associated secretory phenotype and its physiological and pathological implications. *Nat. Rev. Mol. Cell. Biol.* **2024**, *25*, 958–978. [CrossRef] [PubMed]
193. Coryell, P.R.; Diekman, B.O.; Loeser, R.F. Mechanisms and therapeutic implications of cellular senescence in osteoarthritis. *Nat. Rev. Rheumatol.* **2021**, *17*, 47–57. [CrossRef] [PubMed]
194. Lee, K.A.; Flores, R.R.; Jang, I.H.; Saathoff, A.; Robbins, P.D. Immune Senescence, Immunosenescence and Aging. *Front. Aging* **2022**, *3*, 900028. [CrossRef]
195. Perez-Lanzon, M.; Zitvogel, L.; Kroemer, G. Failure of immunosurveillance accelerates aging. *Oncoimmunology* **2019**, *8*, e1575117. [CrossRef]
196. Haga, M.; Okada, M. Systems approaches to investigate the role of NF-κB signaling in aging. *Biochem. J.* **2022**, *479*, 161–183. [CrossRef]
197. Saavedra, D.; Añé-Kourí, A.L.; Barzilai, N.; Caruso, C.; Cho, K.H.; Fontana, L.; Franceschi, C.; Frasca, D.; Ledón, N.; Niedernhofer, L.J.; et al. Aging and chronic inflammation: Highlights from a multidisciplinary workshop. *Immun. Ageing* **2023**, *20*, 25. [CrossRef]
198. Coleman, P.R.; Chang, G.; Hutas, G.; Grimshaw, M.; Vadas, M.A.; Gamble, J.R. Age-associated stresses induce an anti-inflammatory senescent phenotype in endothelial cells. *Aging* **2013**, *5*, 913–924. [CrossRef]
199. Palatella, M.; Guillaume, S.M.; Linterman, M.A.; Huehn, J. The dark side of Tregs during aging. *Front. Immunol.* **2022**, *13*, 940705. [CrossRef]
200. Hou, P.F.; Zhu, L.J.; Chen, X.Y.; Qiu, Z.Q. Age-related changes in CD4+CD25+FOXP3+ regulatory T cells and their relationship with lung cancer. *PLoS ONE* **2017**, *12*, e0173048. [CrossRef]
201. Escrig-Larena, J.I.; Delgado-Pulido, S.; Mittelbrunn, M. Mitochondria during T cell aging. *Semin. Immunol.* **2023**, *69*, 101808. [CrossRef] [PubMed]
202. Gill, J.S.; Bansal, B.; Guo, K.; Huang, F.; Singh, H.; Hur, J.; Khan, N.; Mathur, R. Mitochondrial Oxidative Stress Regulates FOXP3+ T-Cell Activity and CD4-Mediated Inflammation in Older Adults with Frailty. *Int. J. Mol. Sci.* **2024**, *25*, 6235. [CrossRef] [PubMed]
203. Liu, Z.; Liang, Q.; Ren, Y.; Guo, C.; Ge, X.; Wang, L.; Cheng, Q.; Luo, P.; Zhang, Y.; Han, X. Immunosenescence: Molecular mechanisms and diseases. *Signal Transduct. Target Ther.* **2023**, *8*, 200. [CrossRef] [PubMed]
204. Shaw, A.C.; Goldstein, D.R.; Montgomery, R.R. Age-dependent dysregulation of innate immunity. *Nat. Rev. Immunol.* **2013**, *13*, 875–887. [CrossRef]
205. Kasler, H.; Verdin, E. How inflammaging diminishes adaptive immunity. *Nat. Aging* **2021**, *1*, 24–25. [CrossRef]
206. Quiros-Roldan, E.; Sottini, A.; Natali, P.G.; Imberti, L. The Impact of Immune System Aging on Infectious Diseases. *Microorganisms* **2024**, *12*, 775. [CrossRef]
207. Wilkinson, G.S.; Adams, D.M.; Haghani, A.; Lu, A.T.; Zoller, J.; Breeze, C.E.; Arnold, B.D.; Ball, H.C.; Carter, G.G.; Cooper, L.N.; et al. DNA methylation predicts age and provides insight into exceptional longevity of bats. *Nat. Commun.* **2021**, *12*, 1615. [CrossRef]

208. Nardini, C.; Moreau, J.F.; Gensous, N.; Ravaioli, F.; Garagnani, P.; Bacalini, M.G. The epigenetics of inflammaging: The contribution of age-related heterochromatin loss and locus-specific remodelling and the modulation by environmental stimuli. *Semin. Immunol.* **2018**, *40*, 49–60. [CrossRef]
209. Alimohammadi, M.; Makaremi, S.; Rahimi, A.; Asghariazar, V.; Taghadosi, M.; Safarzadeh, E. DNA methylation changes and inflammaging in aging-associated diseases. *Epigenomics* **2022**, *14*, 965–986. [CrossRef]
210. de Lima Camillo, L.P.; Asif, M.H.; Horvath, S.; Larschan, E.; Singh, R. Histone mark age of human tissues and cell types. *Sci. Adv.* **2025**, *11*, eadk9373. [CrossRef]
211. Mrabti, C.; Yang, N.; Desdín-Micó, G.; Alonso-Calleja, A.; Vílchez-Acosta, A.; Pico, S.; Parras, A.; Piao, Y.; Schoenfeldt, L.; Luo, S.; et al. Loss of H3K9 trimethylation leads to premature aging. *bioRxiv* **2024**, 604929.
212. Zhang, X.; Liu, H.; Zhou, J.Q.; Krick, S.; Barnes, J.W.; Thannickal, V.J.; Sanders, Y.Y. Modulation of H4K16Ac levels reduces pro-fibrotic gene expression and mitigates lung fibrosis in aged mice. *Theranostics* **2022**, *12*, 530–541. [CrossRef] [PubMed]
213. Borja-Gonzalez, M.; Casas-Martinez, J.C.; McDonagh, B.; Goljanek-Whysall, K. Inflamma-miR-21 Negatively Regulates Myogenesis during Ageing. *Antioxidants* **2020**, *9*, 345. [CrossRef] [PubMed]
214. Wang, L.; Lee, A.Y.; Wigg, J.P.; Peshavariya, H.; Liu, P.; Zhang, H. miR-126 Regulation of Angiogenesis in Age-Related Macular Degeneration in CNV Mouse Model. *Int. J. Mol. Sci.* **2016**, *17*, 895. [CrossRef]
215. Malenfant, J.H.; Batsis, J.A. Obesity in the geriatric population—A global health perspective. *J. Glob. Health Rep.* **2019**, *3*, e2019045. [CrossRef]
216. Nieva-Vazquez, A.; Pérez-Fuentes, R.; Torres-Rasgado, E.; López-López, J.G.; Romero, J.R. Serum resistin levels are associated with adiposity and insulin sensitivity in obese Hispanic subjects. *Metab. Syndr. Relat. Disord.* **2014**, *12*, 143–148. [CrossRef]
217. Obradovic, M.; Sudar-Milovanovic, E.; Soskic, S.; Essack, M.; Arya, S.; Stewart, A.J.; Gojobori, T.; Isenovic, E.R. Leptin and Obesity: Role and Clinical Implication. *Front. Endocrinol.* **2021**, *12*, 585887. [CrossRef]
218. Mancuso, P. The role of adipokines in chronic inflammation. *Immunotargets Ther.* **2016**, *5*, 47–56. [CrossRef]
219. Mirabelli, M.; Misiti, R.; Sicilia, L.; Brunetti, F.S.; Chiefari, E.; Brunetti, A.; Foti, D.P. Hypoxia in Human Obesity: New Insights from Inflammation towards Insulin Resistance-A Narrative Review. *Int. J. Mol. Sci.* **2024**, *25*, 9802. [CrossRef]
220. Nerstedt, A.; Smith, U. The impact of cellular senescence in human adipose tissue. *J. Cell Commun. Signal.* **2023**, *17*, 563–573. [CrossRef]
221. Pérez, L.M.; Pareja-Galeano, H.; Sanchis-Gomar, F.; Emanuele, E.; Lucia, A.; Gálvez, B.G. 'Adipaging': Ageing and obesity share biological hallmarks related to a dysfunctional adipose tissue. *J. Physiol.* **2016**, *594*, 3187–3207. [CrossRef] [PubMed]
222. Kadry, H.; Noorani, B.; Cucullo, L. A blood-brain barrier overview on structure, function, impairment, and biomarkers of integrity. *Fluids Barriers CNS* **2020**, *17*, 69. [CrossRef] [PubMed]
223. Sweeney, M.D.; Zhao, Z.; Montagne, A.; Nelson, A.R.; Zlokovic, B.V. Blood-Brain Barrier: From Physiology to Disease and Back. *Physiol. Rev.* **2019**, *99*, 21–78. [CrossRef] [PubMed]
224. Knopp, R.C.; Erickson, M.A.; Rhea, E.M.; Reed, M.J.; Banks, W.A. Cellular senescence and the blood-brain barrier: Implications for aging and age-related diseases. *Exp. Biol. Med.* **2023**, *248*, 399–411. [CrossRef]
225. Knox, E.G.; Aburto, M.R.; Clarke, G.; Cryan, J.F.; O'Driscoll, C.M. The blood-brain barrier in aging and neurodegeneration. *Mol. Psychiatry* **2022**, *27*, 2659–2673. [CrossRef]
226. Rempe, R.G.; Hartz, A.M.S.; Bauer, B. Matrix metalloproteinases in the brain and blood-brain barrier: Versatile breakers and makers. *J. Cereb. Blood Flow Metab.* **2016**, *36*, 1481–1507. [CrossRef]
227. Varatharaj, A.; Galea, I. The blood-brain barrier in systemic inflammation. *Brain Behav. Immun.* **2017**, *60*, 1–12. [CrossRef]
228. Li, Q.; Barres, B.A. Microglia and macrophages in brain homeostasis and disease. *Nat. Rev. Immunol.* **2018**, *18*, 225–242. [CrossRef]
229. Anilkumar, S.; Wright-Jin, E. NF-κB as an Inducible Regulator of Inflammation in the Central Nervous System. *Cells* **2024**, *13*, 485. [CrossRef]
230. Peng, J.; Xie, Y.; Pang, J.; Wu, Y.; Zhou, J.; Gu, L.; Guo, K.; Zhang, L.; Xie, B.; Yin, S.; et al. Toll-like receptor 4-mediated microglial inflammation exacerbates early white matter injury following experimental subarachnoid hemorrhage. *J. Neurochem.* **2023**, *166*, 280–293. [CrossRef]
231. Moraes, C.A.; Hottz, E.D.; Dos Santos Ornellas, D.; Adesse, D.; de Azevedo, C.T.; d'Avila, J.C.; Zaverucha-do-Valle, C.; Maron-Gutierrez, T.; Barbosa, H.S.; Bozza, P.T.; et al. Microglial NLRP3 Inflammasome Induces Excitatory Synaptic Loss Through IL-1β-Enriched Microvesicle Release: Implications for Sepsis-Associated Encephalopathy. *Mol. Neurobiol.* **2023**, *60*, 481–494. [CrossRef] [PubMed]
232. Nieman, A.N.; Li, G.; Zahn, N.M.; Mian, M.Y.; Mikulsky, B.N.; Hoffman, D.A.; Wilcox, T.M.; Kehoe, A.S.; Luecke, I.W.; Poe, M.M.; et al. Targeting Nitric Oxide Production in Microglia with Novel Imidazodiazepines for Nonsedative Pain Treatment. *ACS Chem. Neurosci.* **2020**, *11*, 2019–2030. [CrossRef] [PubMed]
233. Di Filippo, M.; de Iure, A.; Giampà, C.; Chiasserini, D.; Tozzi, A.; Orvietani, P.L.; Ghiglieri, V.; Tantucci, M.; Durante, V.; Quiroga-Varela, A.; et al. Persistent activation of microglia and NADPH oxidase [corrected] drive hippocampal dysfunction in experimental multiple sclerosis. *Sci. Rep.* **2016**, *6*, 20926.

234. Smith, A.N.; Shaughness, M.; Collier, S.; Hopkins, D.; Byrnes, K.R. Therapeutic targeting of microglia mediated oxidative stress after neurotrauma. *Front. Med.* **2022**, *9*, 1034692. [CrossRef]
235. Kishi, S.; Nagasu, H.; Kidokoro, K.; Kashihara, N. Oxidative stress and the role of redox signalling in chronic kidney disease. *Nat. Rev. Nephrol.* **2024**, *20*, 101–119. [CrossRef]
236. Zhu, G.; Wang, X.; Chen, L.; Lenahan, C.; Fu, Z.; Fang, Y.; Yu, W. Crosstalk Between the Oxidative Stress and Glia Cells After Stroke: From Mechanism to Therapies. *Front. Immunol.* **2022**, *13*, 852416. [CrossRef]
237. Nicosia, N.; Giovenzana, M.; Misztak, P.; Mingardi, J.; Musazzi, L. Glutamate-Mediated Excitotoxicity in the Pathogenesis and Treatment of Neurodevelopmental and Adult Mental Disorders. *Int. J. Mol. Sci.* **2024**, *25*, 6521. [CrossRef]
238. Troy, C.M.; Akpan, N.; Jean, Y.Y. Regulation of caspases in the nervous system implications for functions in health and disease. *Prog. Mol. Biol. Transl Sci.* **2011**, *99*, 265–305.
239. Zaman, V.; Drasites, K.P.; Myatich, A.; Shams, R.; Shields, D.C.; Matzelle, D.; Haque, A.; Banik, N.L. Inhibition of Calpain Attenuates Degeneration of Substantia Nigra Neurons in the Rotenone Rat Model of Parkinson's Disease. *Int. J. Mol. Sci.* **2022**, *23*, 13849. [CrossRef]
240. Görlach, A.; Bertram, K.; Hudecova, S.; Krizanova, O. Calcium and ROS: A mutual interplay. *Redox Biol.* **2015**, *6*, 260–271. [CrossRef]
241. Li, Y.; Li, Y.J.; Zhu, Z.Q. To re-examine the intersection of microglial activation and neuroinflammation in neurodegenerative diseases from the perspective of pyroptosis. *Front. Aging Neurosci.* **2023**, *15*, 1284214. [CrossRef]
242. Verkhratsky, A.; Nedergaard, M. Physiology of Astroglia. *Physiol. Rev.* **2018**, *98*, 239–389. [CrossRef]
243. Jong Huat, T.; Camats-Perna, J.; Newcombe, E.A.; Onraet, T.; Campbell, D.; Sucic, J.T.; Martini, A.; Forner, S.; Mirzaei, M.; Poon, W.; et al. The impact of astrocytic NF-κB on healthy and Alzheimer's disease brains. *Sci. Rep.* **2024**, *14*, 14305. [CrossRef] [PubMed]
244. Rothhammer, V.; Kenison, J.E.; Li, Z.; Tjon, E.; Takenaka, M.C.; Chao, C.C.; Alves de Lima, K.; Borucki, D.M.; Kaye, J.; Quintana, F.J. Aryl Hydrocarbon Receptor Activation in Astrocytes by Laquinimod Ameliorates Autoimmune Inflammation in the CNS. *Neurol. Neuroimmunol. Neuroinflamm.* **2021**, *8*, e946. [CrossRef] [PubMed]
245. Patani, R.; Hardingham, G.E.; Liddelow, S.A. Functional roles of reactive astrocytes in neuroinflammation and neurodegeneration. *Nat. Rev. Neurol.* **2023**, *19*, 7395–7409. [CrossRef] [PubMed]
246. Ageeva, T.; Rizvanov, A.; Mukhamedshina, Y. NF-κB and JAK/STAT Signaling Pathways as Crucial Regulators of Neuroinflammation and Astrocyte Modulation in Spinal Cord Injury. *Cells* **2024**, *13*, 581. [CrossRef]
247. Simpson, D.S.A.; Oliver, P.L. ROS Generation in Microglia: Understanding Oxidative Stress and Inflammation in Neurodegenerative Disease. *Antioxidants* **2020**, *9*, 743. [CrossRef]
248. Chen, Y.; Qin, C.; Huang, J.; Tang, X.; Liu, C.; Huang, K.; Xu, J.; Guo, G.; Tong, A.; Zhou, L. The role of astrocytes in oxidative stress of central nervous system: A mixed blessing. *Cell Prolif.* **2020**, *53*, e12781. [CrossRef]
249. Qin, P.; Sun, Y.; Li, L. Mitochondrial dysfunction in chronic neuroinflammatory diseases (Review). *Int. J. Mol. Med.* **2024**, *53*, 47. [CrossRef]
250. Olmos, G.; Lladó, J. Tumor necrosis factor alpha: A link between neuroinflammation and excitotoxicity. *Mediat. Inflamm.* **2014**, *2014*, 861231. [CrossRef]
251. Mottahedin, A.; Ardalan, M.; Chumak, T.; Riebe, I.; Ek, J.; Mallard, C. Effect of Neuroinflammation on Synaptic Organization and Function in the Developing Brain: Implications for Neurodevelopmental and Neurodegenerative Disorders. *Front. Cell Neurosci.* **2017**, *11*, 190. [CrossRef] [PubMed]
252. Golia, M.T.; Poggini, S.; Alboni, S.; Garofalo, S.; Ciano Albanese, N.; Viglione, A.; Ajmone-Cat, M.A.; St-Pierre, A.; Brunello, N.; Limatola, C.; et al. Interplay between inflammation and neural plasticity: Both immune activation and suppression impair LTP and BDNF expression. *Brain Behav. Immun.* **2019**, *81*, 484–494. [CrossRef]
253. Jurcau, M.C.; Jurcau, A.; Cristian, A.; Hogea, V.O.; Diaconu, R.G.; Nunkoo, V.S. Inflammaging and Brain Aging. *Int. J. Mol. Sci.* **2024**, *25*, 10535. [CrossRef] [PubMed]
254. Akhgari, A.; Michel, T.M.; Vafaee, M.S. Dendritic spines and their role in the pathogenesis of neurodevelopmental and neurological disorders. *Rev. Neurosci.* **2024**, *35*, 489–502. [CrossRef] [PubMed]
255. Rodríguez-Martín, T.; Cuchillo-Ibáñez, I.; Noble, W.; Nyenya, F.; Anderton, B.H.; Hanger, D.P. Tau phosphorylation affects its axonal transport and degradation. *Neurobiol. Aging* **2013**, *34*, 2146–2157. [CrossRef]
256. Gaikwad, S.; Senapati, S.; Haque, M.A.; Kayed, R. Senescence, brain inflammation, and oligomeric tau drive cognitive decline in Alzheimer's disease: Evidence from clinical and preclinical studies. *Alzheimers Dement.* **2024**, *20*, 709–727. [CrossRef]
257. Mehra, S.; Sahay, S.; Maji, S.K. α-Synuclein misfolding and aggregation: Implications in Parkinson's disease pathogenesis. *Biochim. Biophys. Acta Proteins Proteom.* **2019**, *1867*, 890–908. [CrossRef]
258. Sepúlveda, D.; Cisternas-Olmedo, M.; Arcos, J.; Nassif, M.; Vidal, R.L. Contribution of Autophagy-Lysosomal Pathway in the Exosomal Secretion of Alpha-Synuclein and Its Impact in the Progression of Parkinson's Disease. *Front. Mol. Neurosci.* **2022**, *15*, 805087. [CrossRef]

259. Koszła, O.; Sołek, P. Misfolding and aggregation in neurodegenerative diseases: Protein quality control machinery as potential therapeutic clearance pathways. *Cell Commun. Signal.* **2024**, *22*, 421. [CrossRef]
260. Sabirzhanov, B.; Makarevich, O.; Barrett, J.; Jackson, I.L.; Faden, A.I.; Stoica, B.A. Down-Regulation of miR-23a-3p Mediates Irradiation-Induced Neuronal Apoptosis. *Int. J. Mol. Sci.* **2020**, *21*, 3695. [CrossRef]
261. Yang, L.; Guttman, L.; Dawson, V.L.; Dawson, T.M. Parthanatos: Mechanisms, modulation, and therapeutic prospects in neurodegenerative disease and stroke. *Biochem. Pharmacol.* **2024**, *228*, 116174. [CrossRef] [PubMed]
262. Mustafa, M.; Ahmad, R.; Tantry, I.Q.; Ahmad, W.; Siddiqui, S.; Alam, M.; Abbas, K.; Moinuddin; Hassan, M. I.; Habib, S.; Islam, S. Apoptosis: A Comprehensive Overview of Signaling Pathways, Morphological Changes, and Physiological Significance and Therapeutic Implications. *Cells* **2024**, *13*, 1838. [CrossRef] [PubMed]
263. Huang, P.; Chen, G.; Jin, W.; Mao, K.; Wan, H.; He, Y. Molecular Mechanisms of Parthanatos and Its Role in Diverse Diseases. *Int. J. Mol. Sci.* **2022**, *23*, 7292. [CrossRef] [PubMed]
264. Zhang, J.; Hu, X.; Geng, Y.; Xiang, L.; Wu, Y.; Li, Y.; Yang, L.; Zhou, K. Exploring the role of parthanatos in CNS injury: Molecular insights and therapeutic approaches. *J. Adv. Res.* **2024**, *S2090-1232*, 00174–00177. [CrossRef]
265. Rana, A.; Samtiya, M.; Dhewa, T.; Mishra, V.; Aluko, R.E. Health benefits of polyphenols: A concise review. *J. Food Biochem.* **2022**, *46*, e14264. [CrossRef]
266. Zhang, W.; Qi, S.; Xue, X.; Al Naggar, Y.; Wu, L.; Wang, K. Understanding the Gastrointestinal Protective Effects of Polyphenols using Foodomics-Based Approaches. *Front. Immunol.* **2021**, *12*, 671150. [CrossRef]
267. Wilson, S.M.; Oliver, A.; Larke, J.A.; Naveja, J.J.; Alkan, Z.; Awika, J.M.; Stephensen, C.B.; Lemay, D.G. Fine-Scale Dietary Polyphenol Intake Is Associated with Systemic and Gastrointestinal Inflammation in Healthy Adults. *J. Nutr.* **2024**, *154*, 3286–3297. [CrossRef]
268. Sun, S.; Liu, Z.; Lin, M.; Gao, N.; Wang, X. Polyphenols in health and food processing: Antibacterial, anti-inflammatory, and antioxidant insights. *Front. Nutr.* **2024**, *11*, 1456730. [CrossRef]
269. Sharma, R.; Padwad, Y. Probiotic bacteria as modulators of cellular senescence: Emerging concepts and opportunities. *Gut Microbes* **2020**, *11*, 335–349. [CrossRef]
270. Hamamah, S.; Lobiuc, A.; Covasa, M. Antioxidant Role of Probiotics in Inflammation-Induced Colorectal Cancer. *Int. J. Mol. Sci.* **2024**, *25*, 9026. [CrossRef]
271. Boyajian, J.L.; Ghebretatios, M.; Schaly, S.; Islam, P.; Prakash, S. Microbiome and Human Aging: Probiotic and Prebiotic Potentials in Longevity, Skin Health and Cellular Senescence. *Nutrients* **2021**, *13*, 4550. [CrossRef] [PubMed]
272. Li, Q.; Zhang, Q.; Wang, M.; Zhao, S.; Xu, G.; Li, J. n-3 polyunsaturated fatty acids prevent disruption of epithelial barrier function induced by proinflammatory cytokines. *Mol. Immunol.* **2008**, *45*, 1356–1365. [CrossRef] [PubMed]
273. Xiong, Y.; Li, X.; Liu, J.; Luo, P.; Zhang, H.; Zhou, H.; Ling, X.; Zhang, M.; Liang, Y.; Chen, Q.; et al. Omega-3 PUFAs slow organ aging through promoting energy metabolism. *Pharmacol. Res.* **2024**, *208*, 107384. [CrossRef] [PubMed]
274. Ahmadi, A.R.; Shirani, F.; Abiri, B.; Siavash, M.; Haghighi, S.; Akbari, M. Impact of omega-3 fatty acids supplementation on the gene expression of peroxisome proliferator activated receptors-γ, α and fibroblast growth factor-21 serum levels in patients with various presentation of metabolic conditions: A GRADE assessed systematic review and dose-response meta-analysis of clinical trials. *Front. Nutr.* **2023**, *10*, 1202688.
275. See, X.Z.; Yeo, W.S.; Saptoro, A. A comprehensive review and recent advances of vitamin C: Overview, functions, sources, applications, market survey and processes. *Process Saf. Environ. Prot.* **2024**, *206*, 108–129. [CrossRef]
276. Moore, A.; Khanna, D. The Role of Vitamin C in Human Immunity and Its Treatment Potential Against COVID-19: A Review Article. *Cureus* **2023**, *15*, e33740. [CrossRef]
277. Chu, C.C.; Chew, S.C.; Liew, W.C.; Nyam, K.L. Review article vitamin E: A multi-functional ingredient for health enhancement and food preservation. *J. Food Meas. Charact.* **2023**, *17*, 6144–6156. [CrossRef]
278. Jin, D.; Wei, X.; He, Y.; Zhong, L.; Lu, H.; Lan, J.; Wei, Y.; Liu, Z.; Liu, H. The nutritional roles of zinc for immune system and COVID-19 patients. *Front. Nutr.* **2024**, *11*, 1385591. [CrossRef]
279. Wessels, I.; Maywald, M.; Rink, L. Zinc as a Gatekeeper of Immune Function. *Nutrients* **2017**, *9*, 1286. [CrossRef]
280. Narayan, S.; Dalal, R.; Rizvi, Z.A.; Awasthi, A. Zinc dampens antitumor immunity by promoting Foxp3+ regulatory T cells. *Front. Immunol.* **2024**, *15*, 1389387. [CrossRef]
281. Sergi, C.M. The Role of Zinc in the T-Cell Metabolism in Infection Requires Further Investigation—An Opinion. *Front. Immunol.* **2022**, *13*, 865504. [CrossRef]
282. Di Lorito, C.; Long, A.; Byrne, A.; Harwood, R.H.; Gladman, J.R.F.; Schneider, S.; Logan, P.; Bosco, A.; van der Wardt, V. Exercise interventions for older adults: A systematic review of meta-analyses. *J. Sport Health Sci.* **2021**, *10*, 29–47. [CrossRef] [PubMed]
283. Hernández-Álvarez, D.; Rosado-Pérez, J.; Gavia-García, G.; Arista-Ugalde, T.L.; Aguiñiga-Sánchez, I.; Santiago-Osorio, E.; Mendoza-Núñez, V.M. Aging, Physical Exercise, Telomeres, and Sarcopenia: A Narrative Review. *Biomedicines* **2023**, *11*, 598. [CrossRef] [PubMed]

284. Mathur, S.; Ardestani, A.; Parker, B.; Cappizzi, J.; Polk, D.; Thompson, P.D. Telomere length and cardiorespiratory fitness in marathon runners. *J. Investig. Med.* **2013**, *61*, 613–615. [CrossRef] [PubMed]
285. Knez, W.L.; Coombes, J.S.; Jenkins, D.G. Ultra-endurance exercise and oxidative damage: Implications for cardiovascular health. *Sports Med.* **2006**, *36*, 429–441. [CrossRef]
286. Samjoo, I.A.; Safdar, A.; Hamadeh, M.J.; Raha, S.; Tarnopolsky, M.A. The effect of endurance exercise on both skeletal muscle and systemic oxidative stress in previously sedentary obese men. *Nutr. Diabetes* **2013**, *3*, e88. [CrossRef]
287. Tang, F.; Wang, W.; Wang, Y.; Lee, Y.; Lou, Q. Moderate resistance training reduces intermuscular adipose tissue and risk factors of atherosclerotic cardiovascular disease for elderly patients with type 2 diabetes. *Diabetes Obes. Metab.* **2024**, *26*, 3418–3428. [CrossRef]
288. Goodenough, C.G.; Wogksch, M.D.; Kundu, M.; Lear, M.; Thomas, P.G.; Srivastava, D.K.; Wang, Z.; Armstrong, G.T.; Hudson, M.M.; Robison, L.L.; et al. Associations between exercise capacity, p16INK4a expression and inflammation among adult survivors of childhood cancer. *Front. Oncol.* **2022**, *12*, 1014661. [CrossRef]
289. Justice, J.N.; Gregory, H.; Tchkonia, T.; LeBrasseur, N.K.; Kirkland, J.L.; Kritchevsky, S.B.; Nicklas, B.J. Cellular Senescence Biomarker p16INK4a+ Cell Burden in Thigh Adipose is Associated With Poor Physical Function in Older Women. *J. Gerontol. A Biol. Sci. Med. Sci.* **2018**, *73*, 939–945. [CrossRef]
290. Thanapairoje, K.; Junsiritrakhoon, S.; Wichaiyo, S.; Osman, M.A.; Supharattanasitthi, W. Anti-ageing effects of FDA-approved medicines: A focused review. *J. Basic Clin. Physiol. Pharmacol.* **2023**, *34*, 277–289. [CrossRef]
291. Mohammed, I.; Hollenberg, M.D.; Ding, H.; Triggle, C.R. A Critical Review of the Evidence That Metformin Is a Putative Anti-Aging Drug That Enhances Healthspan and Extends Lifespan. *Front. Endocrinol.* **2021**, *12*, 718942. [CrossRef] [PubMed]
292. Moiseeva, O.; Deschênes-Simard, X.; St-Germain, E.; Igelmann, S.; Huot, G.; Cadar, A.E.; Bourdeau, V.; Pollak, M.N.; Ferbeyre, G. Metformin inhibits the senescence-associated secretory phenotype by interfering with IKK/NF-κB activation. *Aging Cell* **2013**, *12*, 489–498. [CrossRef] [PubMed]
293. Fang, J.; Yang, J.; Wu, X.; Zhang, G.; Li, T.; Wang, X.; Zhang, H.; Wang, C.C.; Liu, G.H.; Wang, L. Metformin alleviates human cellular aging by upregulating the endoplasmic reticulum glutathione peroxidase 7. *Aging Cell* **2018**, *17*, e12765. [CrossRef] [PubMed]
294. Liu, L.; Patnana, P.K.; Nimmagadda, S.C. Low-dose metformin and PEN2-dependent lysosomal AMPK activation: Benefits outnumber side effects. *Signal Transduct. Target Ther.* **2022**, *7*, 178. [CrossRef]
295. Lushchak, O.; Piskovatska, V.; Strilbytska, O.; Kindrat, I.; Stefanyshyn, N.; Koliada, A.; Bubalo, V.; Storey, K.B.; Vaiserman, A. Aspirin as a Potential Geroprotector: Experimental Data and Clinical Evidence. *Adv. Exp. Med. Biol.* **2021**, *1286*, 145–161.
296. Lelarge, V.; Capelle, R.; Oger, F.; Mathieu, T.; Le Calvé, B. Senolytics: From pharmacological inhibitors to immunotherapies, a promising future for patients' treatment. *NPJ Aging* **2024**, *10*, 12. [CrossRef]
297. Luís, C.; Maduro, A.T.; Pereira, P.; Mendes, J.J.; Soares, R.; Ramalho, R. Nutritional senolytics and senomorphics: Implications to immune cells metabolism and aging—From theory to practice. *Front. Nutr.* **2022**, *9*, 958563. [CrossRef]
298. Martel, J.; Ojcius, D.M.; Wu, C.Y.; Peng, H.H.; Voisin, L.; Perfettini, J.L.; Ko, Y.F.; Young, J.D. Emerging use of senolytics and senomorphics against aging and chronic diseases. *Med. Res. Rev.* **2020**, *40*, 2114–2131. [CrossRef]
299. Lagoumtzi, S.M.; Chondrogianni, N. Senolytics and senomorphics: Natural and synthetic therapeutics in the treatment of aging and chronic diseases. *Free Radic. Biol. Med.* **2021**, *171*, 169–190. [CrossRef]
300. Power, H.; Valtchev, P.; Dehghani, F.; Schindeler, A. Strategies for senolytic drug discovery. *Aging Cell* **2023**, *22*, e13948. [CrossRef]
301. Islam, M.T.; Tuday, E.; Allen, S.; Kim, J.; Trott, D.W.; Holland, W.L.; Donato, A.J.; Lesniewski, L.A. Senolytic drugs, dasatinib and quercetin, attenuate adipose tissue inflammation, and ameliorate metabolic function in old age. *Aging Cell* **2023**, *22*, e13767. [CrossRef] [PubMed]
302. Suvakov, S.; Cubro, H.; White, W.M.; Butler Tobah, Y.S.; Weissgerber, T.L.; Jordan, K.L.; Zhu, X.Y.; Woollard, J.R.; Chebib, F.T.; Milic, N.M.; et al. Targeting senescence improves angiogenic potential of adipose-derived mesenchymal stem cells in patients with preeclampsia. *Biol. Sex Differ.* **2019**, *10*, 49. [CrossRef] [PubMed]
303. Dagher, O.; Mury, P.; Thorin-Trescases, N.; Noly, P.E.; Thorin, E.; Carrier, M. Therapeutic Potential of Quercetin to Alleviate Endothelial Dysfunction in Age-Related Cardiovascular Diseases. *Front. Cardiovasc. Med.* **2021**, *8*, 658400. [CrossRef] [PubMed]
304. Saccon, T.D.; Nagpal, R.; Yadav, H.; Cavalcante, M.B.; Nunes, A.D.C.; Schneider, A.; Gesing, A.; Hughes, B.; Yousefzadeh, M.; Tchkonia, T.; et al. Senolytic Combination of Dasatinib and Quercetin Alleviates Intestinal Senescence and Inflammation and Modulates the Gut Microbiome in Aged Mice. *J. Gerontol. A Biol. Sci. Med. Sci.* **2021**, *76*, 1895–1905. [CrossRef]
305. Ota, H.; Kodama, A. Dasatinib plus quercetin attenuates some frailty characteristics in SAMP10 mice. *Sci. Rep.* **2022**, *12*, 2425. [CrossRef]
306. Zhao, J.; Zheng, L.; Dai, G.; Sun, Y.; He, R.; Liu, Z.; Jin, Y.; Wu, T.; Hu, J.; Cao, Y.; et al. Senolytics cocktail dasatinib and quercetin alleviate chondrocyte senescence and facet joint osteoarthritis in mice. *Spine J.* **2025**, *25*, 184–198. [CrossRef]

307. Nogueira-Recalde, U.; Lorenzo-Gómez, I.; Blanco, F.J.; Loza, M.I.; Grassi, D.; Shirinsky, V.; Shirinsky, I.; Lotz, M.; Robbins, P.D.; Domínguez, E.; et al. Fibrates as drugs with senolytic and autophagic activity for osteoarthritis therapy. *EBioMedicine* **2019**, *45*, 588–605. [CrossRef]
308. Sheikh, A.; Iqbal, M. Taurine as a potential anti-ageing therapy: The key to reversing the ageing process? Short communication. *Ann. Med. Surg.* **2023**, *85*, 3759–3760. [CrossRef]
309. Ni, Y.Q.; Liu, Y.S. New Insights into the Roles and Mechanisms of Spermidine in Aging and Age-Related Diseases. *Aging Dis.* **2021**, *12*, 1948–1963. [CrossRef]

biomolecules MDPI

Article

Possible Anti-Aging and Anti-Stress Effects of Long-Term Transcendental Meditation Practice: Differences in Gene Expression, EEG Correlates of Cognitive Function, and Hair Steroids

Supaya Wenuganen [1,2,*,†], Kenneth G. Walton [2,3,*,†], Frederick T. Travis [1], Tobias Stalder [4], R. Keith Wallace [2], Meera Srivastava [5] and John Fagan [2,6]

1 Center for Brain, Cognition and Consciousness, Maharishi International University, Fairfield, IA 52557, USA; ftravis@miu.edu
2 Department of Physiology and Health, Maharishi International University, Fairfield, IA 52557, USA; kwallace@miu.edu (R.K.W.); john.fagan@hrilabs.org (J.F.)
3 Institute for Prevention Research, Maharishi International University, Fairfield, IA 52557, USA
4 Department of Psychology, University of Siegen, 57076 Siegen, Germany; tobias.stalder@uni-siegen.de
5 Department of Anatomy, Physiology, and Genetics, Uniformed Services University of the Health Sciences, Bethesda, MD 20814, USA; meera.srivastava@usuhs.edu
6 Health Research Institute, Fairfield, IA 52556, USA
* Correspondence: wenuganen@miu.edu (S.W.); kwalton@miu.edu (K.G.W.); Tel.: +1-(641)-472-8370 (K.G.W.)
† These authors contributed equally to this work.

Abstract: Background: Our previous comparison of peripheral blood mononuclear cells (PBMCs) from long-term Transcendental Meditation® (TM®) practitioners and matched non-practitioner controls found 200 differentially expressed (DE) genes. Bioinformatics analyses of these DE genes suggested a reduced risk of diseases associated with stress and aging in the TM group. Here we assessed additional signs of reduced stress and aging. **Methods:** A sample of 15 of the 200 DE genes was studied using qPCR in PBMCs from 40-year TM practitioners ("Old TM", $n = 23$) compared to a "Young Control" group ($n = 19$) and an "Old Control" group ($n = 21$) of non-meditators. In these three groups, plus a "Young TM", 12-year practitioner group ($n = 26$), we also studied EEG-based parameters of cognitive function (the Brain Integration Scale (BIS), and latency of three components of the event-related potential (ERP)). Finally, using LC/MS/MS, we compared persistent levels of cortisol (F) and its inactive congener, cortisone (E), in hair. **Results**: qPCR analysis showed that 13 of the 15 genes were more highly expressed in Old Controls than in Young Controls. In the Old TM group, 7 of these 13 were lower than in Old Controls. Both TM groups had higher BIS scores than their age-matched controls. The Old TM group had shorter N2, P3a, and P3b latencies than the Old Control group, and latencies in the Old TM group were not longer than in the Young Control group. The Hair F/Hair E ratio was higher in the control subgroups than in their age-matched TM subgroups, and Hair F was higher in the Young Control and combined control groups than in the Young TM and combined TM groups. **Conclusions**: These results are consistent with reductions in biomarkers of chronic stress and biological age in long-term TM meditators. They are also consistent with results from the previous study suggesting that TM practice lowers energy consumption or leads to more efficient energy metabolism.

Keywords: human gene expression; cortisol; public health; meditation; stress reduction; cognitive function; allostatic load

1. Introduction

Many age-related diseases and disorders involve molecular mechanisms associated with stress [1–4]. Several recent reviews have attempted to systematize stressor types, different responses to stressors, and the connections of these to diseases and aging [1,5,6]. For the past several decades, the concept of "allostatic load"—i.e., the cumulative total of adaptations to stress that may be damaging to an individual—has proven useful in determining stress-related factors that predispose to disease and aging [7–9].

The term "allostasis", succinctly defined as "stability through change", is a concept applied to adaptations to potentially stressful internal or external stimuli or environments [7,9,10]. It involves the nervous, endocrine, and immune systems and is more anticipatory than the momentary perturbations characterizing homeostasis [10]. Allostasis requires energy, and allostatic load is associated with an increase in the rate of energy use [11,12]. This may have particular significance for aging, as shown in a recent study of cellular allostatic load [12], and for understanding the wider implications of the current study [13].

Evidence from our recent exploratory study of transcriptional differences in peripheral blood mononuclear cells (PBMCs) in long-term practitioners of Transcendental Meditation (TM) technologies vs. matched non-practitioner controls (i.e., the parent study) suggested that energy use by the TM group was chronically lower or more efficient than energy use by the control group [14]. Because all study participants were selected to be free of known life-threatening diseases, this suggested the demographically matched control group in the genome-wide microarray component of the study was higher in allostatic load. The finding of transcriptional differences in other stress-related pathways and genes in that study supports this conclusion.

Identification of glucocorticoid receptor signaling as one of the canonical pathways arising from pathway analysis of the microarray data in that study is consistent with prior physiologic studies [15–18]. Together, these studies suggest that reduction of long-lasting effects of stress on the hypothalamic–pituitary–adrenal (HPA) axis, especially reduction of chronically elevated cortisol, is a likely mechanism mediating some of the beneficial effects of TM technologies. Altered function of the HPA axis, particularly alterations leading to chronically elevated cortisol, has long been regarded as a prime biomarker of allostatic load or overload. Recent findings in relation to cardiovascular disease (CVD) [19] and metabolic syndrome (MetS) [20,21] are consistent with this conclusion.

Additionally, based on the genome-wide microarray data in the parent study, it was suggested, and later confirmed [22], that expression of other genes related to the damaging effects of stress, including the conserved transcriptional response to adversity (CTRA) profile, a multi-gene indicator of exposure to chronic threat [23,24], was reduced in this sample of long-term TM practitioners. The CTRA profile, which appears to involve excessive activity of the sympathetic nervous system [25], can be considered another molecular biomarker of allostatic load. It is associated with elevated inflammation and increased susceptibility to infectious diseases [23–25]. Increased inflammation may contribute to failure of the immune system to effectively clear pathogens and dysfunctional host cells, causing changes in mitochondrial function and increased abundance of reactive oxygen species, major contributors to aging [26,27].

Cognitive decline is common with advancing age, even in the absence of disease [28,29], and is strongly linked to stress and cortisol levels [3,4,30,31]. In the current study, along with further investigating transcriptional differences that may be due to long-term meditation practice, we evaluated two EEG measures of cognitive function. One of these is the latency of three components of the event-related potential (ERP), i.e., two frontal–central components, N200 (N2, stimulus evaluation) and P300a (P3a, context updating of novel stimuli), and a parietal component, P300b (P3b, context updating of infrequent stimuli—oddballs). The latency of all three components increases with age [32,33] and may increase under conditions of elevated cortisol [34], especially in subjects with poor glycemic control [35,36]. The gradual increase in these 3 latencies after age 20 suggests that this measure is a direct indicator of age-related decline in neural processing speed or efficiency [33,37].

We also examined a comprehensive scale of healthy brain functioning, the Brain Integration Scale (BIS). The BIS includes three EEG measures identified through a stepwise multiple discriminant analysis of EEG measures that distinguished three groups with different levels of expanded experiences: naïve, novice, and expert meditation. These three measures are the brain preparatory response during simple and choice paired reaction-time tasks, broadband frontal EEG coherence, and alpha/gamma power ratios during a vigilance task [38,39]. The BIS has been reported to relate to higher performance in non-meditators. Examples are: world-class athletes [40] and business leaders [41], as well as correlations with faster habituation to stressful stimuli in college students [39]. BIS levels also correlated positively with creativity in a study of product development engineers [42].

As a long-term indicator of endocrine consequences of chronic stress that may reflect allostatic load [19,43,44], the concentrations of the glucocorticoids cortisol (F) and cortisone (E) in hair were also measured in all subjects. F, the active glucocorticoid, and E, the inactive form, are enzymatically interconverted both in specific tissues and systemically [45,46]. The concentrations in hair of both F (Hair F) and E (Hair E) correlate with the risk of MetS [20], a condition previously reported to be reduced by TM practice [47].

The microarray analysis in the parent study focused on a comparison of a small group of long-term (mean, 38 years) practitioners of TM technologies with demographically well-matched controls [14]. The present study focuses on two larger, less well-matched groups of long-term (means of 12 and 40 years) TM practitioners and age-matched controls. We hypothesized that these comparisons also would show signs consistent with anti-stress and anti-aging effects of meditation, namely, 1) altered expression of specific genes related to aging or stress, 2) beneficial effects on cognitive function, as indicated by EEG tests, and 3) signs of lower systemic cortisol or better adaptation of HPA axis function.

2. Materials and Methods

2.1. Research Design and Participants

This cross-sectional, observational study compared the following 4 groups of approximately 25 subjects each: 1. Young Control, 2. Young TM, 3. Old Control, and 4. Old TM. An overview of subject characteristics and N for each of the three approaches is presented in Table 1. Additional demographics for EEG subjects are available in Table 2. The research design and methods were approved by the Institutional Review Board of Maharishi International University. Following both written and oral explanations of the study, participants gave signed written consent prior to participation.

Table 1. Overview of the Subject Characteristics in Each Group Comparison.

	Subject Characteristics	Young Controls	Young TM	Old Controls	Old TM
Gene Expression Analyses	N (males)	19 (8)	N/A	21 (9)	23 (14)
	Age (mean ± SD)	24.05 ± 3.06	N/A	62.43 ± 4.69	64.2 ± 4.2
	Months TM (mean ± SD)	N/A	N/A	N/A	477 ± 55
	Months TM-Sidhi (mean ± SD)	N/A	N/A	N/A	399 ± 42
Cognitive Performance Analyses	N (males)	25 (13)	30 (15)	22 (10)	26 (17)
	Age (mean ± SD)	24.2 ± 3.0	24.4 ± 2.86	62.7 ± 4.45	64.8 ± 3.5
	Months TM (mean ± SD)	N/A	146.5 ± 64.8	N/A	474.3 ± 33.21
	Months TM-Sidhi (mean ± SD)	N/A	47.1 ± 42.8	N/A	401.3 ± 27.0

Table 1. *Cont.*

	Subject Characteristics	Young Controls	Young TM	Old Controls	Old TM
Hair Glucocorticoid Analyses	N (males)	29 (15)	26 (18)	23 (10)	30 (18)
	Age (mean ± SD)	24.1 ± 2.9	24.0 ± 2.6	62.4 ± 4.7	63.8 ± 3.5
	Months TM (mean ± SD)	N/A	157.3 ± 60.1	N/A	478.8 ± 35.5
	Months TM-Sidhi (mean ± SD)	N/A	49.7 ± 39.2	N/A	399.6 ± 37.4

N/A, not applicable.

Table 2. Demographics for Participants in EEG Recording.

Subject Characteristics	Young Controls	Young TM	Old Controls	Old TM
N (males)	25 (13)	30 (15)	22 (10)	23 (14)
Age (mean ± SD)	24.2 ± 3.0	24.4 ± 2.86	62.7 ± 4.45	64.8 ± 3.5
Vegetarians	2	12	2	15
Smokers	12	13	5	0
Drinkers	8	12	5	2
Moderate Exercise	18	23	18	23
Professional, Skilled Work	3	7	14	17
Retired	0	0	6	10
Student	22	23	0	2
High School Only	6	4	5	3
College Degree	19	26	17	20
Right-Handed	23	28	19	20
Loss of Consciousness (Injury)	2	2	3	4
Non-English-Speaking 1st 20 Years	0	7	0	3
Married	2	1	12	13

Participants from the two age ranges ("Young", aged 20–30; "Old", aged 55–72) were recruited through distribution of advertising brochures on the campus of Maharishi International University and in other public places within a 40-mile radius of Fairfield, Iowa. Interested individuals were prescreened based on a questionnaire regarding their health status and other demographic and lifestyle variables. Note that all qPCR runs were performed at one time, i.e., as part of the original study, and that the EEG studies were performed in the same timeframe and the same subjects as the blood sampling for transcriptomic comparisons. qPCR data for the Old TM group and the Old Control group were presented in the original study [14]. For comparison with the previously unpublished results in the Young Control group, those data are presented in a different format in the second figure of this study. Note also that individuals in the Old TM group averaged having practiced the TM technique for 39.5 ± 2.8 years whereas those in the Young TM group averaged 12.2 ± 5.4 years. In most contexts, both groups would be considered long-term practitioners. Due to restricted resources, qPCR analyses were not performed in the Young TM group.

Both genders were accepted into the study. To minimize genetic variation, only one ethnic group (Caucasian) was admitted. Prospective participants were excluded if they reported having had a doctor-identified history of diabetes, nerve damage, heart attack, coronary heart disease, stroke, kidney failure, cancer, any other life-threatening illness, major psychiatric disorder, or substance abuse. Candidates for control participation were excluded if they had ever been instructed in the TM program. Practitioners of the TM technique, with or without the advancement known as the "TM-Sidhi program", were excluded if they did not report being "regular" in their practice (i.e., usually twice a day).

2.2. Procedures

2.2.1. PBMC Preparation, RNA Extraction, Measurement, and Integrity Check

Blood samples were collected from the Young Control group, the Old Control group, and the Old TM group as described in our previous publication [14]. Likewise, the methods of RNA extraction, measurement, and integrity determination were the same as before, as was the qPCR methodology. In brief, the buffy coat was isolated using BD Vacutainer® CPT™ Mononuclear Cell Preparation Tubes (Franklin Lakes, NJ, USA). The supernatant and the cell pellet were then separated and stored at −80 °C. RNA was extracted from the buffy coat using the RNA-Bee™ Kit (Tel-Test, INC, Friendswood, TX, USA), with concentration estimated using a UV absorption ratio method [48]. Ratios above 1.7 were considered sufficiently pure. RNA integrity was analyzed by automated electrophoresis on microfluidic labchips using the Experion™ System (Bio-Rad, Hercules, CA, USA), with acceptable cutoff values being in the range $7 < RQI \leq 10$. Samples meeting the criterion for integrity were then ready for use as input for cDNA labeling employing the MessageAmp™ II Kit (Ambion® by Life Technologies, Carlsbad, CA, USA).

2.3. qPCR Methodology

Target genes for comparison among the different groups were chosen more or less randomly from the DE genes identified in our previous study of global gene expression [14]. The list and design of primers, construction of cDNAs, and qPCR procedures are all detailed in the previous paper [14].

2.3.1. EEG Recording

Event-Related Potentials

EEG was recorded from 32 active sensors in the 10-10 system using the BIOSEMI Two amplifier and acquisition software system (ActiView version 9.02, Amsterdam, The Netherlands) [49–51]. Two additional sensors were applied to the left and right earlobes for re-referencing offline [39]. One sensor was attached on the non-dominant wrist for measuring heart rate [52], and a respiration belt was applied around the waist to measure breathing rate. All signals were digitized at 256 samples/s with no high- or low-frequency cutoff, and were stored for later analysis offline. Table 2 contains demographic information most relevant to EEG recording participants.

An oddball auditory event-related potential task with rare unexpected and rare expected targets was used to elicit ERPs [53]. Subjects received 240 stimuli with 1 sec inter-stimuli intervals. Twenty-four of the stimuli (10%) were rare, unexpected, 85 dB white noise bursts (*novel* stimuli). Twenty-four of the stimuli (10%) were rare, expected, 85 dB 1000 Hz tones (*oddball* stimuli). The other 80% of the stimuli (192) were standard, 85 dB 500 Hz tones, which subjects were told to ignore.

Recording Continuous EEG to Calculate the Brain Integration Scale

EEG was also recorded while subjects performed two paired reaction-time tasks. The first was a 2 min simple reaction-time task when participants were presented an asterisk on a computer screen (S1) that was followed 1.5 s later by a continuous computer-generated tone (S2, 1200 Hz, 85 dB). They were asked to press the space bar as soon as they heard the tone. The second was a 2 min choice reaction-time task when participants were presented a whole number on a computer screen (S1) that was followed 1.5 s later by another whole number. They were asked to press a right- or left-hand button to indicate which of the two numbers was larger in value.

2.3.2. Hair Steroid Concentrations

Following the protocol described in Gao et al. [54], 4 steroids (F, E, testosterone, and progesterone) were present in sufficient quantities for analysis in most samples. In brief, bundles of hair strands (~3 mm diameter) were cut as close as possible to the scalp from a posterior vertex position using fine scissors. The proximal 3 cm hair segment was used for analyses. Based on an average hair growth rate of 1 cm/month [55], this segment is assumed to reflect a period of about 3 months prior to sampling. In brief, samples were washed in 2.5 mL isopropanol for 3 min and cortisol was extracted from 7.5 mg of whole, non-pulverized hair using 1.6 mL methanol in the presence of 20 μL cortisol-d_4 as an internal standard for 18 h at room temperature. Samples were centrifuged at 15,200× g relative centrifugal force for 2 min, and 1 mL of the clear supernatant was transferred into a new 2 mL tube. The alcohol was evaporated at 50 °C under a constant stream of nitrogen and reconstituted with 175 μL double-distilled water. Samples were analyzed using a Shimadzu HPLC–tandem mass spectrometry system (Shimadzu, Canby, OR, USA) coupled to an AB Sciex API 5000 Turbo-ion-spray triple quadrupole tandem mass spectrometer (AB Sciex, Foster City, CA, USA) with purification by online solid-phase extraction. This method has been shown to achieve excellent sensitivity, specificity, and reliability (intra- and inter-assay CVs between 4.7 and 8.8%) [54] and is considered the current "gold standard" for hair steroid analysis [56].

2.4. *Data and Statistical Analysis*

2.4.1. qPCR Ct (Threshold Cycle) Values

qPCR threshold cycle (Ct) values were normalized to generate ΔCt by subtracting the Ct value of the target genes from the Ct value of the reference gene, using the BioRad CFX Maestro™ software 2.1 intended for this purpose. Then, the average of the ΔCt of each group was calculated to obtain the ΔΔCt for the expression-rate comparison of the Old Control group vs. the Old TM group and the Young Control group vs. the Old Control group. The expression rate of each gene was calculated using a formula ($2^{(\Delta\Delta Ct)}$). Then p values were determined using SPSS 13.0 (Chicago, IL, USA). Outliers were adjusted using the Winsorization procedure, which replaces the outlier with the next closest value, and then kurtosis and skewness < 1 for normal distribution were observed.

Next, MANOVA tests were conducted for the Old Control group vs. the Old TM group to find the meditation effect, the Young Control group vs. the Old Control group to find the aging effect, and the Young Control group vs. the Old TM group to find the anti-aging effect. An alpha level of $p \leq 0.05$ was adopted for statistical significance and $p < 0.1$ for statistical trend.

2.4.2. Event-Related Potentials (ERPs)

The raw data records were manually scanned for artifacts, and artifacts were removed from the analyses. The remaining data were analyzed with the BrainVision Analyzer 1.3 (Munich, Germany) (http://www.brainproducts.com/downloads.php?kid=9, accessed on 29 March 2013). Data were binned by onset of the novel and oddball stimuli in 900 ms windows—100 ms before stimulus onset and 800 ms after the stimulus [53]. The amplitude and latency were recorded for the N2 (stimulus evaluation in a 200–350 ms window in frontal–central sensors), P3a (context updating to novel stimuli in a 300–500 ms window in frontal–central sensors), and P3b (context updating to rare, expected stimuli in a 300–500 ms window in the parietal sensors).

2.4.3. Brain Integration Scale (BIS)

To calculate the brain preparatory response, 2 s epochs were extracted from the data stream beginning 100 ms pre-S1 and ending 400 ms post-S2 during both the simple and choice reaction-time tasks. For both the simple and the choice trials, the late CNV was measured in microvolts as the average amplitude in the 200 ms window before the second stimulus, relative to the 100 ms baseline. Simple-choice difference scores were calculated (CNVsimple minus CNVchoice) to assess the impact of the additional cognitive load of the choice trials independent of possible group differences in the simple trials.

Power and coherence estimates were calculated while the subjects performed Connor's Performance Task—Identical Pairs (CPT-IP), a measure of frontal executive functioning. Participants were presented a capital letter every 0.9 s for 2 min and were instructed to press the left button every time a letter appeared that was different from the preceding one. They were instructed to press the right-hand button every time the current letter was the same as the preceding letter. This task is challenging because the letters come quickly, and 80% of the letters require a left-hand button press. Participants develop a response bias for left-hand responses. When the rare right-hand response is required, the frontal executive system has to inhibit the left-hand response and execute the correct response.

Data from the CPT-IP were visually scanned for epochs with physical movement, including electrode and eye-movement artifacts. These were manually marked and were not included in the spectral analysis. The artifact-free data were then digitally filtered with a 2–50 Hz band pass filter with 48 dB roll off and fast-Fourier-transformed in 2 s epochs. Absolute power ($\mu V^2/Hz$) was calculated from 2–50 Hz in alpha, beta, and gamma bands at the 32 recording sites. Coherence was calculated for the 496 possible combination pairs of 32 recording sites in the same 3 bands. EEG coherence was taken as the absolute value of the cross-correlation function in the frequency domain, reflecting the number and strength of connections between spatially distant brain areas.

The values of broadband frontal coherence, the alpha/beta absolute power ratios from the CPT-IP, and the CNV difference scores were converted to z-scores and compared to the values in the normative database from our earlier studies [38,39]. This database included non-TM, short-term TM, and long-term TM participants. To prevent any subgroup from having a mean BIS value below zero, a value of 2 was added to the score of each subject.

Final analyses of the three ERP latencies and the BIS values were conducted with an omnibus MANOVA with age and TM status as between factors. Individual ANOVAs were used to test individual contrasts. An alpha level of $p \leq 0.05$ was adopted for statistical significance.

2.4.4. Hair Steroids

In the statistical analysis of hair steroids, the overall pattern of glucocorticoid data was similar to that observed in a previous study of 1258 employees of a large German aerospace company, and a similar statistical approach was used [20]. Both Hair F and Hair E data were skewed to the positive side. Therefore, data were tested in raw form as well as $\log_{10}$ transformed in parametric analyses. Two subjects with extremely high Hair F values (>3 SD from the mean) were removed unless otherwise stated. In paired comparisons of TM and Control groups when the homogeneity of variances assumption was not met, either the unpaired t-test with separate variance estimates or the non-parametric Mann–Whitney U test was used. An alpha level of $p \leq 0.05$ was adopted for statistical significance.

3. Results

3.1. Comparisons of Gene Expression Using qPCR

Relative expression of 15 genes randomly selected from the 200 DE genes found in the genome-wide parent study was compared in three groups, namely, the Young Control,

Old Control, and Old TM groups. Most of the genes (13 of the 15) were up-regulated in the Old Controls compared to the Young Controls (Figure 1; a positive gene expression ratio indicates up-regulation relative to the comparison group).

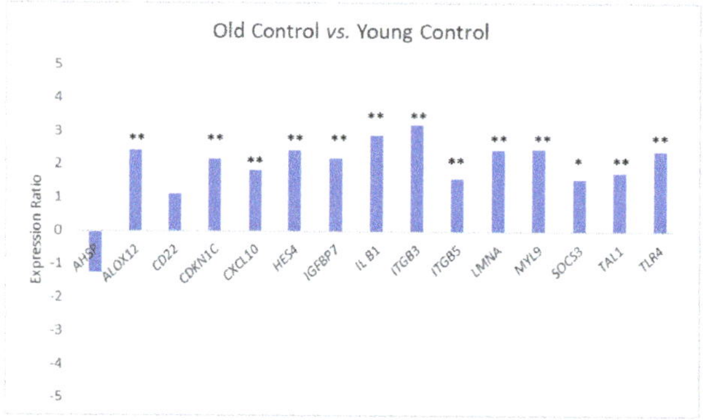

Figure 1. Expression ratio and statistical significance for the qPCR comparison of expression of 15 selected genes in the Old Control group (OC, $n = 21$) relative to the Young Control group (YC, $n = 19$). * $p \leq 0.10$; ** $p \leq 0.05$.

In contrast, expression of these genes in the Old TM group compared to the Old Control group (Figure 2) showed that only 1 of the genes (*CXCL10*) was significantly up-regulated in the TM group while 7 of 15 were down-regulated (3 significantly, 4 trending). Six were highly expressed in the Old Control group compared to the Young Control group and were considered likely age-associated genes. Ct data for subjects in each group are available in Table S1.

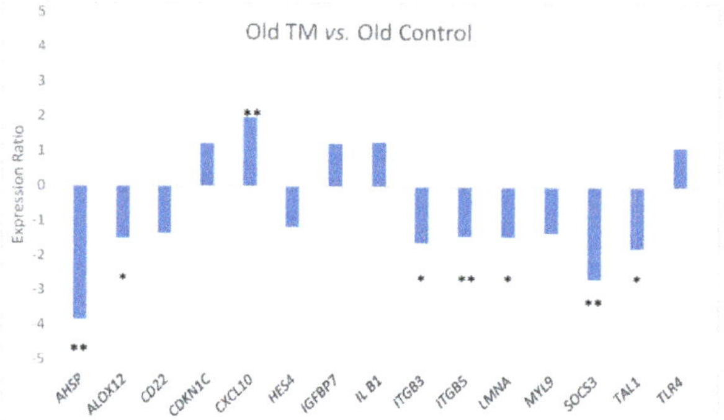

Figure 2. Expression ratio and statistical significance for the qPCR comparison of expression of 15 selected genes in the Old TM group (OTM, $n = 23$) relative to the Old Control group (OC, $n = 21$). (Some of these data appeared in the parent study) [14]. * $p \leq 0.10$; ** $p \leq 0.05$.

Figure 3 shows the relative expression patterns among the three groups, as indicated by threshold cycle values (Ct), for the six DE genes that appeared to be age-associated. Except for *SOCS3*, expression in the Old TM group was intermediate between that of the Young Control group and the Old Control group. Expression of *SOCS3*, *ITGB5*, and *TAL1* in the Old TM group was statistically the same as in the Young Control group (see Table 3). Expression of two other genes, *ITGB3* and *LMNA*, in the Old TM group was significantly higher than in the Young Control group, and expression of *ALOX12* trended higher (Figure 3).

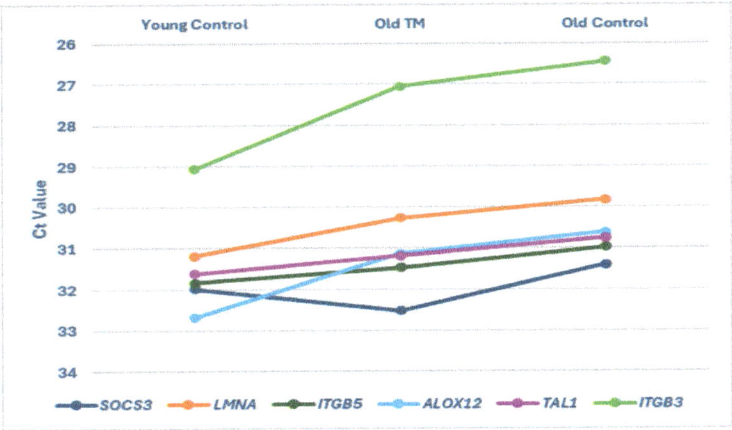

Figure 3. Comparison of average threshold cycle values (Ct) for 6 genes in the Young Control group (YC, n = 19), Old TM group (OTM, n = 23), and Old Control group (OC, n = 21).

Table 3. Expression Ratios and p-Values for Possible Anti-Aging Effect on Gene Expression.

Gene	Expression Ratio of Old Control/Young Control	p-Value	Expression Ratio of Old Control/Old TM	p-Value	Expression Ratio of Old TM/Young Control	p-Value
SOCS3	1.6	0.079	2.15	0.001	1.35 [a]	0.219
ITGB5	1.56	0.009	1.38	0.022	1.13	0.518
TAL1	1.65	0.006	1.25	0.076	1.33	0.262
ITGB3	2.95	0.001	1.44	0.06	2.05	0.002
LMNA	2.33	0.001	1.3	0.074	1.8	0.009
ALOX12	2.28	0.001	1.33	0.086	1.71	0.074

[a] Old TM was insignificantly lower than Young Control.

3.2. Comparison of N2, P3a, and P3b Latencies

An omnibus MANOVA conducted with two between factors (Age Group and TM Status) and N2, P3a, and P3b latencies as variates resulted in highly significant main effects for both factors, with latencies for the Young Group (collapsing across TM Status) and the TM Group (collapsing across Age Group) being shorter. The interaction was not significant ($F(3, 61) = 1.151$, $p = 0.287$). Table 2 (in Section 2) lists demographic data of special importance for EEG data collection. Table 4 presents the N in each group, mean (SEM) of latencies of the three components, F-statistics, and effect sizes. ERP and BIS data for each subject are available in Supplementary Information Table S2: Brain Integration Scale, Grouping Factors, and N2, P3a, and P3b Latencies.

Table 4. Two-Factor MANOVA Comparison of N2, P3a, and P3b Latencies by TM Status and Age Group.

Comparison	N	ERP	Mean ± SEM (ms)	F-Statistic and Effect Size
TM vs. Control				
TM	41	N2 P3a P3b	226.2 (4.2) 322.8 (4.0) 331.5 (6.0)	**F (3,61) = 4.4, p = 0.007, d = 0.84**
Control	42	N2 P3a P3b	247.8 (4.9) 342.6 (7.4) 361.4 (9.8)	

Table 4. *Cont.*

Comparison	N	ERP	Mean ± SEM (ms)	*F*-Statistic and Effect Size
Young vs. Old				
Young	39	N2	217.7 (4.7)	**F (3,62) = 10.9, p < 0.001, d = 0.96**
		P3a	310.4 (4.0)	
		P3b	331.5 (6.1)	
Old	44	N2	256.9 (5.1)	
		P3a	355.6 (6.5)	
		P3b	366.3 (8.5)	

Figures 4 and 5 show the grand averages ± SD of the ERP recordings for the novel and oddball stimuli at the appropriate recording sites. Figure 4 shows the tracings for N2 and P3a from novel stimuli. Figure 5 shows the tracings with P3b from oddball stimuli.

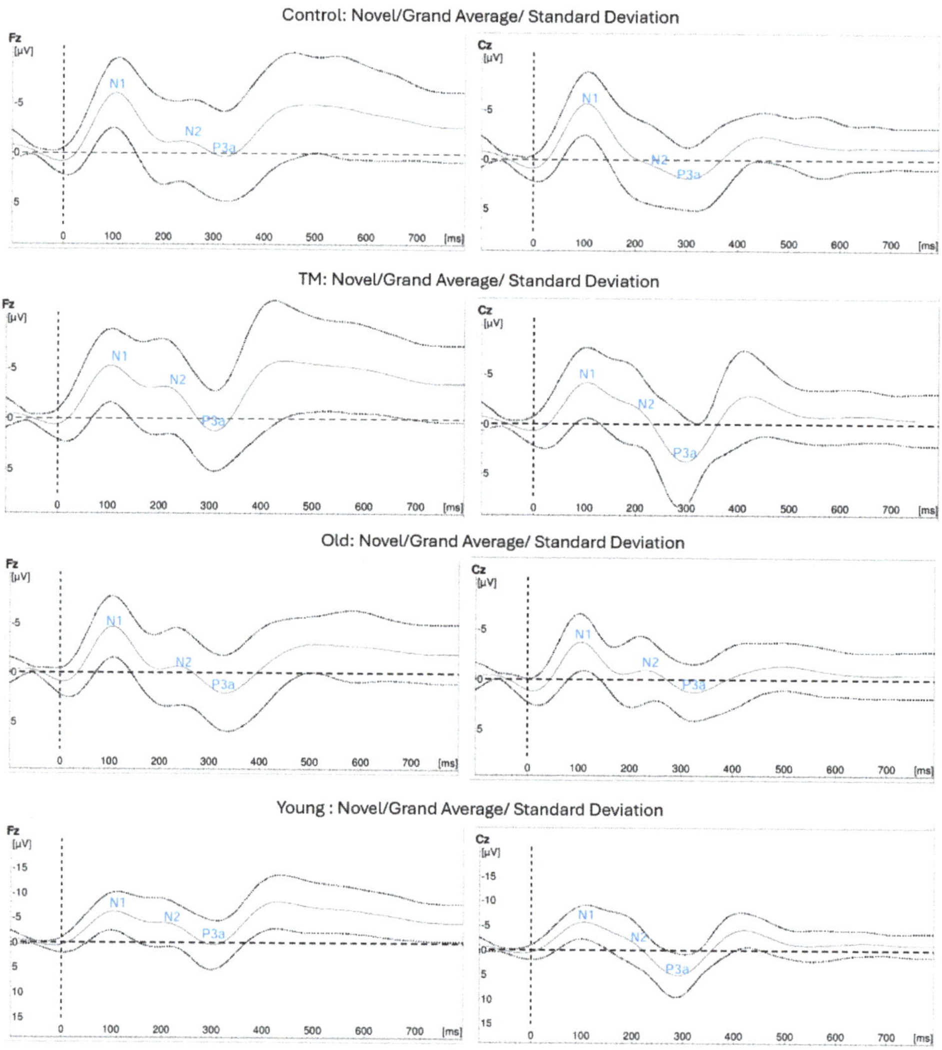

Figure 4. ERP Tracings showing N2 and P3a from Novel Stimuli.

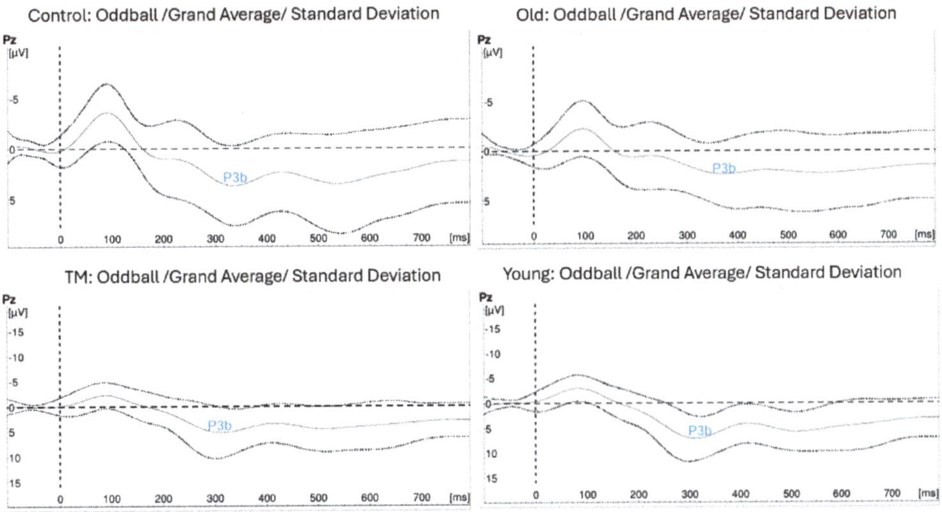

Figure 5. ERP Tracings showing P3b from Oddball Stimuli.

One-way ANOVAs were used to compare the latencies of the three components in two subgroup comparisons: Old TM vs. Old Control and Old TM vs. Young Control. Table 5 presents the results of this analysis. The Old TM group had significantly shorter latencies than those of the Old Control group. The N2 and P3b latencies were not significantly different from those of the Young Control group. P3a latencies were shorter in the Young Control group.

Table 5. Individual Subgroup ANOVAs of N2, P3a, P3b Latencies: Old TM vs. Old Control, and Old TM vs. Young Control.

	N	ERP	Mean ± SEM (ms)	*F*-Statistic and Effect Size
	Old TM vs. Old Control			
Old TM	22	N2	244.1 (3.3)	**F(1,42) = 6.4, p = 0.015,** $\eta 2$ = 0.13
		P3a	340.4 (4.9)	**F(1,37) = 5.5, p = 0.024,** $\eta 2$ = 0.13
		P3b	345.8 (6.7)	**F(1,38) = 5.6, p = 0.023,** $\eta 2$ = 0.13
Old Control	22	N2	269.7 (5.8)	
		P3a	373.5 (7.3)	
		P3b	394.5 (11.0)	
	Old TM vs. Young Control			
Old TM	22	N2	244.1 (3.3)	F(1,42) = 3.6, p = 0.07, $\eta 2$ = 0.08
		P3a	340.4 (4.9)	**F(1,40) = 5.4, p = 0.024,** $\eta 2$ = 0.12
		P3b	345.8 (6.7)	F(1,39) < 1.0, ns, $\eta 2$ = 0.005
Young Control	22	N2	226.0 (5.6)	
		P3a	316.0 (4.9)	
		P3b	339.5 (6.9)	

Note: Significant differences are bolded for easy identification. Note 2: $\eta 2$ stands for eta squared. It is a measure of effect size that indicates the proportion of variance in a dependent variable that is associated with different groups. Small effect size is $\eta 2$ = 0.01, medium effect size is $\eta 2$ = 0.06, and a large effect size is $\eta 2$ = 0.14.

3.3. Comparison of BIS Scores

BIS scores were determined for the three groups. Table 6 shows results of the initial two-factor omnibus ANOVA comparison of scores with the two categorical variables being TM Status (TM, Control) and Age Group (Young, Old). There were significant group effects for higher BIS scores for TM Status, collapsing across Age Group, and the younger group, collapsing across TM Status. There were no significant interactions. Individual BIS scores are available in Supplementary Information Table S2: Brain Integration Scale, Grouping Factors, and N2, P3a, and P3b Latencies.

Table 6. Two-Factor ANOVA Comparison of BIS Scores: TM Status and Age Group.

Comparison	*N*	Mean ± SE	Significance Test	*P* Value, Effect Size (Eta Squared)
TM Status				
TM	**55**	4.63 ± 0.226	***F* (1,92) = 9.46**	**<0.001, $\eta2$ = 0.11**
Control	45	3.62 ± 0.238		
Age Group				
Young	**52**	4.71 ± 0.224	***F* (1,92) = 12.81**	**<0.001, $\eta2$ = 0.12**
Old	48	3.54 ± 0.240		

Table 7 shows the subgroup ANOVA comparisons of BIS scores for the Old TM–Old Control and the Old TM–Young Control comparisons. As seen in the table, the mean BIS score for the Old TM group was significantly higher than for the Old Control group and not significantly different from that of the Young Control group.

Table 7. Individual Subgroup ANOVAs of BIS Scores: Old TM vs. Old Control, and Old TM vs. Young Control.

	N	Mean ± SEM	*F*-Statistic and Effect Size (Eta Squared)
Old TM vs. Old Control			***F* (1,46) = 1.9, *p* = 0.002, $\eta2$= 0.19**
Old TM	26	4.2 (1.3)	
Old Control	22	2.9 (1.4)	
Old TM vs. Young Control			*F* (1,47) < 1.0, ns, $\eta2$ < 0.001
Old TM	26	4.2 (1.3)	
Young Control	23	4.3 (2.0)	

Within the separate young and old groups, correlations between BIS and chronological age were not significant. This can be seen in the long-term Old TM group, for example, where the small negative correlation between BIS and chronological age was far from significance: $r(26) = -0.065$, $p = 0.751$. In this same subgroup, the correlation of BIS with months of TM was positive but fell short of significance: $r(26) = 0.256$, $p = 0.207$.

Exploratory comparison of ERP latencies and BIS scores for the Young TM versus the Young Control showed no significant differences between the two younger groups in ERP latencies ($F(3,31) = 1.9$, $p = 0.15$, but the mean BIS score for the Young TM group (5.3 ± 0.3) was significantly higher than for the Young Control group (4.3 ± 0.4) ($F(1,49) = 4.5$, $p = 0.038$).

3.4. Comparison of Hair Glucocorticoid Concentrations

Hair F values were significantly lower for the combined long-term TM practitioner subgroups (Total TM group) than for the combined Control subgroups (Total Control group) and for the Young TM group than for the Young Control group but were not significantly lower for the Old TM vs. Old Control comparison. The Total TM value was (M ± SE) 3.22 ± 0.466 pg/mg (min = 0.46, max = 22.3) while that of Total Control was 8.61 ± 2.50 pg/mg (min = 0.60, max = 114.6), $t_{\text{separate variances}}(53.5) = 2.132$, $p < 0.04$, Glass's *delta* = 0.30. In the Young TM group, Hair F was 2.88 ± 0.457 pg/mg (min = 0.46, max = 11.16) while in the Young Control group it was 7.64 ± 2.11 pg/mg (min = 0.60, max = 42.0), $t_{\text{separate variances}}(30.6) = 2.19$, $p < 0.04$, Glass's *delta* = 0.42. Hair F in the Old TM group was 3.59 ± 0.783 pg/mg, (min = 0.67, max = 22.3), while for the Old Control group it was 9.88 ± 5.09 pg/mg (min = 0.72, max = 114.6), $t_{\text{separate variances}}$ (22.0) = 1.23, $p = 0.23$, Glass's *delta* = 0.27. Hair F values (see above) for the Old TM group trended lower than for the Young Control group, $t_{\text{separate variances}}$ (35.5) = 1.82, $p = 0.077$, Glass's *delta* = 0.36. Data for individual subjects are shown in Supplementary Information Table S1: Hair Steroids, Grouping Factors, and Gene Expression Ct Values.

Raw Hair E values did not significantly differ between long-term TM practitioner groups and the age-matched Control groups. Hair E in the Old TM group was (M ± SE) 8.72 ± 1.33 pg/mg, (min = 1.05, max = 37.08), while for the Old Control group it was 10.11 ± 1.97 pg/mg (min = 0.45, max = 30.06). In the Young TM group, Hair E was 8.36 ± 0.98 pg/mg (min = 1.94, max = 23.2) while in the Young Control group it was 7.71 ± 1.26 pg/mg (min = 1.95, max = 38.64). Hair E values for the young and old groups combined were: Total TM = 8.56 ± 0.839 pg/mg (min = 1.05, max = 37.1), Total Control = 8.74 ± 1.12 pg/mg (min = 0.45, max = 38.6).

Because of the enzymatic interconversion (also known as "recycling") of F and E, we reasoned that the Hair F/Hair E ratio might more accurately reflect glucocorticoid status than either steroid alone. Figure 6 shows comparisons of the mean Hair F/Hair E ratios.

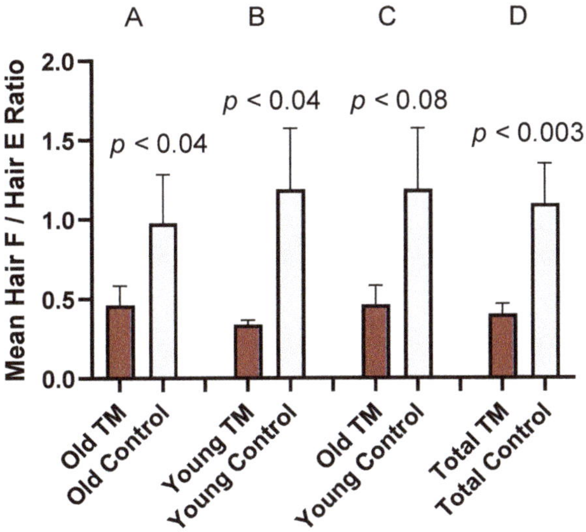

Figure 6. Comparison of Hair F/Hair E Ratios in TM and Control Subgroups. Statistical comparisons were made with the non-parametric Mann–Whitney U test. (A) Old TM vs. Old Control, Z = 2.10, $p < 0.04$, Glass's *delta* = 0.36; (B) Young TM vs. Young Control, Z = 2.10, $p < 0.04$, Glass's *delta* = 0.41; (C) Old TM vs. Young Control, Z = 1.77, $p < 0.08$, Glass's *delta* = 0.35; (D) Total TM vs. Total Control, Z = 2.98, $p < 0.003$, Glass's *delta* = 0.38. Error bars represent SE.

These results for ratio differences were similar to those for Hair F alone, but they appeared to show more robust differences and gave greater effect sizes. In an attempt to better understand the quantitative relationship between Hair F and Hair E, we conducted an analysis of regression of the $\log_{10}$ values of dependent variable Hair F on the explanatory variable $\log_{10}$ Hair E, examining all cases with no missing data (N = 108, after exclusion of two subjects with missing values). Figure 7 shows the scatter plot of the data with the ordinary least squares (OLS) regression line. The estimate of the slope regression coefficient for the explanatory variable $\log_{10}$ E is 1.2389 ($t(106) = 9.56$, $p < 0.001$), with beta coefficient 0.6959. R-square is 0.4631, with adjusted R-square 0.4581, indicating that $\log_{10}$ E explains approximately 46% of the variation in $\log_{10}$ F. The analysis was performed with Stata 18 software (StataCorp LLC, College Station, TX, USA, 2023).

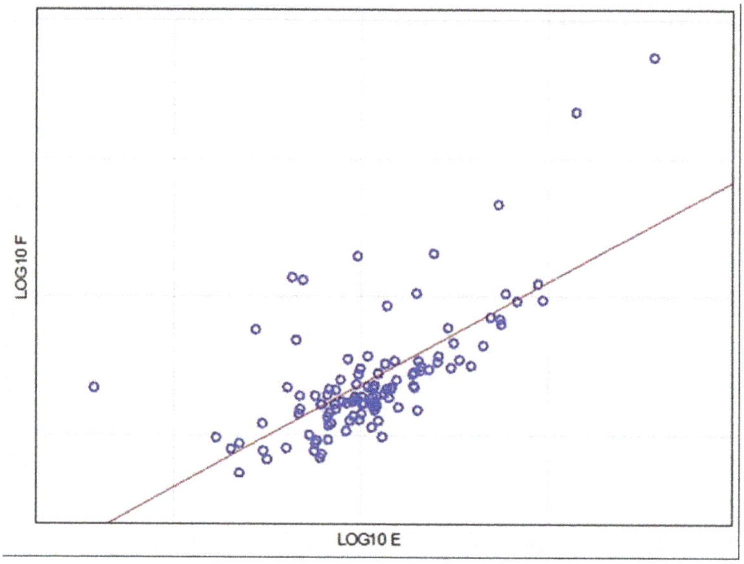

Figure 7. Scatterplot of Log_{10} Hair F Values vs. Log_{10} Hair E Values for all Study Participants.

However, a formal analysis (Doornik–Hansen test) rejected the null hypothesis that the regression errors are drawn from a normal (Gaussian) distribution ($p < 0.001$). This finding violates a key assumption of OLS regression. Formal testing [57] also indicated violation of the OLS assumption of no heteroskedasticity of the regression errors ($p < 0.001$). Heteroskedasticity is defined as non-constant conditional variance of the regression errors given the predictor variables.

The rejection of normal regression errors and visual inspection of Figure 7 suggest the possible presence of a few extreme or outlying positive observations. These potential outliers may simply reflect inherent variability of $\log_{10}$ F or may indicate that at least some of the outliers are observations drawn from a different data distribution.

Re-analysis after sequential exclusion of 12 positive outliers and 1 negative outlier by removing observations that were more than 3.5 standard errors from the regression line yielded a data set that meets all assumptions for OLS regression. After deleting these postulated outliers, the estimated slope coefficient decreased to 1.0836 ($t(94) = 19.750$, $p < 0.001$). The R-square was 0.9132, with adjusted R-square 0.9012, indicating that $\log_{10}$ E explained approximately 90% of the variation in $\log_{10}$ F.

To further characterize this high-Hair-F "Positive Outlier" subgroup, comparisons were made with the remaining "Ordinary" subgroup. Inspection of the raw data confirmed that all members of the Positive Outlier subgroup showed Hair F/Hair E ratios greater than 1.0. All members of the Ordinary group showed Hair F/Hair E ratios less than 1.0.

Because this ratio is a measure of the balance between the active glucocorticoid F and its inactive congener E, we further explored these two groups by comparing their differences in the main variables using the non-parametric Mann–Whitney U test (see Table 8). Mean Hair F in the Positive Outlier group was 8.9 times higher than in the Ordinary group. Mean Hair E in the Positive Outlier group, while numerically 2.3 times higher than that of the mean for the Ordinary group, was not statistically different. The Hair F/Hair E ratio showed the greatest difference between groups (12-fold).

Table 8. Glucocorticoid Comparisons in the Positive Outlier and Ordinary Groups.

	Positive Outlier Group Mean (SE)	Ordinary Group Mean (SE)	Mann–Whitney Z	*p*-Value
Hair F (pg/mg)	29.6 (9.42)	3.33 (0.383)	−4.53	6×10^{-6}
Hair E (pg/mg)	20.57 (9.71)	8.82 (0.727)	−0.989	0.32
Hair F/Hair E	4.35 (0.64)	0.361 (0.015)	−5.18	$<1 \times 10^{-6}$

The number of cases falling in the Positive Outlier group differed between TM and Control subjects. Two of the Old TM cases and none of the Young TM cases were in the Positive Outlier group. The 10 Control cases in the Positive Outlier group fell equally into Young Control and Old Control subgroups. Overall, five male and seven female cases fell into the Positive Outlier group.

In the Ordinary group, Hair F, Hair E, and the Hair F/Hair E ratio were compared between TM and Control groups. No significant differences in Hair F or Hair E were found between Total TM and Total Control in the Ordinary group: Hair F: $t(94) = 0.631$, $p = 0.53$; Hair E: $t(94) = 0.022$, $p = 0.98$. For the Hair F/Hair E ratio in the Ordinary group, there was a numerical tendency for lower mean (± SE) in the Total TM subgroup (0.3399 ± 0.020 pg/mg) compared to the Total Control subgroup (0.3873 ± 0.024 pg/mg), $t(94) = 1.54$, $p = 0.13$.

4. Discussion

The main findings of the study appear to support the stated hypotheses. These findings are: (relating to hypothesis 1) lower expression of 13 of the 15 selected genes in the Young Control group than in the Old Control group; lower expression of 7 of these 15 in the Old TM group than in the Old Control group; identification of specific DE genes that appear to reflect an anti-aging, anti-stress state in the Old TM group, (relating to hypothesis 2) shorter latencies of N2, P3a, and P3b (indicating higher cognitive processing speed) and higher BIS score (indicating higher cognitive performance) in the TM groups, and (relating to hypothesis 3) lower Hair F/Hair E ratio in TM groups in all comparisons of TM and control, along with lower mean Hair F in three of the four group comparisons. These outcomes are discussed in detail in the following subsections because of their relevance not only to effects of this meditation practice but also because they may help lay groundwork for new biomarkers of stress and aging.

4.1. Aging-Related Differences in Gene Expression

Participants in the present study were more numerous ($n = 23$ and 21 for the Old TM and Old Control groups, respectively) and were not as well matched as in the microarray comparison [14]. Nevertheless, qPCR analysis in this more diverse group generally validated the microarray findings. Furthermore, in the present analysis, 13 of the 15 genes that were randomly selected from the DE genes in that microarray also showed an apparent age-related difference, i.e., all 13 of these genes were more expressed in the Old Control group vs. the Young Control group. Six of these showed lower expression in the Old TM group than in the Old Control group, suggesting possible anti-aging and/or anti-stress effects of long-term meditation practice. Other researchers report that, among the small percentage of genes that differ with chronologic age, most are overexpressed in older age groups [58–60]. In one large-scale transcriptomic study, for example, 2% of the probes were associated with age; 82% of these were overexpressed in the older group [59].

4.2. *Differential Expression of Specific Genes Indicating Anti-Aging and Anti-Stress Effects*

The first four of the six genes identified here as reflecting possible anti-aging, anti-stress effects of TM (*SOCS3, ITGB3, ITGB5, LMNA, TAL1*, and *ALOX12*) were also reported by others to show age-related increases in expression in PBMCs [60]. The following subsections describe their known functions and other evidence associating them with aging and stress.

4.2.1. *SOCS3* Expression—Chronic Stress, Energy Metabolism, and CVD

Among the 15 genes studied, *SOCS3* expression provides the clearest links to possible anti-stress and anti-aging effects. Expression of *SOCS3* in the Old TM group was significantly down-regulated compared to the Old Control group and somewhat lower compared to the Young Control group.

Suppressor of cytokine signaling 3 (SOCS3) is a signaling protein induced by pro-inflammatory cytokines, including IL-6 and IL-10, and is increased in the stress response [61]. SOCS3 plays a major role in regulating inflammation and infections [62,63]. SOCS3 also plays a vital role in a complex of energy regulatory mechanisms involving interactions with AMPK, JAK-STAT, and leptin [64]. In both these roles, *SOCS3* down-regulation is consistent with a healthier state.

The SOCS3 effect on AMPK is particularly important. AMPK is an energy-status sensor for maintaining energy homeostasis [65,66]. Its full activation is prevented by SOCS3 [64]. Not only does AMPK directly affect energy metabolism through its influence on oxidative phosphorylation, but it also regulates mitochondrial generation and disposal, being responsible for ensuring there are enough functional mitochondria for normal energy metabolism [65]. Elevated level of SOCS3, by inhibiting full activation of AMPK, interferes with mitochondrial energy metabolism [64]. The parent study [14] provided evidence for reduced erythropoiesis in a long-term TM group compared with the matched control group, consistent with reduced or more efficient energy metabolism in the TM group, that could derive from reduced SOCS3.

Increased level of SOCS3, then, appears to provide a key molecular mechanism linking chronic stress to reduced mitochondrial function and aging. This is consistent with the recent recognition of mitochondrial energy metabolism as a key aspect of stress responses and adaptation [11] and may explain mitochondrial sensitivity to chronic stress. Decreased mitochondrial number and function due to chronic stress are known to contribute to aging [67]. In addition, mitochondrial health is sensitive to mood and caregiver stress [68], an effect modifiable by effective stress-reduction approaches.

A recent study of cellular allostatic load and energy metabolism in fibroblast cell lines may reinforce this interpretation of the *SOCS3* results and also connects importantly with our results on glucocorticoid activity, i.e., the Hair F/Hair E ratio, in this paper. Bobba-Alves et al. [12] studied effects of growing fibroblasts in a medium with elevated glucocorticoid concentrations comparable to those found in an organism with high allostatic load. This chronic glucocorticoid exposure caused a 60% increase in energy use in the fibroblasts along with a shift from glycolysis to oxidative phosphorylation. These signs of increased energy metabolism were accompanied by multiple signs of increased cellular aging [12]. In the present study, increased systemic glucocorticoid, as reflected by increased Hair F and increased Hair F/Hair E ratio, provides a parallel situation and probable explanation for the elevated oxygen demand evidenced in the control group in the parent study [14].

Clinical trials have found TM practice to prevent or reverse insulin resistance [47], a precursor to diabetes and obesity. Increased SOCS3 level mediates the inhibitory effects of IL-6 on insulin signaling and glucose metabolism and is documented to lead to insulin resistance in skeletal muscle, liver, and adipose tissue [64]. Other researchers have verified that chronic JAK-STAT3-SOCS3 signaling induced by leptin and IL-6 is implicated in obesity [69,70], cancer [71], and aging [60,72].

This reduced expression of *SOCS3* aligns with other research on effects of the TM program. Specifically, randomized controlled trials (RCTs) indicate this program reduces insulin resistance [47], atherosclerosis [73], other risk factors for CVD (see, for reviews, [74,75]), and cardiac events [76]. Studies also suggest practice of the technique has beneficial effects in diabetes [47,77]. Evidence for novel effects of the TM technique on energy metabolism was among the earliest findings of research on this technique (see, for review, [78]), and *SOCS3* expression provides a possible mechanistic link.

4.2.2. Expression of *ITGB5* and *ITGB3*—Inflammaging, Platelet Aggregation

Two members of the integrin family of genes that were studied in this qPCR analysis (*ITGB5* and *ITGB3*), and another member (*ITGA2B*) observed in our microarray comparison [14], were down-regulated in the Old TM group compared with the Old Control group. The integrin beta 5 (ITGB5) gene product, integrin, is important in platelet aggregation. Its related super-pathways are integrin-mediated cell adhesion and extracellular matrix–receptor interaction [79,80]. ITGB5 also is associated with extracellular matrix stiffness, one of the hallmarks of aging [81].

In addition to their adhesive functions, integrins can activate intracellular signaling pathways that control growth, differentiation, apoptosis, cell motility, cell migration, and cell survival [79,80,82]. Up-regulation of these genes is associated with increased production of inflammatory mediators through their roles as signaling molecules (e.g., IL-8 signaling, tight-junction signaling, and ILK signaling) [60]. Other researchers, studying the up-regulation of *ITGB3* with age, have suggested that the integrin β3 subunit is a marker for and regulator of cell senescence [83].

Besides inflammation, elevated expression of *ITGB* genes increases cell-to-cell aggregation and cell-to-extracellular matrix adhesion. This characteristic can lead to low mobility of cells in the circulatory system. Increased cell aggregation and adhesion in the blood contributes to disorders such as atherosclerosis, heart attack, and stroke, which are more common in old age [84]. Recently, integrins have been mentioned as novel therapeutic targets for treating cardiovascular disease [85].

4.2.3. *TAL1* and *ALOX12* Expression—Links to Leukemia, Telomerase Activity, and Oxidative Stress

Expression levels of *TAL1* and *ALOX12* were low in the Young Control group and in the Old TM group but higher in the Old Control group than in either of these. T-cell acute lymphoblastic leukemia 1 (*TAL1*, also known as "stem cell leukemia" or *SCL*) is a protein-coding gene that encodes a transcription factor regulating both embryonic and adult hematopoiesis [86]. In human adults, *TAL1* is up-regulated in erythropoiesis and down-modulated in granulopoiesis [87]. Increased level of TAL1 leads to hypersensitivity to erythropoietin, resulting in excessive erythrocytosis [88], and is the most common molecular abnormality found in human T-cell leukemia [89,90]. The elevated *TAL1* expression found here in the Old Control group is consistent with the mechanism described previously (i.e., increased erythropoiesis, see [14]) that may help compensate for the loss of energy efficiency due to elevated SOCS3. TAL1/SCL is also known to be important in vascular development and inflammation [91].

The possible anti-aging effect of reduced expression of *TAL1* suggested by our data may also relate in part to the role of TAL1 as a negative regulator of the promoter of hTERT, the protein subunit of the human telomerase gene. Because increased TAL1 leads to a decrease in hTERT mRNA abundance and hence to reduced telomerase activity [92], which is expected to accelerate aging [93], the relative down-regulation of *TAL1* in the TM group is consistent with an anti-aging effect.

ALOX12 codes for the enzyme arachidonate 12-lipoxygenase, which metabolizes arachidonic acid to generate potent inflammatory mediators and plays an important role in inflammation-associated diseases. ALOX12 and inducible nitric oxide synthase (iNOS) are well-known mediators of inflammation [94] and oxidative stress [95] in advancing age. Furthermore, ALOX12 has an important function in obesity-related complex phenotypes, including hypertension [96], atherosclerosis [97], diabetes and insulin secretion [98], and obesity itself [99].

4.2.4. *LMNA* Expression—Chromatin Structure and Nuclear Stability

LMNA, coding for the lamin proteins, is involved in nuclear stability and chromatin structure. Inflammation appears to be a part of the accelerated aging phenotype caused by *LMNA* mutations [100]. Additional reports have suggested that reduced *LMNA* expression is involved in healthy human aging [101,102]. Others have reported up-regulated expression of this gene in old age in a large meta-analysis [58] and in a study of PBMCs in nonagenarians [60].

To summarize the data on reduced gene expression in the TM group, the association of these six genes with healthy aging through their roles in controlling inflammation, energy metabolism and mitochondrial function, stability of nuclear DNA, and other key cell functions is clear. Increased expression of these genes is connected with a number of age-related diseases. Other transcriptomic studies of PBMCs and muscle are consistent with our findings concerning the key pathways that increase with age and may reflect a common aging signature [103]. The age-related down-regulation of adult cell proliferation pathways, e.g., through TAL1, is a natural strategy to prevent unregulated hematopoietic cell growth and for maintenance of telomerase activity. A separate study of hypertensive patients reported that the TM program increased telomerase gene expression [104]. Other techniques of meditation also may increase telomerase activity [15,105].

It should be noted that in certain tissues other than blood, up-regulation (instead of the down-regulation found in PBMCs) of one or more of these six genes with age may represent a healthier state. For example, in brain tissue, integrin signaling (controlled by *ITGB* genes) is needed because of its important roles in modulating neuronal survival by protecting against oxidative stress and apoptosis [106] as well as in neurogenesis, axonal outgrowth, and axonal pathfinding [107]. Also, continual *LMNA* expression is needed in many tissues for the integrity of nuclear DNA [102].

4.2.5. Up-Regulation of *CXCL10*

The *CXCL10* gene codes for C-X-C motif chemokine 10, also known as interferon gamma-induced protein 10 (IP-10) or small-inducible cytokine B10, a small cytokine belonging to the CXC chemokine family. In the current study, it is the one gene up-regulated in the Old TM group relative to the Old Control group. The significance of this difference is unclear without more information. Up-regulation of *CXCL10*, in conjunction with a unique structural domain, Glu-Leu-Arg, within the protein, can have tumor suppressor and tissue repair roles rather than pro-inflammatory and tumorigenic roles. This unique structural domain allows it to attenuate angiogenesis and have anti-tumor action [108]. An in vitro study found that a considerably higher concentration of the specific variant of CXCL10 with this structural domain is required to exhibit anti-inflammatory and anti-tumor actions [109]. If this variant is present in the older TM group, then the up-regulation of CXCL10 would have tumor suppressor actions. Further studies are required.

4.3. Latency of N2, P3a, and P3b

Our data indicate that the N2, P3a, and P3b latencies for the Young TM and Young Control collapsed together were shorter than for the Old TM and Old Control subjects collapsed together. This is consistent with a major association of latency with age. Further, the N2 and the P3b latencies of the Old TM group were not significantly different than those in the Young Control and significantly shorter than those in the Old Control subjects, consistent with possible anti-aging effect of this meditation technique.

N2, P3a, and P3b latencies index different steps of cognitive processing. The N2 component indexes response inhibition during stimulus evaluation—irrelevant stimuli are dampened so that the object in attention can be processed [110]. The P3a originates from stimulus-driven disruption of frontal attention engagement during task processing [111]; this occurred in this study when the novel white noise stimulus interrupted the counting task. The P3b originates when parietal mechanisms take the result from the stimulus evaluation processes and update the context [112]; this occurred in this study when subjects updated the mental sum of the number of oddball stimuli they saw. These three components contribute to the relative timing of cognitive processing [113].

These findings of the effect of age on N2, P3a, and P3b latencies are consistent with results of other studies and with our gene expression data. For example, the oxidative stress-related gene *ALOX12* is down-regulated in the Old TM group; oxidative stress as well as ALOX12 are associated with reduced cerebral cortical thickness [114]. These findings are also consistent with shorter P3 latencies in both auditory and visual oddball tests for long-term meditation practitioners compared to non-practitioner controls [115,116].

The latency of these components gives a cortical index of the maintenance of a more youthful cognitive processing speed with aging. Further supporting this evidence of improved cognitive function in the TM group, a random-assignment, prospective study on aging in individuals in their 80s under residential care found marked improvements in several measures of cognitive function as well as in longevity after beginning TM practice [117].

4.4. BIS Score

The BIS scores in the Old TM group were significantly higher than those in the Old Control subjects and not significantly different than those in the Young Control subjects. This suggests that Transcendental Meditation practice may help to maintain higher levels of brain integration as we age.

Previous studies of BIS support the significance of BIS scores for cognitive function. Groups of non-meditators at different levels of achievement exhibited higher BIS scores in mid-level managers [41] and in winning athletes [40] relative to those who seldom placed at the top. In other studies, the BIS was found to correlate positively with moral reasoning, inner directedness, and emotional stability and to correlate negatively with anxiety [118]. Prospective, randomized, controlled trials provide direct evidence that introducing TM practice into one's daily routine can increase BIS scores after as few as 10 weeks of 20 min twice-daily practice [39,119]. Higher BIS scores, along with shorter N2, P3a, and P3b latencies, support the possibility that TM practice leads to a more youthful state of brain function into older chronological age.

4.5. Hair Glucocorticoid Concentrations

Hair F/Hair E ratios were significantly higher in controls in all comparisons of control subgroups with TM subgroups. Hair F alone also was higher in all control subgroup comparisons except for the Old Control vs. Old TM subgroup, which fell short of significance. Close examination of the hair steroid results revealed that most of the difference between

TM and Control groups was due to a small subgroup with extremely high Hair F and Hair F/Hair E ratio. This "Positive Outlier" subgroup represented 19% of all control subjects but only 3.6% of TM subjects. It was characterized by a 12-fold higher F/E ratio than the remaining "Ordinary" subgroup. All members of this Positive Outlier subgroup showed F/E ratios less than 1.0 (mean, 0.36). The mean ratio for the Ordinary subgroup was 4.4, suggesting that if part of the body's F comes from a pool of E, as suggested by others [45,46], then this pool is greatly reduced in the high-Hair-F, Positive Outlier subgroup and might represent a suboptimal reserve for normal HPA axis function.

Previous studies in blood have reported that F levels are reduced both acutely during an individual TM practice session [120] and chronically, in a prospective, random assignment, active-control study of male medical students after 4 months of twice-daily practice of TM [18]. More relevant to the current study, lower urinary excretion rates of F were found in a previous study of long-term (avg. 8.5 y) TM-practitioner college students compared with non-meditating controls [16]. That study also reported evidence for two classes of F excreters. The excessively high-F class, termed the "sodium-loaders", showed markedly higher urinary excretion rates of not only F but also sodium, other minerals, and (during waking hours) vanillylmandelic acid (VMA), a metabolite of norepinephrine and epinephrine, than did the remaining controls or the TM group. These extra-high-F controls also showed a higher waking/sleeping ratio of F excretion. Excretion rates of aldosterone and dehydroepiandrosterone sulfate (DS), the other steroids measured, were not different between sodium-loaders and the remaining control subjects.

None of these other measures were examined in the present study, leaving uncertainty as to whether the high Hair F Outlier group observed here is comparable to the high-F subgroup in that study, which represented 33% of controls (vs. 0% in the TM group). It may be noteworthy that an exploratory aspect of the large hair steroid study involving 1258 factory employees, mentioned earlier, identified a subclass with extra-high Hair F that represented 16.6% of the total [20], possibly matching the Positive Outlier group in the current study.

It is unknown whether the observed high-Hair-F subgroup reflects genuine variability in long-term systemic F secretion not captured by Hair E or, alternatively, arises from differences in local enzymatic conversion of F to E in the hair follicle, as suggested previously [20]. However, more recent studies have directly shown cycling between F and E on the systemic level. For example, studies using pulses of F and E that were deuterium labeled in different carbon atoms, allowing the cycling rates between F and E to be directly determined, showed that not only does cycling between F and E occur within specific organs and tissues but also on a whole-body level under conditions such as hyperinsulinemia and therapeutic steroid administration [45,46]. In the current study, the large (12-fold) difference in Hair F/Hair E ratio between the Outlier and Ordinary groups suggests the possibility of different states of glucocorticoid regulation, perhaps different set-points in the balance between the active and inactive forms. Large shifts of the F/E ratio occur in the adrenal glands because this is where most circulating catabolic steroids (glucocorticoids) as well as anabolic steroids (e.g., DHEA) are synthesized [121]. The F/E ratio should serve as a better indicator of the balance of F and E than the concentration of either F or E alone. The possibility that a subgroup of individuals with high Hair F and high Hair F/Hair E ratio may reflect a distinctive allostatic load or overload condition deserves further investigation.

4.6. Glucocorticoid Activity, Allostatic Load, and Energy Usage

Our initial transcriptional comparison of long-term TM practitioners and matched controls found evidence consistent with lower allostatic load and lower, or more efficient, energy consumption [14]. The present results support these findings. Not only was

SOCS3, whose product is known to suppress full activation of the energy regulator AMPK, confirmed to be less expressed in the Old TM group but the Hair F/Hair E ratio was lower as well, suggesting lower persistent glucocorticoid activity. This may relate to the recent study of energy use in cultured fibroblasts, mentioned earlier, which showed that cortisol in the culture medium increased energy usage [12]. The authors viewed this system as an example of "cellular allostatic load" and reported that fibroblasts cultured in the presence of cortisol also had a lower Hayflick limit, i.e., more rapid aging. The authors concluded that cellular allostatic load increases energy consumption and contributes to more rapid aging [12]. This finding, along with other research, has led to a "brain–body energy conservation" model of aging [122]. This model is consistent with other research indicating that biological age undergoes a rapid increase in response to diverse forms of stress but that this change in biological age is reversed following recovery from stress [123]. The transcriptional part of our study was conducted in blood cells (PBMCs) which, according to the hair steroid results, likely experienced chronically elevated levels of cortisol, making the comparison with this study of fibroblasts in culture a reasonable one.

Other evidence that TM practice is capable of reversing allostatic load or overload comes from studies of post-traumatic stress disorder (PTSD). A systematic review and meta-analysis of 61 studies using meditation techniques to treat PTSD reports that all four categories of techniques that have been studied produced significant benefits [124]. Analysis of the 16 studies employing TM found this technique to have a large effect size (Hedges's $g = -1.13$) in reducing PTSD symptoms compared to moderate effect sizes with the three other types of meditation.

Finally, another recent article highlights the central but complex role of glucocorticoids in adaptation to stress and discusses the possibility of reversing the long-lasting effects of allostatic load contributing to disease and aging [125]. The authors elaborate on the role of the two receptors for glucocorticoids—GR and MR—in the brain and how their sequence, level, and duration of activation appear to underlie physiological, psychological, and behavioral effects, including effects on cognitive functions. These observations do not contradict the notion of age reversal by correcting effects of stress, alluded to in an earlier paragraph, and may help to understand the wide range of effects observed.

4.7. Limitations

This study lacked a suitable placebo, i.e., an active-control condition, for either the 12-year, Young-TM or the 40-year, Old-TM group. The study also was conducted at a single research site and at a single time point, i.e., without a baseline or known challenge condition. Further, the number of participants was relatively small and group sizes were sometimes unequal. Restricted resources prevented transcriptional analysis in the Young TM group. Where differences in paired outcomes were observed, these design features limit causal inferences. It is primarily the result of pre-existing, short-term RCTs of TM in the areas studied that, when coupled with the present results, provide a basis for proposing a causal role of long-term meditation practice.

5. Summary and Conclusions

The main findings of this study are consistent with the role of meditation in reducing effects of stress and aging. These are: (1) reduced expression of specific aging-related genes, including *SOCS3*, importantly related to stress and energy metabolism; (2) benefits for cognitive functions that decline with age; (3) reduced level of the Hair F/Hair E ratio and Hair F in the Young TM and the Old TM subgroups, suggesting more optimal and more stable HPA axis regulation. The prior existence of short-term RCTs and other studies whose results align with the present observations make a causal role for meditation likely.

These results, when combined with other evidence from the parent study (see [14,22]), are consistent with the conclusion that long-term practice of TM technologies can reduce or reverse allostatic load. This conclusion is supported by other studies, such as those on their efficacy in treating PTSD [124]. It is also consistent with results from a study in fibroblasts in which cellular allostatic load produced by elevated glucocorticoid level was found to cause a shift from glycolysis to oxidative phosphorylation as the main energy supply [12]. Early research on TM indicates it may foster the reverse shift, i.e., from oxidative phosphorylation to glycolysis during meditation (see, for review, [78]). Results in the parent study suggest that such a shift may pertain outside of meditation as well. From this knowledge and the known ability of SOCS3 to reduce the activity of AMPK [64], a key regulator of mitochondrial energy metabolism [65,66], a deeper understanding of the role of energetics in the benefits of meditation may be emerging [13,126].

Supplementary Materials: The following supporting information can be downloaded at: https://www.mdpi.com/article/10.3390/biom15030317/s1, Table S1: Hair Steroids, Grouping Factors, and Gene Expression Ct Values; Table S2: Brain Integration Scale, Grouping Factors, and N2, P3a, and P3b Latencies.

Author Contributions: Conceptualization, S.W., K.G.W., F.T.T. and R.K.W.; Methodology, S.W., K.G.W., F.T.T., T.S. and M.S.; Validation, M.S.; Formal analysis, S.W., K.G.W. and F.T.T.; Investigation, S.W., K.G.W., F.T.T., T.S. and M.S.; Data curation, M.S.; Writing—original draft, S.W. and K.G.W.; Writing—review and editing, K.G.W., T.S., R.K.W. and J.F.; Supervision, R.K.W. and J.F.; Project administration, J.F. All authors have read and agreed to the published version of the manuscript.

Funding: This research received on external funding.

Institutional Review Board Statement: The study was conducted in accordance with the Declaration of Helsinki, and the protocol was approved by the Ethics Committee of Maharishi International University (Project Identification Code: 06012010W) completed on 1 June 2010.

Informed Consent Statement: All participants gave written informed consent for inclusion before they participated in the study.

Data Availability Statement: The original contributions presented in this study are included in the article/Supplementary Materials.

Acknowledgments: This study represents partial fulfillment of doctoral degree requirements in physiology and health at Maharishi International University (S.W.). We thank Clemens Kirschbaum, Department of Psychology, Technical University of Dresden, Dresden, Germany, for encouragement and for suggesting hair cortisol as long-term indicator of cortisol levels. We thank Shiv Srivastava, Department of Biochemistry and Molecular & Cell Biology, Georgetown University School of Medicine, Washington, DC, USA, along with Clifton Dalgard and Gauthaman Sukumar, Department of Anatomy, Physiology and Genetics, Uniformed Services University of the Health Sciences, Maryland, USA, for assistance in conducting the qPCR analysis, Kenneth L. Cavanaugh, Tony Nader Institute for Research on Consciousness and Its Applied Technologies, Maharishi International University, Fairfield, IA 52557, USA, for assistance in regression analysis of Hair F and Hair E, and Niyazi Parim from the Center for Brain, Consciousness and Cognition at MIU for assistance with EEG recording.

Conflicts of Interest: The authors declare no conflict of interest.

References

1. Gilgoff, R.; Mengelkoch, S.; Elbers, J.; Kotz, K.; Radin, A.; Pasumarthi, I.; Murthy, R.; Sindher, S.; Burke Harris, N.; Slavich, G.M. The Stress Phenotyping Framework: A multidisciplinary biobehavioral approach for assessing and therapeutically targeting maladaptive stress physiology. *Stress* **2024**, *27*, 2327333. [CrossRef]
2. Yegorov, Y.E.; Poznyak, A.V.; Nikiforov, N.G.; Sobenin, I.A.; Orekhov, A.N. The Link between Chronic Stress and Accelerated Aging. *Biomedicines* **2020**, *8*, 198. [CrossRef]

3. Lupien, S.J.; de Leon, M.; de Santi, S.; Convit, A.; Tarshish, C.; Nair, N.P.; Thakur, M.; McEwen, B.S.; Hauger, R.L.; Meaney, M.J. Cortisol levels during human aging predict hippocampal atrophy and memory deficits. *Nat. Neurosci.* **1998**, *1*, 69–73. [CrossRef] [PubMed]
4. Lupien, S.J.; Maheu, F.; Tu, M.; Fiocco, A.; Schramek, T.E. The effects of stress and stress hormones on human cognition: Implications for the field of brain and cognition. *Brain Cogn.* **2007**, *65*, 209–237. [CrossRef]
5. McEwen, B.S. Biomarkers for assessing population and individual health and disease related to stress and adaptation. *Metabolism* **2015**, *64* (Suppl. S1), S2–S10. [CrossRef] [PubMed]
6. Epel, E.S. The geroscience agenda: Toxic stress, hormetic stress, and the rate of aging. *Ageing Res. Rev.* **2020**, *63*, 101167. [CrossRef]
7. McEwen, B.S. Stress, adaptation, and disease. Allostasis and allostatic load. *Ann. N. Y. Acad. Sci.* **1998**, *840*, 33–44. [CrossRef] [PubMed]
8. Guidi, J.; Lucente, M.; Sonino, N.; Fava, G.A. Allostatic Load and Its Impact on Health: A Systematic Review. *Psychother. Psychosom.* **2021**, *90*, 11–27. [CrossRef] [PubMed]
9. Juster, R.P.; McEwen, B.S.; Lupien, S.J. Allostatic load biomarkers of chronic stress and impact on health and cognition. *Neurosci. Biobehav. Rev.* **2010**, *35*, 2–16. [CrossRef] [PubMed]
10. Danese, A.; McEwen, B.S. Adverse childhood experiences, allostasis, allostatic load, and age-related disease. *Physiol. Behav.* **2012**, *106*, 29–39. [CrossRef] [PubMed]
11. Picard, M.; McEwen, B.S.; Epel, E.S.; Sandi, C. An energetic view of stress: Focus on mitochondria. *Front. Neuroendocrinol.* **2018**, *49*, 72–85. [CrossRef] [PubMed]
12. Bobba-Alves, N.; Sturm, G.; Lin, J.; Ware, S.A.; Karan, K.R.; Monzel, A.S.; Bris, C.; Procaccio, V.; Lenaers, G.; Higgins-Chen, A.; et al. Cellular allostatic load is linked to increased energy expenditure and accelerated biological aging. *Psychoneuroendocrinology* **2023**, *155*, 106322. [CrossRef]
13. Monzel, A.S.; Levin, M.; Picard, M. The energetics of cellular life transitions. *Life Metab.* **2024**, *3*, load051. [CrossRef] [PubMed]
14. Wenuganen, S.; Walton, K.G.; Katta, S.; Dalgard, C.L.; Sukumar, G.; Starr, J.; Travis, F.T.; Wallace, R.K.; Morehead, P.; Lonsdorf, N.K.; et al. Transcriptomics of Long-Term Meditation Practice: Evidence for Prevention or Reversal of Stress Effects Harmful to Health. *Medicina* **2021**, *57*, 218. [CrossRef] [PubMed]
15. Jacobs, T.L.; Epel, E.S.; Lin, J.; Blackburn, E.H.; Wolkowitz, O.M.; Bridwell, D.A.; Zanesco, A.P.; Aichele, S.R.; Sahdra, B.K.; MacLean, K.A.; et al. Intensive meditation training, immune cell telomerase activity, and psychological mediators. *Psychoneuroendocrinology* **2011**, *36*, 664–681. [CrossRef]
16. Walton, K.G.; Pugh, N.D.; Gelderloos, P.; Macrae, P. Stress reduction and preventing hypertension: Preliminary support for a psychoneuroendocrine mechanism. *J. Altern. Complement. Med.* **1995**, *1*, 263–283. [CrossRef]
17. Klimes-Dougan, B.; Chong, L.S.; Samikoglu, A.; Thai, M.; Amatya, P.; Cullen, K.R.; Lim, K.O. Transcendental meditation and hypothalamic-pituitary-adrenal axis functioning: A pilot, randomized controlled trial with young adults. *Stress* **2020**, *23*, 105–115. [CrossRef] [PubMed]
18. MacLean, C.R.; Walton, K.G.; Wenneberg, S.R.; Levitsky, D.K.; Mandarino, J.P.; Waziri, R.; Hillis, S.L.; Schneider, R.H. Effects of the Transcendental Meditation program on adaptive mechanisms: Changes in hormone levels and responses to stress after 4 months of practice. *Psychoneuroendocrinology* **1997**, *22*, 277–295. [CrossRef] [PubMed]
19. Iob, E.; Steptoe, A. Cardiovascular Disease and Hair Cortisol: A Novel Biomarker of Chronic Stress. *Curr. Cardiol. Rep.* **2019**, *21*, 116. [CrossRef] [PubMed]
20. Stalder, T.; Kirschbaum, C.; Alexander, N.; Bornstein, S.R.; Gao, W.; Miller, R.; Stark, S.; Bosch, J.A.; Fischer, J.E. Cortisol in hair and the metabolic syndrome. *J. Clin. Endocrinol. Metab.* **2013**, *98*, 2573–2580. [CrossRef]
21. Mazgelyte, E.; Mazeikiene, A.; Burokiene, N.; Matuzeviciene, R.; Linkeviciute, A.; Kucinskiene, Z.A.; Karciauskaite, D. Association between hair cortisol concentration and metabolic syndrome. *Open Med.* **2021**, *16*, 873–881. [CrossRef] [PubMed]
22. Walton, K.G.; Wenuganen, S.; Cole, S.W. Transcendental Meditation practitioners show reduced expression of the Conserved Transcriptional Response to Adversity. *Brain Behav. Immun. Health* **2023**, *32*, 100672. [CrossRef] [PubMed]
23. Cole, S.W. Social regulation of human gene expression: Mechanisms and implications for public health. *Am. J. Public Health* **2013**, *103* (Suppl. S1), S84–S92. [CrossRef] [PubMed]
24. Cole, S.W. The Conserved Transcriptional Response to Adversity. *Curr. Opin. Behav. Sci.* **2019**, *28*, 31–37. [CrossRef] [PubMed]
25. Powell, N.D.; Sloan, E.K.; Bailey, M.T.; Arevalo, J.M.; Miller, G.E.; Chen, E.; Kobor, M.S.; Reader, B.F.; Sheridan, J.F.; Cole, S.W. Social stress up-regulates inflammatory gene expression in the leukocyte transcriptome via beta-adrenergic induction of myelopoiesis. *Proc. Natl. Acad. Sci. USA* **2013**, *110*, 16574–16579. [CrossRef] [PubMed]
26. Salminen, A.; Kaarniranta, K.; Kauppinen, A. Inflammaging: Disturbed interplay between autophagy and inflammasomes. *Aging* **2012**, *4*, 166–175. [CrossRef]
27. Franceschi, C.; Campisi, J. Chronic inflammation (inflammaging) and its potential contribution to age-associated diseases. *J. Gerontol. A Biol. Sci. Med. Sci.* **2014**, *69* (Suppl. S1), S4–S9. [CrossRef] [PubMed]

28. Andrews-Hanna, J.R.; Snyder, A.Z.; Vincent, J.L.; Lustig, C.; Head, D.; Raichle, M.E.; Buckner, R.L. Disruption of large-scale brain systems in advanced aging. *Neuron* **2007**, *56*, 924–935. [CrossRef] [PubMed]
29. Lee, B.K.; Glass, T.A.; McAtee, M.J.; Wand, G.S.; Bandeen-Roche, K.; Bolla, K.I.; Schwartz, B.S. Associations of salivary cortisol with cognitive function in the Baltimore memory study. *Arch. Gen. Psychiatry* **2007**, *64*, 810–818. [CrossRef] [PubMed]
30. Lupien, S.; Lecours, A.R.; Lussier, I.; Schwartz, G.; Nair, N.P.; Meaney, M.J. Basal cortisol levels and cognitive deficits in human aging. *J. Neurosci.* **1994**, *14 Pt 1*, 2893–2903. [CrossRef]
31. Ouanes, S.; Popp, J. High Cortisol and the Risk of Dementia and Alzheimer's Disease: A Review of the Literature. *Front. Aging Neurosci.* **2019**, *11*, 43. [CrossRef] [PubMed]
32. Tarawneh, H.Y.; Mulders, W.; Sohrabi, H.R.; Martins, R.N.; Jayakody, D.M.P. Investigating Auditory Electrophysiological Measures of Participants with Mild Cognitive Impairment and Alzheimer's Disease: A Systematic Review and Meta-Analysis of Event-Related Potential Studies. *J. Alzheimers Dis.* **2021**, *84*, 419–448. [CrossRef]
33. Yamaguchi, S.; Knight, R.T. Age effects on the P300 to novel somatosensory stimuli. *Electroencephalogr. Clin. Neurophysiol.* **1991**, *78*, 297–301. [CrossRef]
34. Siripornpanich, V.; Rachiwong, S.; Ajjimaporn, A. A pilot study on salivary cortisol secretion and auditory P300 event-related potential in patients with physical disability-related stress. *Int. J. Neurosci.* **2020**, *130*, 170–175. [CrossRef]
35. Hamed, S.A.; Youssef, A.H.; Elserogy, Y.; Herdan, O.; Abd-Elaal, R.F.; Metwaly, N.A.-H.; Hassan, M.M.; Mohamad, H.O. Cognitive function in patients with type 2 diabetes mellitus: Relationship to stress hormone (cortisol). *J. Neurol. Neurosci.* **2013**, *4*, e335. [CrossRef]
36. do Vale, S.; Selinger, L.; Martins, J.M.; Bicho, M.; do Carmo, I.; Escera, C. Dehydroepiandrosterone (DHEA) and dehydroepiandrosterone-sulfate (DHEAS) and emotional processing—A behavioral and electrophysiological approach. *Horm. Behav.* **2015**, *73*, 94–103. [CrossRef]
37. van Dinteren, R.A.M.; Jongsma, M.L.A.; Kessels, R.P. P300 development across the lifespan: A systematic review and meta-analysis. *PLoS ONE* **2014**, *9*, e87347. [CrossRef]
38. Travis, F.; Tecce, J.; Arenander, A.; Wallace, R.K. Patterns of EEG coherence, power, and contingent negative variation characterize the integration of transcendental and waking states. *Biol. Psychol.* **2002**, *61*, 293–319. [CrossRef] [PubMed]
39. Travis, F.; Haaga, D.A.; Hagelin, J.; Tanner, M.; Nidich, S.; Gaylord-King, C.; Grosswald, S.; Rainforth, M.; Schneider, R.H. Effects of Transcendental Meditation practice on brain functioning and stress reactivity in college students. *Int. J. Psychophysiol.* **2009**, *71*, 170–176. [CrossRef]
40. Harung, H.S.; Travis, F.; Pensgaard, A.M.; Boes, R.; Cook-Greuter, S.; Daley, K. Higher psycho-physiological refinement in world-class Norwegian athletes: Brain measures of performance capacity. *Scand. J. Med. Sci. Sports* **2011**, *21*, 32–41. [CrossRef]
41. Harung, H.S.; Travis, F. Higher mind-brain development in successful leaders: Testing a unified theory of performance. *Cogn. Process.* **2012**, *13*, 171–181. [CrossRef] [PubMed]
42. Travis, F.; Lagrosen, Y. Creativity and Brain-Functioning in Product Development Engineers: A Canonical Correlation Analysis. *Creat. Res. J.* **2014**, *26*, 239–243. [CrossRef]
43. Stalder, T.; Kirschbaum, C. Analysis of cortisol in hair--state of the art and future directions. *Brain Behav. Immun.* **2012**, *26*, 1019–1029. [CrossRef]
44. Stalder, T.; Steudte-Schmiedgen, S.; Alexander, N.; Klucken, T.; Vater, A.; Wichmann, S.; Kirschbaum, C.; Miller, R. Stress-related and basic determinants of hair cortisol in humans: A meta-analysis. *Psychoneuroendocrinology* **2017**, *77*, 261–274. [CrossRef] [PubMed]
45. Anderson, A.J.; Andrew, R.; Homer, N.Z.M.; Hughes, K.A.; Boyle, L.D.; Nixon, M.; Karpe, F.; Stimson, R.H.; Walker, B.R. Effects of Obesity and Insulin on Tissue-Specific Recycling Between Cortisol and Cortisone in Men. *J. Clin. Endocrinol. Metab.* **2021**, *106*, e1206–e1220. [CrossRef]
46. Tsai, K.H.; Cooper, M.S. Recycling glucocorticoids: Therapeutic implications of the 11beta-HSD1 enzyme system. *J. Endocrinol.* **2023**, *258*, e220289. [CrossRef]
47. Paul-Labrador, M.; Polk, D.; Dwyer, J.H.; Velasquez, I.; Nidich, S.; Rainforth, M.; Schneider, R.; Merz, C.N. Effects of a randomized controlled trial of transcendental meditation on components of the metabolic syndrome in subjects with coronary heart disease. *Arch. Intern. Med.* **2006**, *166*, 1218–1224. [CrossRef]
48. Okamoto, T.; Okabe, S. Ultraviolet absorbance at 260 and 280 nm in RNA measurement is dependent on measurement solution. *Int. J. Mol. Med.* **2000**, *5*, 657–659. [CrossRef] [PubMed]
49. Travis, F. Comparison of coherence, amplitude, and eLORETA patterns during Transcendental Meditation and TM-Sidhi practice. *Int. J. Psychophysiol.* **2011**, *81*, 198–202. [CrossRef] [PubMed]
50. Fein, G.; Raz, J.; Brown, F.F.; Merrin, E.L. Common reference coherence data are confounded by power and phase effects. *Electroencephalogr. Clin. Neurophysiol.* **1988**, *69*, 581–584. [CrossRef]

51. Travis, F. A second linked-reference issue: Possible biasing of power and coherence spectra. *Int. J. Neurosci.* **1994**, *75*, 111–117. [CrossRef] [PubMed]
52. Fowles, D.C.; Christie, M.J.; Edelberg, R.; Grings, W.W.; Lykken, D.T.; Venables, P.H. Committee report. Publication recommendations for electrodermal measurements. *Psychophysiology* **1981**, *18*, 232–239. [CrossRef]
53. Travis, F.; Harung, H.S.; Lagrosen, Y. Moral development, executive functioning, peak experiences and brain patterns in professional and amateur classical musicians: Interpreted in light of a Unified Theory of Performance. *Conscious. Cogn.* **2011**, *20*, 1256–1264. [CrossRef] [PubMed]
54. Gao, W.; Stalder, T.; Foley, P.; Rauh, M.; Deng, H.; Kirschbaum, C. Quantitative analysis of steroid hormones in human hair using a column-switching LC-APCI-MS/MS assay. *J. Chromatogr. B Analyt. Technol. Biomed. Life Sci.* **2013**, *928*, 1–8. [CrossRef] [PubMed]
55. Wennig, R. Potential problems with the interpretation of hair analysis results. *Forensic. Sci. Int.* **2000**, *107*, 5–12. [CrossRef]
56. Gao, W.; Kirschbaum, C.; Grass, J.; Stalder, T. LC-MS based analysis of endogenous steroid hormones in human hair. *J. Steroid. Biochem. Mol. Biol.* **2016**, *162*, 92–99. [CrossRef]
57. White, H. A heteroskedastic-consistent covariance matrix estimator and a direct test for hetereoskedasticity. *Econometrica* **1980**, *48*, 817–838. [CrossRef]
58. Peters, M.J.; Joehanes, R.; Pilling, L.C.; Schurmann, C.; Conneely, K.N.; Powell, J.; Reinmaa, E.; Sutphin, G.L.; Zhernakova, A.; Schramm, K.; et al. The transcriptional landscape of age in human peripheral blood. *Nat. Commun.* **2015**, *6*, 8570. [CrossRef]
59. Harries, L.W.; Hernandez, D.; Henley, W.; Wood, A.R.; Holly, A.C.; Bradley-Smith, R.M.; Yaghootkar, H.; Dutta, A.; Murray, A.; Frayling, T.M.; et al. Human aging is characterized by focused changes in gene expression and deregulation of alternative splicing. *Aging Cell* **2011**, *10*, 868–878. [CrossRef] [PubMed]
60. Kuparinen, T.; Marttila, S.; Jylhava, J.; Tserel, L.; Peterson, P.; Jylha, M.; Hervonen, A.; Hurme, M. Cytomegalovirus (CMV)-dependent and -independent changes in the aging of the human immune system: A transcriptomic analysis. *Exp. Gerontol.* **2013**, *48*, 305–312. [CrossRef] [PubMed]
61. Niemand, C.; Nimmesgern, A.; Haan, S.; Fischer, P.; Schaper, F.; Rossaint, R.; Heinrich, P.C.; Muller-Newen, G. Activation of STAT3 by IL-6 and IL-10 in primary human macrophages is differentially modulated by suppressor of cytokine signaling 3. *J. Immunol.* **2003**, *170*, 3263–3272. [CrossRef]
62. Carow, B.; Rottenberg, M.E. SOCS3, a Major Regulator of Infection and Inflammation. *Front. Immunol.* **2014**, *5*, 58. [CrossRef] [PubMed]
63. Rastmanesh, M.M.; Bluyssen, H.A.; Joles, J.A.; Boer, P.; Willekes, N.; Braam, B. Increased expression of SOCS3 in monocytes and SOCS1 in lymphocytes correlates with progressive loss of renal function and cardiovascular risk factors in chronic kidney disease. *Eur. J. Pharmacol.* **2008**, *593*, 99–104. [CrossRef]
64. Sarvas, J.L.; Khaper, N.; Lees, S.J. The IL-6 Paradox: Context dependent interplay of SOCS3 and AMPK. *J. Diabetes. Metab.* **2013**, *813* (Suppl. S13), 003. [CrossRef] [PubMed]
65. Hardie, D.G. AMP-activated protein kinase: An energy sensor that regulates all aspects of cell function. *Genes Dev.* **2011**, *25*, 1895–1908. [CrossRef] [PubMed]
66. Herzig, S.; Shaw, R.J. AMPK: Guardian of metabolism and mitochondrial homeostasis. *Nat. Rev. Mol. Cell Biol.* **2018**, *19*, 121–135. [CrossRef] [PubMed]
67. Sun, N.; Youle, R.J.; Finkel, T. The Mitochondrial Basis of Aging. *Mol. Cell* **2016**, *61*, 654–666. [CrossRef] [PubMed]
68. Picard, M.; Prather, A.A.; Puterman, E.; Cuillerier, A.; Coccia, M.; Aschbacher, K.; Burelle, Y.; Epel, E.S. A Mitochondrial Health Index Sensitive to Mood and Caregiving Stress. *Biol Psychiatry* **2018**, *84*, 9–17. [CrossRef]
69. Wunderlich, C.M.; Hovelmeyer, N.; Wunderlich, F.T. Mechanisms of chronic JAK-STAT3-SOCS3 signaling in obesity. *JAKSTAT* **2013**, *2*, e23878. [CrossRef]
70. Steinberg, G.R.; McAinch, A.J.; Chen, M.B.; O'Brien, P.E.; Dixon, J.B.; Cameron-Smith, D.; Kemp, B.E. The suppressor of cytokine signaling 3 inhibits leptin activation of AMP-kinase in cultured skeletal muscle of obese humans. *J. Clin. Endocrinol. Metab.* **2006**, *91*, 3592–3597. [CrossRef]
71. Evans, M.K.; Yu, C.R.; Lohani, A.; Mahdi, R.M.; Liu, X.; Trzeciak, A.R.; Egwuagu, C.E. Expression of SOCS1 and SOCS3 genes is differentially regulated in breast cancer cells in response to proinflammatory cytokine and growth factor signals. *Oncogene* **2007**, *26*, 1941–1948. [CrossRef]
72. McKay, B.R.; Ogborn, D.I.; Baker, J.M.; Toth, K.G.; Tarnopolsky, M.A.; Parise, G. Elevated SOCS3 and altered IL-6 signaling is associated with age-related human muscle stem cell dysfunction. *Am. J. Physiol. Cell Physiol.* **2013**, *304*, C717–C728. [CrossRef] [PubMed]
73. Castillo-Richmond, A.; Schneider, R.H.; Alexander, C.N.; Cook, R.; Myers, H.; Nidich, S.; Haney, C.; Rainforth, M.; Salerno, J. Effects of stress reduction on carotid atherosclerosis in hypertensive African Americans. *Stroke* **2000**, *31*, 568–573. [CrossRef]
74. Barnes, V.A.; Orme-Johnson, D.W. Prevention and treatment of cardiovascular disease in adolescents and adults through the Transcendental Meditation((R)) Program: A research review update. *Curr. Hypertens. Rev.* **2012**, *8*, 227–242. [CrossRef]

75. Anderson, J.W.; Liu, C.; Kryscio, R.J. Blood pressure response to transcendental meditation: A meta-analysis. *Am. J. Hypertens.* **2008**, *21*, 310–316. [CrossRef] [PubMed]
76. Schneider, R.H.; Grim, C.E.; Rainforth, M.V.; Kotchen, T.; Nidich, S.I.; Gaylord-King, C.; Salerno, J.W.; Kotchen, J.M.; Alexander, C.N. Stress reduction in the secondary prevention of cardiovascular disease: Randomized, controlled trial of transcendental meditation and health education in Blacks. *Circ. Cardiovasc. Qual. Outcomes* **2012**, *5*, 750–758. [CrossRef]
77. Elder, C.; Aickin, M.; Bauer, V.; Cairns, J.; Vuckovic, N. Randomized trial of a whole-system ayurvedic protocol for type 2 diabetes. *Altern. Ther. Health Med.* **2006**, *12*, 24–30.
78. Jevning, R.; Wallace, R.K.; Beidebach, M. The physiology of meditation: A review. A wakeful hypometabolic integrated response. *Neurosci. Biobehav. Rev.* **1992**, *16*, 415–424. [CrossRef] [PubMed]
79. Shaw, L.M. Integrin function in breast carcinoma progression. *J. Mammary Gland Biol. Neoplasia* **1999**, *4*, 367–376. [CrossRef] [PubMed]
80. Taddei, I.; Faraldo, M.M.; Teuliere, J.; Deugnier, M.A.; Thiery, J.P.; Glukhova, M.A. Integrins in mammary gland development and differentiation of mammary epithelium. *J. Mammary Gland Biol. Neoplasia* **2003**, *8*, 383–394. [CrossRef]
81. Pun, F.W.; Leung, G.H.D.; Leung, H.W.; Liu, B.H.M.; Long, X.; Ozerov, I.V.; Wang, J.; Ren, F.; Aliper, A.; Izumchenko, E.; et al. Hallmarks of aging-based dual-purpose disease and age-associated targets predicted using PandaOmics AI-powered discovery engine. *Aging* **2022**, *14*, 2475–2506. [CrossRef] [PubMed]
82. Hood, J.D.; Bednarski, M.; Frausto, R.; Guccione, S.; Reisfeld, R.A.; Xiang, R.; Cheresh, D.A. Tumor regression by targeted gene delivery to the neovasculature. *Science* **2002**, *296*, 2404–2407. [CrossRef] [PubMed]
83. Rapisarda, V.; Borghesan, M.; Miguela, V.; Encheva, V.; Snijders, A.P.; Lujambio, A.; O'Loghlen, A. Integrin Beta 3 Regulates Cellular Senescence by Activating the TGF-beta Pathway. *Cell Rep.* **2017**, *18*, 2480–2493. [CrossRef]
84. Libby, P.; Ridker, P.M.; Maseri, A. Inflammation and atherosclerosis. *Circulation* **2002**, *105*, 1135–1143. [CrossRef]
85. Lal, H.; Guleria, R.S.; Foster, D.M.; Lu, G.; Watson, L.E.; Sanghi, S.; Smith, M.; Dostal, D.E. Integrins: Novel therapeutic targets for cardiovascular diseases. *Cardiovasc. Hematol. Agents Med. Chem.* **2007**, *5*, 109–132. [CrossRef] [PubMed]
86. Porcher, C.; Swat, W.; Rockwell, K.; Fujiwara, Y.; Alt, F.W.; Orkin, S.H. The T cell leukemia oncoprotein SCL/tal-1 is essential for development of all hematopoietic lineages. *Cell* **1996**, *86*, 47–57. [CrossRef]
87. Condorelli, G.; Vitelli, L.; Valtieri, M.; Marta, I.; Montesoro, E.; Lulli, V.; Baer, R.; Peschle, C. Coordinate expression and developmental role of Id2 protein and TAL1/E2A heterodimer in erythroid progenitor differentiation. *Blood* **1995**, *86*, 164–175. [CrossRef] [PubMed]
88. Rogers, H.; Wang, L.; Yu, X.; Alnaeeli, M.; Cui, K.; Zhao, K.; Bieker, J.J.; Prchal, J.; Huang, S.; Weksler, B.; et al. T-cell acute leukemia 1 (TAL1) regulation of erythropoietin receptor and association with excessive erythrocytosis. *J. Biol. Chem.* **2012**, *287*, 36720–36731. [CrossRef]
89. Mansour, M.R.; Sanda, T.; Lawton, L.N.; Li, X.; Kreslavsky, T.; Novina, C.D.; Brand, M.; Gutierrez, A.; Kelliher, M.A.; Jamieson, C.H.; et al. The TAL1 complex targets the FBXW7 tumor suppressor by activating miR-223 in human T cell acute lymphoblastic leukemia. *J. Exp. Med.* **2013**, *210*, 1545–1557. [CrossRef] [PubMed]
90. Robb, L.; Begley, C.G. The SCL/TAL1 gene: Roles in normal and malignant haematopoiesis. *Bioessays* **1997**, *19*, 607–613. [CrossRef]
91. Bian, X.W.; Chen, J.H.; Jiang, X.F.; Bai, J.S.; Wang, Q.L.; Zhang, X. Angiogenesis as an immunopharmacologic target in inflammation and cancer. *Int. Immunopharmacol.* **2004**, *4*, 1537–1547. [CrossRef]
92. Terme, J.M.; Mocquet, V.; Kuhlmann, A.S.; Zane, L.; Mortreux, F.; Wattel, E.; Duc Dodon, M.; Jalinot, P. Inhibition of the hTERT promoter by the proto-oncogenic protein TAL1. *Leukemia* **2009**, *23*, 2081–2089. [CrossRef] [PubMed]
93. Olovnikov, A.M. Telomeres, telomerase, and aging: Origin of the theory. *Exp. Gerontol.* **1996**, *31*, 443–448. [CrossRef]
94. Tersey, S.A.; Bolanis, E.; Holman, T.R.; Maloney, D.J.; Nadler, J.L.; Mirmira, R.G. Minireview: 12-Lipoxygenase and Islet beta-Cell Dysfunction in Diabetes. *Mol. Endocrinol.* **2015**, *29*, 791–800. [CrossRef] [PubMed]
95. Almeida, M.; Ambrogini, E.; Han, L.; Manolagas, S.C.; Jilka, R.L. Increased lipid oxidation causes oxidative stress, increased peroxisome proliferator-activated receptor-gamma expression, and diminished pro-osteogenic Wnt signaling in the skeleton. *J. Biol. Chem.* **2009**, *284*, 27438–27448. [CrossRef]
96. Quintana, L.F.; Guzman, B.; Collado, S.; Claria, J.; Poch, E. A coding polymorphism in the 12-lipoxygenase gene is associated to essential hypertension and urinary 12(S)-HETE. *Kidney Int.* **2006**, *69*, 526–530. [CrossRef] [PubMed]
97. Huo, Y.; Zhao, L.; Hyman, M.C.; Shashkin, P.; Harry, B.L.; Burcin, T.; Forlow, S.B.; Stark, M.A.; Smith, D.F.; Clarke, S.; et al. Critical role of macrophage 12/15-lipoxygenase for atherosclerosis in apolipoprotein E-deficient mice. *Circulation* **2004**, *110*, 2024–2031. [CrossRef] [PubMed]
98. Bleich, D.; Chen, S.; Zipser, B.; Sun, D.; Funk, C.D.; Nadler, J.L. Resistance to type 1 diabetes induction in 12-lipoxygenase knockout mice. *J. Clin. Investig.* **1999**, *103*, 1431–1436. [CrossRef] [PubMed]
99. Xiao, W.J.; He, J.W.; Zhang, H.; Hu, W.W.; Gu, J.M.; Yue, H.; Gao, G.; Yu, J.B.; Wang, C.; Ke, Y.H.; et al. ALOX12 polymorphisms are associated with fat mass but not peak bone mineral density in Chinese nuclear families. *Int. J. Obes.* **2011**, *35*, 378–386. [CrossRef]

100. Tran, J.R.; Chen, H.; Zheng, X.; Zheng, Y. Lamin in inflammation and aging. *Curr. Opin. Cell Biol.* **2016**, *40*, 124–130. [CrossRef]
101. Ragnauth, C.D.; Warren, D.T.; Liu, Y.; McNair, R.; Tajsic, T.; Figg, N.; Shroff, R.; Skepper, J.; Shanahan, C.M. Prelamin A acts to accelerate smooth muscle cell senescence and is a novel biomarker of human vascular aging. *Circulation* **2010**, *121*, 2200–2210. [CrossRef]
102. Scaffidi, P.; Misteli, T. Lamin A-dependent nuclear defects in human aging. *Science* **2006**, *312*, 1059–1063. [CrossRef]
103. Zahn, J.M.; Sonu, R.; Vogel, H.; Crane, E.; Mazan-Mamczarz, K.; Rabkin, R.; Davis, R.W.; Becker, K.G.; Owen, A.B.; Kim, S.K. Transcriptional profiling of aging in human muscle reveals a common aging signature. *PLoS Genet.* **2006**, *2*, e115. [CrossRef]
104. Duraimani, S.; Schneider, R.H.; Randall, O.S.; Nidich, S.I.; Xu, S.; Ketete, M.; Rainforth, M.A.; Gaylord-King, C.; Salerno, J.W.; Fagan, J. Effects of lifestyle modification on telomerase gene expression in hypertensive patients: A pilot trial of stress reduction and health education programs in African Americans. *PLoS ONE* **2015**, *10*, e0142689. [CrossRef] [PubMed]
105. Daubenmier, J.; Lin, J.; Blackburn, E.; Hecht, F.M.; Kristeller, J.; Maninger, N.; Kuwata, M.; Bacchetti, P.; Havel, P.J.; Epel, E. Changes in stress, eating, and metabolic factors are related to changes in telomerase activity in a randomized mindfulness intervention pilot study. *Psychoneuroendocrinology* **2012**, *37*, 917–928. [CrossRef]
106. Gary, D.S.; Mattson, M.P. Integrin signaling via the PI3-kinase-Akt pathway increases neuronal resistance to glutamate-induced apoptosis. *J. Neurochem.* **2001**, *76*, 1485–1496. [CrossRef] [PubMed]
107. Myers, J.P.; Santiago-Medina, M.; Gomez, T.M. Regulation of axonal outgrowth and pathfinding by integrin-ECM interactions. *Dev. Neurobiol.* **2011**, *71*, 901–923. [CrossRef] [PubMed]
108. Persano, L.; Crescenzi, M.; Indraccolo, S. Anti-angiogenic gene therapy of cancer: Current status and future prospects. *Mol. Aspects. Med.* **2007**, *28*, 87–114. [CrossRef]
109. Campanella, G.S.; Grimm, J.; Manice, L.A.; Colvin, R.A.; Medoff, B.D.; Wojtkiewicz, G.R.; Weissleder, R.; Luster, A.D. Oligomerization of CXCL10 is necessary for endothelial cell presentation and in vivo activity. *J. Immunol.* **2006**, *177*, 6991–6998. [CrossRef]
110. Hoyniak, C.P.; Petersen, I.T. A meta-analytic evaluation of the N2 component as an endophenotype of response inhibition and externalizing psychopathology in childhood. *Neurosci. Biobehav. Rev.* **2019**, *103*, 200–215. [CrossRef]
111. Polich, J.; Criado, J.R. Neuropsychology and neuropharmacology of P3a and P3b. *Int. J. Psychophysiol.* **2006**, *60*, 172–185. [CrossRef]
112. Polich, J. Updating P300: An integrative theory of P3a and P3b. *Clin. Neurophysiol.* **2007**, *118*, 2128–2148. [CrossRef] [PubMed]
113. Coles, M.G.H.; Rugg, M.D. Event-related brain potentials: An introduction. In *Electrophysiology of Mind*; Rugg, M.D., Coles, M.G.H., Eds.; Oxford University Press: New York, NY, USA, 1995; pp. 1–26.
114. Miller, M.W.; Wolf, E.J.; Sadeh, N.; Logue, M.; Spielberg, J.M.; Hayes, J.P.; Sperbeck, E.; Schichman, S.A.; Stone, A.; Carter, W.C.; et al. A novel locus in the oxidative stress-related gene ALOX12 moderates the association between PTSD and thickness of the prefrontal cortex. *Psychoneuroendocrinology* **2015**, *62*, 359–365. [CrossRef]
115. Polich, J. Meta-analysis of P300 normative aging studies. *Psychophysiology* **1996**, *33*, 334–353. [CrossRef] [PubMed]
116. Cahn, B.R.; Polich, J. Meditation states and traits: EEG, ERP, and neuroimaging studies. *Psychol. Bull.* **2006**, *132*, 180–211. [CrossRef] [PubMed]
117. Alexander, C.N.; Langer, E.J.; Newman, R.I.; Chandler, H.M.; Davies, J.L. Transcendental meditation, mindfulness, and longevity: An experimental study with the elderly. *J. Pers. Soc. Psychol.* **1989**, *57*, 950–964. [CrossRef] [PubMed]
118. Travis, F.; Arenander, A.; DuBois, D. Psychological and physiological characteristics of a proposed object-referral/self-referral continuum of self-awareness. *Conscious. Cogn.* **2004**, *13*, 401–420. [CrossRef] [PubMed]
119. Travis, F.; Valosek, L.; Konrad, A.T.; Link, J.; Salerno, J.; Scheller, R.; Nidich, S. Effect of meditation on psychological distress and brain functioning: A randomized controlled study. *Brain Cogn.* **2018**, *125*, 100–105. [CrossRef]
120. Jevning, R.; Wilson, A.F.; Davidson, J.M. Adrenocortical activity during meditation. *Horm. Behav.* **1978**, *10*, 54–60. [CrossRef] [PubMed]
121. Parker, L.N. Control of adrenal androgen secretion. *Endocrinol. Metab. Clin. N. Am.* **1991**, *20*, 401–421. [CrossRef]
122. Shaulson, E.D.; Cohen, A.A.; Picard, M. The brain-body energy conservation model of aging. *Nat. Aging* **2024**, *4*, 1354–1371. [CrossRef] [PubMed]
123. Poganik, J.R.; Zhang, B.; Baht, G.S.; Tyshkovskiy, A.; Deik, A.; Kerepesi, C.; Yim, S.H.; Lu, A.T.; Haghani, A.; Gong, T.; et al. Biological age is increased by stress and restored upon recovery. *Cell Metab.* **2023**, *35*, 807–820.e5. [CrossRef] [PubMed]
124. Orme-Johnson, D.W.; Barnes, V.A.; Rees, B.; Tobin, J.; Walton, K.G. Effectiveness of meditation techniques in treating post-traumatic stress disorder (PTSD): A systematic review and meta-analysis. *Medicina* **2024**, *60*, 2050. [CrossRef] [PubMed]

125. Daskalakis, N.P.; Meijer, O.C.; de Kloet, E.R. Mineralocorticoid receptor and glucocorticoid receptor work alone and together in cell-type-specific manner: Implications for resilience prediction and targeted therapy. *Neurobiol. Stress* **2022**, *18*, 100455. [CrossRef] [PubMed]
126. Trumpff, C.; Monzel, A.S.; Sandi, C.; Menon, V.; Klein, H.U.; Fujita, M.; Lee, A.; Petyuk, V.A.; Hurst, C.; Duong, D.M.; et al. Psychosocial experiences are associated with human brain mitochondrial biology. *Proc. Natl. Acad. Sci. USA* **2024**, *121*, e2317673121. [CrossRef] [PubMed]

biomolecules

MDPI

Review

Perception and Longevity Control in Invertebrate Model Organisms—A Mini-Review of Recent Advances

Nicholas Pontillo and Yang Lyu *

Department of Molecular Biology and Biochemistry, Rutgers, The State University of New Jersey, Piscataway, NJ 08854-8000, USA; np743@cabm.rutgers.edu
* Correspondence: yang.lyu@rutgers.edu

Abstract: Perception alone can, in some cases, be sufficient to modulate aging and longevity. These influences on aging are perhaps mediated by changes in motivational states that regulate metabolism and physiology to impact health. Simple invertebrate models uniquely enable detailed dissection of integrative pathways linking perceptions to aging and remain the leading systems for advancing this field. Over the past 25 years, studies using the fruit fly *Drosophila melanogaster* and the nematode *Caenorhabditis elegans* have demonstrated that sensory cues, such as those related to food or mating, can influence aging independently of the physical acts associated with them. In this review, we highlight recent advancements in these invertebrate models, focusing on two key areas of progress: (i) the discovery of lifespan modulation driven by novel sensory cues across multiple modalities, including non-sexual social experience, light, and dietary choices; and (ii) the assignment of new aging-regulation functions to specific neurons downstream of sensory perception. The latter offers an exciting first glimpse at the neuronal circuits integrating sensory cues, motivational states, physiology, and aging.

Keywords: perception; longevity; lifespan; life history theory; *Drosophila melanogaster*; *Caenorhabditis elegans*

1. Introduction

According to life history theory, the rate of aging can be modified by the division of scarce resources between short-term and long-term reproduction—the latter including somatic maintenance and behaviors that tend to increase long-term survival. Part of this division, and therefore, the rate of aging, is set genetically, but another part can be modulated in real time by the organism in response to environmental conditions [1–4]. Perhaps the most well-studied manifestation of this phenomenon is the promotion of reproductive investments by dietary resources—particularly total calories and protein—and the converse promotion of longevity by food or protein scarcity—although there has been debate over whether and to what extent longevity gains "must" be traded off with worse reproductive or other types of performance [5,6].

As the organism itself can change its physiology to prioritize either short-term reproduction or longevity, chronic sensory perceptions and associated motivational states can powerfully influence the rate of aging. Somewhat counterintuitively, this subject is most well-documented in comparatively simple invertebrate models—the nematode *Caenorhabditis elegans* and the fruit fly *Drosophila melanogaster*. Sensory systems that detect specific environmental cues have been shown to regulate physiology and longevity through neurosignaling pathways in these species [1–4]. While this biology of aging is well-documented, the precise mechanisms underlying these processes remain poorly understood.

https://doi.org/10.3390/biom15020187

In this mini-review, we examine recent findings on the perceptive regulation of longevity, focusing on (i) critically evaluating the evidence that perception truly mediates at least part of a reported longevity phenotype, especially for newly reported sensory modalities, (ii) highlighting newly discovered mechanisms and pathways, and (iii) discussing the implications of these findings for classical and evolutionary theories of aging. Aging is a deeply studied research topic, and we limit the scope of our discussion only to longevity phenotypes that may be attributed to "external" perceptual cues such as via olfaction, gustation, vision, and somatosensation.

2. Dietary Cues

Food-Related Cues and Dietary Restriction

Diet is one of the most well-studied variables that influences aging. In particular, dietary restriction (DR)—defined as the restriction of nutrient ingestion to the minimum necessary to prevent malnutrition—is known to extend lifespan in multiple animal models [7]. Studies in *D. melanogaster* and *C. elegans* produced some of the earliest evidence that diet-related sensations can modulate the rate of aging [8–10].

One such piece of evidence was the finding that exposure to food odors alone is sufficient to abrogate the pro-longevity effect of DR in *D. melanogaster* [11] and *C. elegans* [12]. In *C. elegans*, two recent studies have delineated the neural pathways responsible in remarkable detail [13,14], though with some unresolved differences. Food odorants partially abrogate the effect of DR via a neural circuit involving serotonergic olfactory neurons and downstream dopaminergic and octopaminergic neurons, which ultimately regulate intestinal expression of either the energy sensor adenosine monophosphate-activated protein kinase (AMPK) [13] or flavin-containing monooxygenase 2 (FMO-2) [14]. Commonalities and discrepancies between the models proposed by the two studies are noted in Figure 1.

Beyond chemosensation, food-related tactile stimuli also regulate lifespan. In *C. elegans*, dopaminergic neurons innervate the epidermis with mechanosensitive cilia, which are sensitive to the light tactile stimuli produced by encountering bacteria, their main food source. Activation of these neurons by tactile stimuli putatively induces dopamine release, which inhibits locomotion (the "slowing response" to touched food), presumably to facilitate feeding [15]. A recent study reported that this type of tactile stimulation partially abrogates the pro-longevity effect of dietary restriction by impairing DR-induced changes in insulin and FMO-2 signaling [16]. Thus, both chemosensory and mechanosensory perception of food (or lack thereof) are critical for DR lifespan phenotypes in *C. elegans*.

In line with the unique roles of various nutrients in modulating lifespan and physiology, the perception of different nutrients may likewise have distinct effects. In *D. melanogaster*, we and colleagues found that merely segregating the primary carbohydrate (sucrose) and protein (yeast) sources into separate portions (termed the "choice" paradigm) can either dramatically [17–19] or moderately [20] reduce lifespan, though with sex differences. Furthermore, there is evidence that differential positioning of the nutrients can reverse the direction of the effect [21], though this requires validation in a well-controlled environment. Strilbytska and colleagues suggest that flies choosing their own diet may live shorter because of differences in nutrient intake, specifically an increase in consumed protein, which was argued to reflect a fitness-maximizing investment in reproduction at the expense of optimal longevity [20,22]. In contrast, our group's data suggest that nutrient intake alone does not fully account for the observed effects and that feeding and longevity phenotypes can be uncoupled, suggesting instead that some aspect of nutrient perception in the choice paradigm may shorten lifespan [19]. Nevertheless, discrepancies in measured food intake across these two research groups (perhaps due to differences in the time allowed for the consumption assay) constitute an unresolved issue for future investigation.

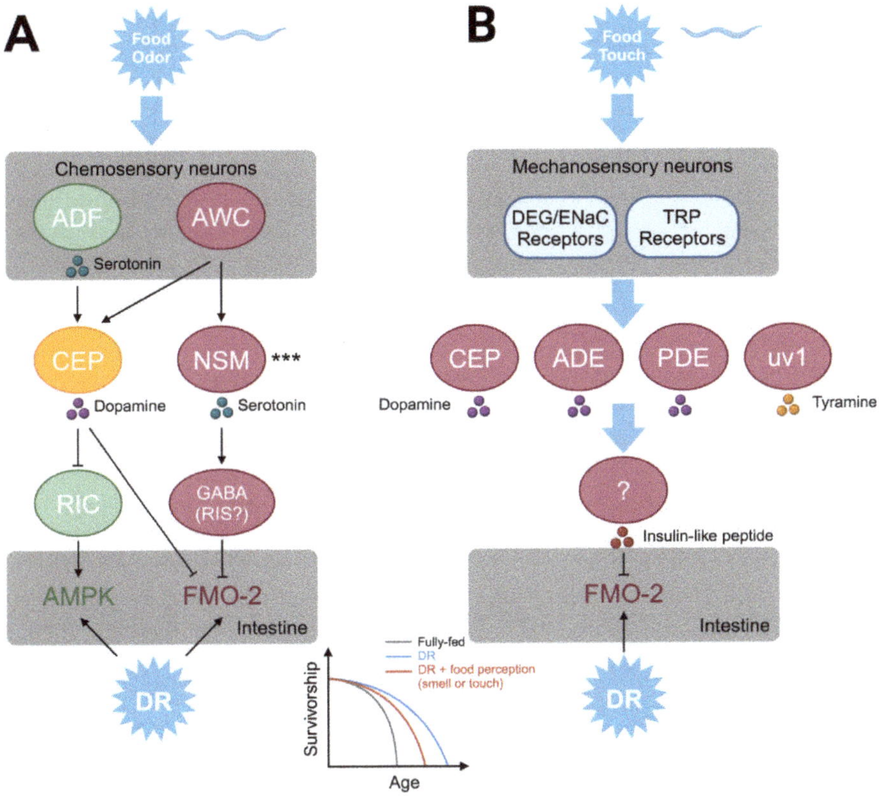

Figure 1. Neural circuits linking food cues and longevity in *C. elegans*. (**A**). Circuit by which food odorant information regulates FMO-2 and/or AMPK activity to abrogate the pro-longevity effect of DR. Pathway components suggested by Zhang et al. [13] are shown in green, while components suggested by Miller et al. [14] are shown in red, and common components are shown in yellow. *** Note that conflicting evidence was presented on whether NSM neurons respond to food odors. (**B**). Summary of the circuit delineated by Kitto et al. [16]. Created in BioRender. Lyu, Y. (2025) https://BioRender.com/k57e037 (accessed on 1 January, 2025).

Shortened lifespan on the choice diet requires neuronal expression of the type 2A serotonin receptor (5-HT2A) [19], and there is preliminary evidence that 5-HT2B plays a similar role [23] (unpublished). However, the identity of the serotonergic neurons and their downstream, 5-HT2R - expressing partners within the Drosophila nervous system that regulate aging in response to choice remain unknown. 5-HT2Rs are expressed by "F cells", neurons that innervate the fan-shaped body, a brain region that receives serotonergic input [24] and is involved in nutrient valence and feeding choice [25,26]. They are also expressed in the olfactory system [27], and the *Drosophila* sensory integration center known as the Ellipsoid Body, which is known to regulate longevity in response to externally perceived cues [28].

Additionally, both 5-HT2Rs are expressed in median neurosecretory cells that produce *Drosophila* insulin-like peptides 2, 3, and 5, known as insulin-producing cells (IPCs) [29]. An important advance in the perception–metabolism relationship was recently provided by

Yao and Scott [30], who demonstrated that IPCs receive serotonergic input via 5-HT2A from second-order sweet-sensing neurons. Because sweet-sensing gustatory receptor neurons scale their sensitivity to sucrose according to the magnitude of previous chronic sweet taste [31], the serotonergic input to IPCs could conceivably be altered by the chronic consumption of unmixed sucrose in the choice paradigm.

Despite pan-neuronal *5-HT2A* knockdown being sufficient to abrogate the choice phenotype [19], it is possible that it is not an example of aging regulation by external cues per se, but instead that nutrient separation results in shorter lifespan by affecting gut motility, nutrient absorption, and/or gut-derived endocrine signals, consilient with the mammalian role of serotonin in gut peristalsis and nutrient sensing [32,33]. Transcriptomic studies suggest 5-HT2B is, in fact, expressed in *D. melanogaster* enteroendocrine cells (EECs) [34]. *D. melanogaster* EECs also express canonical neuronal markers and gustatory receptors, and are known to regulate longevity in response to the gut's nutrient contents via neuropeptide F (the *Drosophila* homolog of mammalian neuropeptide Y) signaling to insulin-secreting neurons, ultimately regulating circulating levels of juvenile hormone titer [35]. However, whether the choice phenotype is attributable to adult serotonergic signaling to EECs or other intestinal cells remains to be directly addressed.

How generalizable or translatable is the regulation of lifespan by dietary choice? In humans, dietary choice is the norm, and an emerging market of meal-replacement products (MRPs) is the closest analog to laboratory "fixed" diets. MRPs are known to be effective temporary regimes for achieving weight loss and ameliorating metabolic syndrome [36,37], but long-term use by healthy individuals is not well-studied. Additional research on longevity benefits and the mechanisms behind them in invertebrates, future research in mammalian models, and additional research on long-term usage in humans, would greatly inform the scientific understanding of their true potential as a longevity intervention, and may conceivably justify more widespread adoption of MRPs for general health across the lifespan rather than solely as a weight loss intervention.

3. Social Cues

3.1. Mating-Related Cues

Various aspects of the social environment are known to influence longevity in *D. melanogaster* and other "non-social" invertebrates [5,38,39]. We focus here on the "external" cue of sex pheromones rather than mating itself and downstream "internal" pathways such as the *Drosophila* Sex Peptide pathway, which have been extensively discussed elsewhere [40].

In *C. elegans*, exposure to male-secreted ascaroside pheromones shortens lifespan of hermaphrodites [41,42] and is especially toxic to other males, although the latter appears to be true only in androdecious species (males and hermaphrodites instead of males and females), possibly as a mechanism to regulate the population sex ratio. Masculinization of the hermaphrodite nervous system is sufficient to sensitize them [43]. The pheromone nacq#1 is synthesized and secreted disproportionately by males, starting at sexual maturity. Nacq#1 is perceived by chemosensory neurons in hermaphrodites, and promotes increased fecundity and reduced lifespan—thus constituting a clear example of a signal of mating opportunity modifying the allocation of resources between reproduction and longevity [44].

In male *D. melanogaster*, exposure to females reduces longevity, and to a greater degree, if it is not followed by mating, which was initially interpreted to reflect a cost of courtship [45]. However, the cost of female pheromone exposure—which also includes reduced triglyceride stores and increased sensitivity to starvation and oxidative stress—can be partially rescued by allowing exposed males to mate either with wild-type [46] or pheromone-masculinized females [47]. Stress and lifespan effects are mediated by neuropeptide F signaling downstream of pickpocket 23-expressing chemosensory foreleg neurons, as well as by insulin-independent changes in *Drosophila* Forkhead Box O (dFOXO) activity and can be modulated by chronic excitation/inhibition of neurons that express the fly homolog of gonadotropin-releasing hormone, Corazonin. Importantly, neither wing damage (suggestive of male-male aggression) nor time spent courting were decreased in males that were exposed but allowed to mate, thus failing to explain the partial rescue of lifespan [47]. These results suggest that in male *D. melanogaster*, the perception of unsatisfied mating opportunities via female pheromone exposure causes physiological stress and shortens lifespan, while actual reproductive resource investments (mating behavior and seminal fluid production) may be largely benign if uncoupled from this stressor.

A recent report provided additional evidence that these stresses are imposed by the nervous system by attributing them to the disinhibition of neuropeptide F target neurons [48]. Additionally, the *Drosophila* homolog of Cholecystokinin (CCK)—Drosulfakinin (DSK)—which is released from a broad set of central neurons including but not limited to IPCs—encodes reward and rescues the effect of female pheromones on lifespan [49]. Interestingly, short-term female pheromone exposure increases subsequent male reproductive competitiveness in mixed-sex housing, suggesting that the costs of exposure are trade-offs for improved fitness [50]. In contrast, the physiological consequences of long-term exposure—if not paired with mating—accentuate the effect of sexual selection by imposing additional fitness costs on unsuccessful males [51].

Thus, female pheromones rather than courtship costs appear to mediate the health effects of female exposure. Pheromones initiate a neuronal response that begins in peripheral chemosensory neurons and likely involves NPF, Corazonin, and DSK signaling, but how pheromone information streams are ultimately integrated into the central brain and affected by these neuropeptides at the circuit level remains unclear.

To what extent are the results in invertebrates extendable to mammals? In mice, neither male nor female odorants affect male lifespan [52], although the presence of female odorants does shorten lifespan when paired with access to additional, different females [53], perhaps reflecting a dosage effect. While mating and/or courtship shorten female lifespan compared to housing with another female [54], exposure to male pheromones does not shorten female lifespan more than living alone [52]. Thus, in "normal" living conditions, opposite-sex pheromones do not have strong effects on lifespan in the mouse, though negative effects on male lifespan reminiscent of invertebrate findings may be present under certain conditions.

3.2. Other Social Cues

We have discussed opposite-sex pheromone effects on longevity, but does perceiving same-sex animals also affect health and longevity? In *D. melanogaster*, social isolation is known to extend lifespan compared to co-housing with same-sex vial-mates [55–58]. In males, this was presumed to result from sperm competition costs. Male *D. melanogaster* react to perceived sperm competition by increasing mating duration [59]. However, it was recently shown that while this behavioral change requires auditory and/or olfactory cues from the rival, they are dispensable for the decrease in lifespan [60]. It should be noted that previously discussed pheromone exposure paradigms compared focal flies to others living in different social environments, but none were alone. Thus, it is possible that isolation extends lifespan by a completely different mechanism (such as increased resource availability) than the relative pro and anti-longevity effects of pheromones from different conspecifics.

Aside from sex, two recent studies have analyzed longevity as a function of conspecific age. Longevity and multiple aspects of health are improved or worsened by co-housing with younger or older same-sex conspecifics, respectively. Co-housing with older conspecifics also decreases resistance to a variety of stresses, though not in all genotypes [61], while the opposite is true in the presence of younger conspecifics [62].

A key question is: What is the mechanism by which conspecifics influence longevity, and to what extent does it involve solely perception? Pheromones from young vial-mates seem to be responsible for extending lifespan, as mutating chemosensory genes in aged flies, or preventing cuticular hydrocarbon (pheromone) production in young flies, both abrogate the pro-longevity effect of young vial-mates [62]. Despite the practice of transferring flies to fresh food every 2–3 days, the fact that *Drosophila* cultures contain a shared food substrate highlights the importance of thoroughly disentangling perceptual or social effects from the transmission of biological factors through feeding. Importantly, a prior study observed that health and longevity improvements can be conferred to otherwise short-lived Cu/Zn superoxide dismutase mutants simply by cohousing with healthier wild-type flies, but pre-exposing the food medium to healthy flies before administering to mutants was not sufficient, validating that transmission of biological factors through food is not necessarily sufficient to explain a robust impact of cohousing on health and aging in fruit flies [63]. Future research could fortify this interpretation by allowing visual and olfactory interaction between separately housed flies.

In summary, co-housing flies with old or young conspecifics exerts largely opposite effects on health and longevity, which may be caused by chemosensory cues rather than transmission of biological factors through shared food. These experiments suggest that, even in non-social insects, social environment may influence health and aging. In contrast, the case for social effects on aging in the context of male–male cohousing (vs isolation) is less strong.

In mice, similar findings on the beneficial effects of younger or healthier conspecifics to the immune function and longevity of otherwise "prematurely aging" individuals have been reported [64], as well as in a Huntington's Disease model [65,66]. Like in flies, the effect of social interaction, per se, is not clearly distinguished from other consequences of cohabitation, and may, for example, be attributable to differences in shared nest quality [65] or transference of microbiota [64].

4. Stress Perception

As stressors in the environment may directly impact an animal's future fitness, they constitute an important class of signals informing the allocation of resources toward different life history traits. In *D. melanogaster*, exposure to dead conspecifics shortens lifespan via serotonin receptor 2A (5-HT2A). Unlike "necromone"-mediated behavioral responses in social insects [67], this longevity phenotype depends on sight, as exposure in the dark or in visually impaired mutants fails to reduce lifespan. The initial study that characterized this effect reported metabolomic changes in death-exposed flies, which the authors tentatively attributed to a depression or anxiety-like state [68], but the mechanism by which death perception shortens lifespan and its evolutionary significance was otherwise unknown.

The terminal investment hypothesis posits that reproductive effort is inversely related to residual reproductive opportunity [69]. One implication of this idea is that investment in reproduction should plastically respond to changes in environmental conditions that may signal changes in long-term survival prospects. Consistent with this hypothesis, female flies exposed to dead conspecifics display a greater reproductive rate during the exposure period, and this effect can be rapidly reversed upon removal of dead flies [70]. Of note, after measuring fecundity, Corbel and Carazo [70] subsequently tracked longevity, and did not observe shorter lifespan in death-exposed female flies. However, they note that their experimental paradigm only exposed flies to dead conspecifics in early adulthood rather than the whole lifespan. Thus, both studies [68,70] are consistent with death perception acutely shifting organismal resource allocation toward immediate reproduction at the expense of somatic maintenance.

Recently, additional insights have also emerged about the neural signals that mediate health changes in the death perception paradigm. The *Drosophila* Ellipsoid Body is a major sensory integration center within the fly brain and is organized in concentric rings of neuropil. The sight of dead conspecifics activates 5-HT2A^{+} neurons that innervate layers R2, R4d, and R4m. Subsequently, the expression of *Drosophila* insulin-like peptide 3 (*dilp* 3) is increased in the IPCs [28]. Thus, the perception of dead conspecifics likely alters organismal insulin signaling, but how Ellipsoid Body neuron activity is connected to IPCs and regulates Dilp production and/or release is not yet clear.

Like *Drosophila*, *C. elegans* display shortened lifespan and an increase in egg-laying rate after exposure to dead conspecifics. Unlike *Drosophila*, *C. elegans* avoid the corpses of conspecifics (*Drosophila* avoid death-exposed living flies but not corpses themselves [68]). Both the behavioral and longevity phenotypes of *C. elegans* require the function of two olfactory neurons, AWB and ASH, and the detection of the "death signatures" adenosine monophosphate and cysteine rather than visual input, while serotonin signaling is dispensable [71]. Thus, while the perception of dead conspecifics consistently drives a terminal investment phenotype of increased reproductive rate concomitant with reduced lifespan, these effects are mediated by different sensory and neurotransmitter systems.

While death exposure seems to reduce lifespan via the promotion of terminal investment, exposure to other stressors can regulate physiology and aging through other mechanisms. In *C. elegans*, olfactory neurons detect volatiles from pathogenic bacteria and initiate the unfolded protein response (UPR) in the intestines, likely to anticipate proteotoxicity challenges associated with infection. UPR activation depends on cue detection by odorant receptor 3 (ODR-3) and not on an immune response. Chronic exposure and activation of this pathway extend lifespan [72]. In contrast, the regular application of an alarming vibratory stimuli to induce a "flight response" reduces longevity when applied over the lifetime, and transient stimulation worsens subsequent resistance to oxidative stress. These effects are mediated by a neural circuit connecting mechanosensory neurons to tyraminergic RIM neurons, which signal to the intestine to promote insulin signaling, likely promoting the fulfillment of the energetic demands of the flight response while simultaneously inhibiting cytoprotective pathways [73]. In summary, stressor-related perception can have opposite effects on longevity, likely reflecting different organismal response strategies (i.e., terminal investment, stress anticipatory responses, and fight-or-flight stress responses).

5. Light

Light exposure has well-established negative effects on invertebrate longevity [74–76], particularly via short (blue/ultraviolet) wavelength-induced photooxidative stress and DNA damage [77]. In *D. melanogaster*, lifespan-shortening effects of blue light are not dependent on vision [74], and studies of "stress response" mutants such as superoxide dismutase and sirtuins established that the deleterious effects of light on fly lifespan are attributable to oxidative stress and DNA damage [78]. It is also well-established that light, particularly blue light, is the primary external cue for the entrainment of organismal circadian rhythms in physiology and behavior. Circadian rhythms are driven by an endogenous, cell-autonomous molecular "clock" consisting of a transcription-translation feedback loop of core clock proteins. The phase of the clock is shifted by light exposure [79], and there is evidence that chronic light disruption of circadian rhythms is also harmful (the "circadian resonance" hypothesis) [80,81]. Although discussions of phototoxicity and circadian resonance are outside the scope of this review, a pertinent question is whether there exist any health and longevity effects of light perception or the exercise of vision that can be disentangled from radiative damage and circadian dysregulation.

Two recent studies dissected the effects of monochromatic light on fly lifespan, but obtained different results. In one, daily 12 h exposure to green light extended *D. melanogaster* lifespan even compared to constant darkness. However, the effect was stronger in the w^{1118} mutant, which lacks eye pigment, and was abrogated by lacing the food with the antibiotic doxycycline, therefore pointing to microbiota and not visual perception as mediators of green light-induced longevity [82]. In contrast, Krittika and Yadav [83], observed the longest lifespan in darkness and progressively decreasing lifespan with 12 h exposure to monochromatic light of decreasing wavelengths, consistent with the notion of direct, wavelength-dependent phototoxicity [83]. Thus, neither study provided evidence that monochromatic light exerts any effect on longevity via a perceptual circuit, and indeed, they obtained contradictory evidence on the effect of green light. Both studies are instead consistent with phototoxicity being the primary driver of effects on health and aging.

Consistent with a previous report [84], a recent study observed an extension of *D. melanogaster* lifespans in constant darkness [85]. However, a surprisingly shorter lifespan could be recapitulated by the daily activation of *Rh1*-expressing photoreceptor neurons to mimic the rhythmic perception of blue light in constant darkness. Circadian rhythm mutants still displayed an extended lifespan in constant darkness, and genetic ablation of the eyes abrogated the effect of light on lifespan. The former shows the independence of the phenotype to circadian rhythms, and the latter to direct phototoxic effects of light. While it is not disputed that at higher light intensities, known phototoxic effects likely dominate, these results nevertheless suggest the existence of uncharacterized pathways linking photoreceptor neuron activity to physiology and aging [85]. Interestingly, it has been argued based on dFOXO mutant research that reduced insulin signaling and increased dFOXO activity occur in constant darkness, and that upregulated dFOXO activity is required for the resulting extension of lifespan [78]. As *D. melanogaster* insulin is released from brain neuroendocrine cells, this hints that the nervous system may regulate a reproduction/longevity trade-off in response to light levels. Directly investigating a potential link between visual exercise and dFOXO should be a goal of future research in this field.

There is evidence that non-circadian, neuronal pathways regulate physiology in response to light in mammals as well as flies, but not in a manner necessarily involving image-forming pathways. In mammals, intrinsically photosensitive retinal ganglion cell (ipRGC) innervation of the suprachiasmatic nucleus (SCN) is the canonical pathway by which light exerts circadian modulation on physiology and behavior. While many non-circadian effects of light on mammalian physiology have been reported (reviewed in [86]), they are mediated either by direct effects on peripheral tissues (such as brown adipose tissue and epidermis) via endogenously expressed opsins, or via non-visual detection by ipRGCs. For example, it was reported that acute thermoregulatory and sleep responses to light in the mouse are mediated by a subset of ipRGCs defined by expression of the *Brn3b* transcription factor, which project widely but not to the SCN, essentially bypassing the circadian master clock [87]. To our knowledge, the exercise of vision per se has not been implicated in longevity regulation in mammals.

6. Conclusions and Outlook

There is abundant evidence that sensory perception can modify the rate of aging in animal models, but the levels of understanding of the mechanistic details are starkly different in different sensory modalities and model organisms. The most extensive progress has been achieved in the area of sensory perception of food-related cues, which interface with known "internal" nutrient sensing systems in *C. elegans*. Other prominent research areas include mating and stress-related cues, which are also well-described in both *Drosophila* and *C. elegans*. While particular neurons, genes, ligands, and receptors have been variously identified as critical mediators of the perception-longevity relationship in different cases, more detailed circuit-level understanding often remains to be fully elucidated. The simple anatomy of invertebrate models should render promising the continued identification and characterization of such circuits.

Sensory experience is universal, and its effects on aging, first uncovered through studies in *Drosophila melanogaster* and *Caenorhabditis elegans*, have yet to be rigorously tested in higher organisms. Research using simple animal models offers advantages such as rapid experimentation, genetic access to specific cells, and the ease of uncovering mechanistic pathways, providing initial evidence that inspires the translation of findings to mammals, including humans. However, while highly conserved molecular signaling pathways associated with sensory effects on longevity have been identified in these systems, the regulatory networks that modulate these effects remain poorly understood. Moving forward, investigating how sensory-driven effects function independently of or interact with established longevity pathways—particularly within specific environmental contexts—could significantly advance our understanding of how aging is regulated in real-world conditions.

Author Contributions: Both N.P. and Y.L. drafted and edited the manuscript. Both N.P. and Y.L. have read and agreed to the published version of the manuscript. All authors have read and agreed to the published version of the manuscript.

Funding: This research was conducted while Yang Lyu was an AFAR Grant for Junior Faculty awardee.

Institutional Review Board Statement: Not applicable.

Informed Consent Statement: Not applicable.

Data Availability Statement: Not applicable.

Acknowledgments: The authors thank Laura Byron for commentary on the manuscript.

Conflicts of Interest: The authors declare that they have no competing interests, whether financial or non-financial, to disclose.

References

1. Miller, H.A.; Dean, E.S.; Pletcher, S.D.; Leiser, S.F. Cell Non-Autonomous Regulation of Health and Longevity. *Elife* **2020**, *9*, e62659. [CrossRef] [PubMed] [PubMed Central]
2. Riera, C.E.; Dillin, A. Emerging Role of Sensory Perception in Aging and Metabolism. *Trends Endocrinol. Metab.* **2016**, *27*, 294–303. [CrossRef] [PubMed]
3. Dixit, A.; Bhattacharya, B. Sensory Perception of Environmental Cues as a Modulator of Aging and Neurodegeneration: Insights from *Caenorhabditis elegans*. *J. Neurosci. Res.* **2021**, *99*, 2416–2426. [CrossRef] [PubMed]
4. Gendron, C.M.; Chakraborty, T.S.; Chung, B.Y.; Harvanek, Z.M.; Holme, K.J.; Johnson, J.C.; Lyu, Y.; Munneke, A.S.; Pletcher, S.D. Neuronal Mechanisms that Drive Organismal Aging Through the Lens of Perception. *Annu. Rev. Physiol.* **2020**, *82*, 227–249. [CrossRef] [PubMed] [PubMed Central]
5. Flatt, T. Survival Costs of Reproduction in *Drosophila*. *Exp. Gerontol.* **2011**, *46*, 369–375. [CrossRef] [PubMed]
6. Partridge, L.; Gems, D.; Withers, D.J. Sex and Death: What Is the Connection? *Cell* **2005**, *120*, 461–472. [CrossRef] [PubMed]
7. Krittika, S.; Yadav, P. An Overview of Two Decades of Diet Restriction Studies Using *Drosophila*. *Biogerontology* **2019**, *20*, 723–740. [CrossRef] [PubMed]
8. Ostojic, I.; Boll, W.; Waterson, M.J.; Chan, T.; Chandra, R.; Pletcher, S.D.; Alcedo, J. Positive and Negative Gustatory Inputs Affect *Drosophila* Lifespan Partly in Parallel to dFOXO Signaling. *Proc. Natl. Acad. Sci. USA* **2014**, *111*, 8143–8148. [CrossRef] [PubMed] [PubMed Central]
9. Apfeld, J.; Kenyon, C. Regulation of Lifespan by Sensory Perception in *Caenorhabditis elegans*. *Nature* **1999**, *402*, 804–809. [CrossRef] [PubMed]

10. Alcedo, J.; Kenyon, C. Regulation of *C. elegans* Longevity by Specific Gustatory and Olfactory Neurons. *Neuron* **2004**, *41*, 45–55. [CrossRef] [PubMed]
11. Libert, S.; Zwiener, J.; Chu, X.; Vanvoorhies, W.; Roman, G.; Pletcher, S.D. Regulation of *Drosophila* Life Span by Olfaction and Food-Derived Odors. *Science* **2007**, *315*, 1133–1137. [CrossRef] [PubMed]
12. Smith, E.D.; Kaeberlein, T.L.; Lydum, B.T.; Sager, J.; Welton, K.L.; Kennedy, B.K.; Kaeberlein, M. Age- and Calorie-Independent Life Span Extension from Dietary Restriction by Bacterial Deprivation in *Caenorhabditis elegans*. *BMC Dev. Biol.* **2008**, *8*, 49. [CrossRef] [PubMed] [PubMed Central]
13. Zhang, B.; Jun, H.; Wu, J.; Liu, J.; Xu, X.Z.S. Olfactory Perception of Food Abundance Regulates Dietary Restriction-Mediated Longevity via a Brain-to-Gut Signal. *Nat. Aging* **2021**, *1*, 255–268. [CrossRef] [PubMed] [PubMed Central]
14. Miller, H.A.; Huang, S.; Dean, E.S.; Schaller, M.L.; Tuckowski, A.M.; Munneke, A.S.; Beydoun, S.; Pletcher, S.D.; Leiser, S.F. Serotonin and Dopamine Modulate Aging in Response to Food Odor and Availability. *Nat. Commun.* **2022**, *13*, 3271. [CrossRef] [PubMed] [PubMed Central]
15. Sawin, E.R.; Ranganathan, R.; Horvitz, H.R. *C. elegans* Locomotory Rate Is Modulated by the Environment through a Dopaminergic Pathway and by Experience through a Serotonergic Pathway. *Neuron* **2000**, *26*, 619–631. [CrossRef] [PubMed]
16. Kitto, E.S.; Beydoun, S.; Leiser, S.F. Food Touch Limits Lifespan Through Bioamine and Neuroendocrine Signaling. *bioRxiv* **2024**.
17. Lyu, Y.; Promislow, D.E.L.; Pletcher, S.D. Serotonin Signaling Modulates Aging-Associated Metabolic Network Integrity in Response to Nutrient Choice in *Drosophila melanogaster*. *Commun. Biol.* **2021**, *4*, 740. [CrossRef] [PubMed] [PubMed Central]
18. Ro, J.; Pak, G.; Malec, P.A.; Lyu, Y.; Allison, D.B.; Kennedy, R.T.; Pletcher, S.D. Serotonin Signaling Mediates Protein Valuation and Aging. *Elife* **2016**, *5*, e16843. [CrossRef] [PubMed] [PubMed Central]
19. Lyu, Y.; Weaver, K.J.; Shaukat, H.A.; Plumoff, M.L.; Tjilos, M.; Promislow, D.E.; Pletcher, S.D. *Drosophila* Serotonin 2A Receptor Signaling Coordinates Central Metabolic Processes to Modulate Aging in Response to Nutrient Choice. *Elife* **2021**, *10*, e59399. [CrossRef] [PubMed] [PubMed Central]
20. Strilbytska, O.; Yurkevych, I.; Semaniuk, U.; Gospodaryov, D.; Simpson, S.J.; Lushchak, O. Life-History Trade-Offs in *Drosophila*: Flies Select a Diet to Maximize Reproduction at the Expense of Lifespan. *J. Gerontol. A Biol. Sci. Med. Sci.* **2024**, *79*, glae057. [CrossRef] [PubMed] [PubMed Central]
21. Vigne, P.; Frelin, C. Food Presentation Modifies Longevity and the Beneficial Action of Dietary Restriction in *Drosophila*. *Exp. Gerontol.* **2010**, *45*, 113–118. [CrossRef] [PubMed]
22. Strilbytska, O.; Semaniuk, U.; Bubalo, V.; Storey, K.B.; Lushchak, O. Dietary Choice Reshapes Metabolism in *Drosophila* by Affecting Consumption of Macronutrients. *Biomolecules* **2022**, *12*, 1201. [CrossRef] [PubMed] [PubMed Central]
23. Pontillo, N.; Lyu, Y. Rutgers, the State University of New Jersey, New Brunswick, NJ, USA. Unpublished work. 2025.
24. Kahsai, L.; Winther, A.M. Chemical Neuroanatomy of the *Drosophila* Central Complex: Distribution of Multiple Neuropeptides in Relation to Neurotransmitters. *J. Comp. Neurol.* **2011**, *519*, 290–315. [CrossRef] [PubMed]
25. Sareen, P.F.; McCurdy, L.Y.; Nitabach, M.N. A Neuronal Ensemble Encoding Adaptive Choice During Sensory Conflict in *Drosophila*. *Nat. Commun.* **2021**, *12*, 4131. [CrossRef] [PubMed] [PubMed Central]
26. Munneke, A.S.; Chakraborty, T.S.; Porter, S.S.; Gendron, C.M.; Pletcher, S.D. The Serotonin Receptor 5-HT2A Modulates Lifespan and Protein Feeding in *Drosophila melanogaster*. *Front. Aging* **2022**, *3*, 1068455. [CrossRef] [PubMed] [PubMed Central]
27. Sizemore, T.R.; Dacks, A.M. Serotonergic Modulation Differentially Targets Distinct Network Elements within the Antennal Lobe of *Drosophila melanogaster*. *Sci. Rep.* **2016**, *6*, 37119. [CrossRef] [PubMed] [PubMed Central]
28. Gendron, C.M.; Chakraborty, T.S.; Duran, C.; Dono, T.; Pletcher, S.D. Ring Neurons in the *Drosophila* Central Complex Act as a Rheostat for Sensory Modulation of Aging. *PLoS Biol.* **2023**, *21*, e3002149. [CrossRef] [PubMed] [PubMed Central]
29. Gnerer, J.P.; Venken, K.J.; Dierick, H.A. Gene-Specific Cell Labeling Using MiMIC Transposons. *Nucleic Acids Res.* **2015**, *43*, e56. [CrossRef] [PubMed] [PubMed Central]
30. Yao, Z.; Scott, K. Serotonergic Neurons Translate Taste Detection into Internal Nutrient Regulation. *Neuron* **2022**, *110*, 1036–1050.e7. [CrossRef] [PubMed] [PubMed Central]
31. Wang, Q.P.; Lin, Y.Q.; Lai, M.L.; Su, Z.; Oyston, L.J.; Clark, T.; Park, S.J.; Khuong, T.M.; Lau, M.T.; Shenton, V.; et al. PGC1α Controls Sucrose Taste Sensitization in *Drosophila*. *Cell Rep.* **2020**, *31*, 107480. [CrossRef] [PubMed]
32. Uneyama, H.; Niijima, A.; San Gabriel, A.; Torii, K. Luminal Amino Acid Sensing in the Rat Gastric Mucosa. *Am. J. Physiol. Gastrointest. Liver Physiol.* **2006**, *291*, G1163–G1170. [CrossRef] [PubMed]
33. Mohammad-Zadeh, L.F.; Moses, L.; Gwaltney-Brant, S.M. Serotonin: A Review. *J. Vet. Pharmacol. Ther.* **2008**, *31*, 187–199. [CrossRef] [PubMed]
34. Guo, X.; Yin, C.; Yang, F.; Zhang, Y.; Huang, H.; Wang, J.; Deng, B.; Cai, T.; Rao, Y.; Xi, R. The Cellular Diversity and Transcription Factor Code of *Drosophila* Enteroendocrine Cells. *Cell Rep.* **2019**, *29*, 4172–4185.e5. [CrossRef] [PubMed]
35. Chen, J.; Nouzová, M.; Noriega, F.G.; Tatar, M. Gut-to-Brain Regulation of *Drosophila* Aging through Neuropeptide F, Insulin, and Juvenile Hormone. *Proc. Natl. Acad. Sci. USA* **2024**, *121*, e2411987121. [CrossRef] [PubMed] [PubMed Central]

36. Edwards-Hampton, S.A.; Ard, J. The Latest Evidence and Clinical Guidelines for Use of Meal Replacements in Very-Low-Calorie Diets or Low-Calorie Diets for the Treatment of Obesity. *Diabetes Obes. Metab.* **2024**, *26* (Suppl. S4), 28–38. [CrossRef] [PubMed]
37. Durbin, C.G.; Hutchison, A.; Colecchi, T.; Mulligan, J.; Bodepudi, S.; Zanni, M.V.; Anekwe, C.V. Meal Replacement Therapy for Obesity and Diabetes Remission: Efficacy Assessment and Considerations of Barriers and Facilitators to Adherence among U.S. Individuals with Low Economic Resources. *Rev. Endocr. Metab. Disord.* **2024**. *ahead of print*. [CrossRef] [PubMed]
38. Shi, C.; Murphy, C.T. Sex and Death. *Curr. Top. Dev. Biol.* **2021**, *144*, 353–375. [CrossRef] [PubMed]
39. Harrison, L.M.; Churchill, E.R.; Fairweather, M.; Smithson, C.H.; Chapman, T.; Bretman, A. Ageing Effects of Social Environments in 'Non-Social' Insects. *Philos. Trans. R. Soc. Lond. B Biol. Sci.* **2024**, *379*, 20220463. [PubMed] [PubMed Central]
40. White, M.A.; Wolfner, M.F. The Effects of Male Seminal Fluid Proteins on Gut/Gonad Interactions in *Drosophila*. *Insects* **2022**, *13*, 623. [CrossRef] [PubMed] [PubMed Central]
41. Aprison, E.Z.; Ruvinsky, I. Sexually Antagonistic Male Signals Manipulate Germline and Soma of *C. elegans* Hermaphrodites. *Curr. Biol.* **2016**, *26*, 2827–2833. [CrossRef] [PubMed]
42. Maures, T.J.; Booth, L.N.; Benayoun, B.A.; Izrayelit, Y.; Schroeder, F.C.; Brunet, A. Males Shorten the Life Span of *C. elegans* Hermaphrodites via Secreted Compounds. *Science* **2014**, *343*, 541–544. [CrossRef] [PubMed] [PubMed Central]
43. Shi, C.; Runnels, A.M.; Murphy, C.T. Mating and Male Pheromone Kill *Caenorhabditis* Males through Distinct Mechanisms. *Elife* **2017**, *6*, e23493. [CrossRef] [PubMed] [PubMed Central]
44. Ludewig, A.H.; Artyukhin, A.B.; Aprison, E.Z.; Rodrigues, P.R.; Pulido, D.C.; Burkhardt, R.N.; Panda, O.; Zhang, Y.K.; Gudibanda, P.; Ruvinsky, I.; et al. An Excreted Small Molecule Promotes *C. elegans* Reproductive Development and Aging. *Nat. Chem. Biol.* **2019**, *15*, 838–845. [CrossRef] [PubMed] [PubMed Central]
45. Cordts, R.; Partridge, L. Courtship Reduces Longevity of Male *Drosophila melanogaster*. *Anim. Behav.* **1996**, *52*, 269–278. [CrossRef]
46. Gendron, C.M.; Kuo, T.H.; Harvanek, Z.M.; Chung, B.Y.; Yew, J.Y.; Dierick, H.A.; Pletcher, S.D. *Drosophila* Life Span and Physiology Are Modulated by Sexual Perception and Reward. *Science* **2014**, *343*, 544–548. [CrossRef] [PubMed] [PubMed Central]
47. Harvanek, Z.M.; Lyu, Y.; Gendron, C.M.; Johnson, J.C.; Kondo, S.; Promislow, D.E.L.; Pletcher, S.D. Perceptive Costs of Reproduction Drive Ageing and Physiology in Male *Drosophila*. *Nat. Ecol. Evol.* **2017**, *1*, 152. [CrossRef] [PubMed] [PubMed Central]
48. Ryvkin, J.; Omesi, L.; Kim, Y.K.; Levi, M.; Pozeilov, H.; Barak-Buchris, L.; Agranovich, B.; Abramovich, I.; Gottlieb, E.; Jacob, A.; et al. Failure to Mate Enhances Investment in Behaviors That May Promote Mating Reward and Impairs the Ability to Cope with Stressors via a Subpopulation of Neuropeptide F receptor neurons. *PLoS Genet.* **2024**, *20*, e1011054. [CrossRef] [PubMed] [PubMed Central]
49. Fedina, T.Y.; Cummins, E.T.; Promislow, D.E.L.; Pletcher, S.D. The Neuropeptide Drosulfakinin Enhances Choosiness and Protects Males from the Aging Effects of Social Perception. *Proc. Natl. Acad. Sci. USA* **2023**, *120*, e2308305120. [CrossRef] [PubMed] [PubMed Central]
50. Corbel, Q.; Serra, M.; García-Roa, R.; Carazo, P. Male Adaptive Plasticity Can Explain the Evolution of Sexual Perception Costs. *Am. Nat.* **2022**, *200*, E110–E123. [CrossRef] [PubMed]
51. García-Roa, R.; Serra, M.; Carazo, P. Ageing via Perception Costs of Reproduction Magnifies Sexual Selection. *Proc. Biol. Sci.* **2018**, *285*, 20182136. [CrossRef] [PubMed] [PubMed Central]
52. Garratt, M.; Erturk, I.; Alonzo, R.; Zufall, F.; Leinders-Zufall, T.; Pletcher, S.D.; Miller, R.A. Lifespan Extension in Female Mice by Early, Transient Exposure to Adult Female Olfactory Cues. *Elife* **2022**, *11*, e84060. [CrossRef] [PubMed] [PubMed Central]
53. Garratt, M.; Try, H.; Neyt, C.; Brooks, R.C. Exposure to Female Olfactory Cues Hastens Reproductive Ageing and Increases Mortality when Mating in Male Mice. *Proc. Biol. Sci.* **2024**, *291*, 20231848. [CrossRef] [PubMed] [PubMed Central]
54. Garratt, M.; Try, H.; Smiley, K.O.; Grattan, D.R.; Brooks, R.C. Mating in the Absence of Fertilization Promotes a Growth-Reproduction Versus Lifespan Trade-Off in Female Mice. *Proc. Natl. Acad. Sci. USA* **2020**, *117*, 15748–15754. [CrossRef] [PubMed] [PubMed Central]
55. Leech, T.; Sait, S.M.; Bretman, A. Sex-Specific Effects of Social Isolation on Ageing in *Drosophila melanogaster*. *J. Insect Physiol.* **2017**, *102*, 12–17. [CrossRef] [PubMed]
56. Bretman, A.; Westmancoat, J.D.; Gage, M.J.; Chapman, T. Costs and Benefits of Lifetime Exposure to Mating Rivals in Male *Drosophila melanogaster*. *Evolution* **2013**, *67*, 2413–2422. [CrossRef] [PubMed]
57. Iliadi, K.G.; Iliadi, N.N.; Boulianne, G.L. Regulation of *Drosophila* Life-Span: Effect of Genetic Background, Sex, Mating and Social Status. *Exp. Gerontol.* **2009**, *44*, 546–553. [CrossRef] [PubMed]
58. Carazo, P.; Green, J.; Sepil, I.; Pizzari, T.; Wigby, S. Inbreeding Removes Sex Differences in Lifespan in a Population of *Drosophila melanogaster*. *Biol. Lett.* **2016**, *12*, 20160337. [CrossRef] [PubMed] [PubMed Central]
59. Bretman, A.; Fricke, C.; Chapman, T. Plastic Responses of Male *Drosophila melanogaster* to the Level of Sperm Competition Increase Male Reproductive Fitness. *Proc. Biol. Sci.* **2009**, *276*, 1705–1711. [CrossRef] [PubMed] [PubMed Central]
60. Smithson, C.H.; Duncan, E.J.; Sait, S.M.; Bretman, A. Sensory Perception of Rivals Has Trait-Dependent Effects on Plasticity in *Drosophila melanogaster*. *Behav. Ecol.* **2024**, *35*, arae031. [CrossRef] [PubMed] [PubMed Central]

61. Lin, Y.C.; Zhang, M.; Wang, S.H.; Chieh, C.W.; Shen, P.Y.; Chen, Y.L.; Chang, Y.C.; Kuo, T.H. The Deleterious Effects of Old Social Partners on *Drosophila* Lifespan and Stress Resistance. *NPJ Aging* **2022**, *8*, 1. [CrossRef] [PubMed] [PubMed Central]
62. Cho, L.C.; Yu, C.C.; Kao, C.F. Social Perception of Young Adults Prolongs the Lifespan of Aged *Drosophila*. *NPJ Aging Mech. Dis.* **2021**, *7*, 21. [CrossRef] [PubMed] [PubMed Central]
63. Ruan, H.; Wu, C.F. Social Interaction-Mediated Lifespan Extension of *Drosophila* Cu/Zn Superoxide Dismutase Mutants. *Proc. Natl. Acad. Sci. USA* **2008**, *105*, 7506–7510. [CrossRef] [PubMed] [PubMed Central]
64. Garrido, A.; Cruces, J.; Ceprián, N.; Corpas, I.; Tresguerres, J.A.; De la Fuente, M. Social Environment Improves Immune Function and Redox State in Several Organs from Prematurely Aging Female Mice and Increases Their Lifespan. *Biogerontology* **2019**, *20*, 49–69. [CrossRef] [PubMed]
65. Story, D.; Gallien, J.; Al-Gharaibeh, A.; Sandstrom, M.; Rossignol, J.; Dunbar, G.L. Housing R6/2 Mice with Wild-Type Littermates Increases Lifespan. *J. Huntingtons Dis.* **2021**, *10*, 455–458. [CrossRef] [PubMed]
66. Carter, R.J.; Hunt, M.J.; Morton, A.J. Environmental Stimulation Increases Survival in Mice Transgenic for Exon 1 of the Huntington's Disease Gene. *Mov. Disord.* **2000**, *15*, 925–937. [CrossRef] [PubMed]
67. Sun, Q.; Zhou, X. Corpse Management in Social Insects. *Int. J. Biol. Sci.* **2013**, *9*, 313–321. [CrossRef] [PubMed] [PubMed Central]
68. Chakraborty, T.S.; Gendron, C.M.; Lyu, Y.; Munneke, A.S.; DeMarco, M.N.; Hoisington, Z.W.; Pletcher, S.D. Sensory Perception of Dead Conspecifics Induces Aversive Cues and Modulates Lifespan through Serotonin in *Drosophila*. *Nat. Commun.* **2019**, *10*, 2365. [CrossRef] [PubMed] [PubMed Central]
69. Clutton-Brock, T.H. Reproductive Effort and Terminal Investment in Iteroparous Animals. *Am. Nat.* **1984**, *123*, 212–229. [CrossRef]
70. Corbel, Q.; Carazo, P. Perception of Dead Conspecifics Increases Reproductive Investment in Fruit Flies. *Funct. Ecol.* **2022**, *36*, 1834–1844. [CrossRef]
71. Hernandez-Lima, M.A.; Seo, B.; Urban, N.D.; Truttmann, M.C. *C. elegans* Behavior, Fitness, and Lifespan, Are Modulated by AWB/ASH-Dependent Death Perception. *bioRxiv* **2024**. [CrossRef] [PubMed] [PubMed Central]
72. De-Souza, E.A.; Thompson, M.A.; Taylor, R.C. Olfactory Chemosensation Extends Lifespan through TGF-β Signaling and UPR Activation. *Nat. Aging* **2023**, *3*, 938–947. [CrossRef] [PubMed] [PubMed Central]
73. De Rosa, M.J.; Veuthey, T.; Florman, J.; Grant, J.; Blanco, M.G.; Andersen, N.; Donnelly, J.; Rayes, D.; Alkema, M.J. The Flight Response Impairs Cytoprotective Mechanisms by Activating the Insulin Pathway. *Nature* **2019**, *573*, 135–138. [CrossRef] [PubMed] [PubMed Central]
74. Nash, T.R.; Chow, E.S.; Law, A.D.; Fu, S.D.; Fuszara, E.; Bilska, A.; Bebas, P.; Kretzschmar, D.; Giebultowicz, J.M. Daily Blue-Light Exposure Shortens Lifespan and Causes Brain Neurodegeneration in *Drosophila*. *NPJ Aging Mech. Dis.* **2019**, *5*, 8. [CrossRef] [PubMed] [PubMed Central]
75. Hori, M.; Shibuya, K.; Sato, M.; Saito, Y. Lethal Effects of Short-Wavelength Visible Light on Insects. *Sci. Rep.* **2014**, *4*, 7383. [CrossRef] [PubMed] [PubMed Central]
76. De Magalhaes Filho, C.D.; Henriquez, B.; Seah, N.E.; Evans, R.M.; Lapierre, L.R.; Dillin, A. Visible Light Reduces *C. elegans* Longevity. *Nat. Commun.* **2018**, *9*, 927. [CrossRef] [PubMed] [PubMed Central]
77. Cadet, J.; Sage, E.; Douki, T. Ultraviolet Radiation-Mediated Damage to Cellular DNA. *Mutat. Res.* **2005**, *571*, 3–17. [CrossRef] [PubMed]
78. Shostal, O.A.; Moskalev, A.A. The Genetic Mechanisms of the Influence of the Light Regime on the Lifespan of *Drosophila melanogaster*. *Front. Genet.* **2013**, *3*, 325. [CrossRef] [PubMed] [PubMed Central]
79. Dubowy, C.; Sehgal, A. Circadian Rhythms and Sleep in *Drosophila melanogaster*. *Genetics* **2017**, *205*, 1373–1397. [CrossRef] [PubMed] [PubMed Central]
80. McLay, L.K.; Green, M.P.; Jones, T.M. Chronic Exposure to Dim Artificial Light at Night Decreases Fecundity and Adult Survival in *Drosophila melanogaster*. *J. Insect Physiol.* **2017**, *100*, 15–20. [CrossRef] [PubMed]
81. Pittendrigh, C.S.; Minis, D.H. Circadian Systems: Longevity as a Function of Circadian Resonance in *Drosophila melanogaster*. *Proc. Natl. Acad. Sci. USA* **1972**, *69*, 1537–1539. [CrossRef] [PubMed] [PubMed Central]
82. Shen, J.; Yang, P.; Luo, X.; Li, H.; Xu, Y.; Shan, J.; Yang, Z.; Liang, B. Green Light Extends *Drosophila* Longevity. *Exp. Gerontol.* **2021**, *147*, 111268. [CrossRef] [PubMed]
83. Krittika, S.; Yadav, P. Alterations in Lifespan and Sleep:Wake Duration Under Selective Monochromes of Visible Light in *Drosophila melanogaster*. *Biol. Open* **2022**, *11*, bio059273. [CrossRef] [PubMed] [PubMed Central]
84. Allemand, R.; Cohet, Y.; David, J. Increase in the Longevity of Adult *Drosophila melanogaster* Kept in Permanent Darkness. *Exp. Gerontol.* **1973**, *8*, 279–283. [CrossRef] [PubMed]
85. Johnson, J.C.; Munneke, A.S.; Richardson, H.M.; Gendron, C.M.; Pletcher, S.D. Light Modulates *Drosophila* Lifespan via Perceptual Systems Independent of Circadian Rhythms. *Aging* **2023**, *15*, 396–420. [CrossRef] [PubMed] [PubMed Central]

86. Rao, F.; Xue, T. Circadian-Independent Light Regulation of Mammalian Metabolism. *Nat. Metab.* **2024**, *6*, 1000–1007. [CrossRef] [PubMed]
87. Rupp, A.C.; Ren, M.; Altimus, C.M.; Fernandez, D.C.; Richardson, M.; Turek, F.; Hattar, S.; Schmidt, T.M. Distinct ipRGC Subpopulations Mediate Light's Acute and Circadian Effects on Body Temperature and Sleep. *Elife* **2019**, *8*, e44358. [CrossRef] [PubMed] [PubMed Central]

biomolecules

MDPI

Article

Bcl-2 Orthologues, *Buffy* and *Debcl*, Can Suppress *Drp1*-Dependent Age-Related Phenotypes in Drosophila

Azra Hasan and Brian E. Staveley *

Department of Biology, Memorial University of Newfoundland, St. John's, NL A1C 5S7, Canada; ahasan@mun.ca
* Correspondence: bestave@mun.ca; Tel.: +1-(709)-864-4317

Abstract: The relationship of Amyotrophic Lateral Sclerosis, Parkinson's disease, and other age-related neurodegenerative diseases with mitochondrial dysfunction has led to our study of the mitochondrial fission gene *Drp1* in *Drosophila melanogaster* and aspects of aging. Previously, the Drp1 protein has been demonstrated to interact with the Drosophila Bcl-2 mitochondrial proteins, and *Drp1* mutations can lead to mitochondrial dysfunction and neuronal loss. In this study, the *Dopa decarboxylase-Gal4* (*Ddc-Gal4*) transgene was exploited to direct the expression of *Drp1* and *Drp1-RNAi* transgenes in select neurons. Here, the knockdown of *Drp1* seems to compromise locomotor function throughout life but does not alter longevity. The co-expression of *Buffy* suppresses the poor climbing induced by the knockdown of the *Drp1* function. The consequences of *Drp1* overexpression, which specifically reduced median lifespan and diminished climbing abilities over time, can be suppressed through the directed co-overexpression of pro-survival *Bcl-2* gene *Buffy* or by the co-knockdown of the pro-cell death *Bcl-2* homologue *Debcl*. Alteration of the expression of *Drp1* acts to phenocopy neurodegenerative disease phenotypes in Drosophila, while overexpression of *Buffy* can counteract or rescue these phenotypes to improve overall health. The diminished healthy aging due to either the overexpression of *Drp1* or the RNA interference of *Drp1* has produced novel Drosophila models for investigating mechanisms underlying neurodegenerative disease.

Keywords: *Drp1*; *Buffy*; *Debcl*; *Drosophila melanogaster*; Amyotrophic Lateral Sclerosis; Parkinson's disease; mitochondrial fission; mitochondrial dysfunction

1. Introduction

The mitochondrial network is essential for many aspects of the subcellular survival mechanisms of organisms. Known to be the "powerhouse of the cell", mitochondria are responsible for various aspects of energy homeostasis, oxidative stress, calcium handling, cell signalling, and, thus, cell survival [1–3]. The dynamic nature of the mitochondria population is critical to the integrity of the subcellular network structures that these organelles maintain and to the control of the quality of mitochondrial proteins and other components [4,5]. For example, the early dynamic events that occur during apoptosis include cristae remodelling, mitochondrial fragmentation, and membrane "blebbing". Inhibition of these processes, either through knockdown directed by RNA interference (RNAi) or by expression of a dominant-negative mutant form of *dynamin-related protein 1 (Drp1)*, can slow the rate of mitochondrial fragmentation and, in turn, the cascade of apoptotic events. The activity of the Drp1 protein promotes caspase-independent mitochondrial fission and cristae remodelling to amplify the process of apoptosis, whether or not cell death is instigated by either the specific activity of the pro-apoptotic protein BH3 interacting-domain death agonist (BID) or by the general consequences of oxidative stress [6]. Overall, the mechanics of mitochondrial fission plays a crucial role in the amplification of aspects of the essential cellular process of apoptosis.

Modifications to the mitochondrial network seem to differentially influence a number of signalling pathways. In response to a series of molecular cues [7,8], the mitochondrial network participates in a delicate balance between the continuous division and fusion processes [9]. Mitochondrial fusion helps compromised mitochondria, likely bearing highly damaged DNA and proteins, to actively exchange components with other more healthy mitochondria. This process acts to decrease the severity of this accumulated heteroplasmy (differing mitochondrial sub-lineages) and help with functional complementation [10]. Recently, much has been understood about mitochondrial fission, a fundamental process in the maintenance of the mitochondrial network, and the requirements for *Drp1* [2,3]. Yet, a complete understanding of the factors that control mitochondrial dynamics during age-related processes [11] remains limited.

The B-cell lymphoma-2 (Bcl-2) family of proteins interact with the Drp1 protein, and expression of *Drp1* can promote apoptosis in both Bcl-2 protein-dependent and independent manners [6]. The two Bcl-2 family homologues in *Drosophila melanogaster* are *Buffy* (anti-apoptotic) and *Debcl* (pro-apoptotic) [12]. The Debcl protein can interact with Drp1 in Drosophila to activate apoptosis via the c-Jun N-terminal kinase (JNK) pathway [13]. *Drp1* is required for a standard rate of Cyt-c release and caspase activation during programmed cell death. The Drp1 protein interacts with other proteins involved in a number of mitochondrial processes, such as the protein product of *Bax* [8]. Diverse stresses can increase the translocation of cytosolic Drp1 to the mitochondria, thus leading to the induction of excessive fragmentation and initiation of apoptosis or mitophagy [14]. In a number of models of ALS, dephosphorylation of Drp1 through the activity of protein phosphatase 1 has been identified as causative of the disease-associated phenotypes [15]. As well, Drp1-mediated mitochondrial fragmentation caused by the administration of rotenone has been identified in a rat model of PD-like changes to the olfactory bulb [16]. As adjustment of *Drp1* gene activity may be key to the inhibition of the pathology of ALS and PD, investigation of such alterations may provide some knowledge helpful in the development of therapies. Excessive activity of Drp1 increases mitochondrial fission and consequently promotes cell death and/or degeneration.

Here, we show that alteration of *Drp1* expression can result in distinct phenotypes over time. Our research group employs "the fruit fly", *Drosophila melanogaster*, to model neurodegenerative disease because it is an excellent model system for studying the genes and proteins affected in ALS, PD, and aging [17]. The anticipated role of mitochondria in pathogenesis has made the study of the interactions of the *Drp1* gene important for modelling these diseases in Drosophila. In these experiments, we exploited the *UAS-Gal4* system to direct the overexpression and knockdown of the genes of interest in selected neuronal tissues, using the *Ddc-Gal4* transgene [18]. We propose that the *Drp1* overexpression phenotype is due to excessive activities related to apoptosis and can be rescued by the appropriate regulation by the anti-apoptotic *Bcl-2* gene, *Buffy*. The knockdown phenotype produced in response to the expression of *Drp1-RNAi* may be due to the diminishment of mitochondrial integrity and may be rescued through modification of the responsible signalling pathway. The careful regulation of mitochondrial dynamics must be important in the control of age-related mitochondrial-induced defects. Overall, our strategy is to identify the basic mechanisms in the fly model to encourage further validation in mammalian model organisms.

2. Materials and Methods

2.1. Bioinformatic Analysis

Protein sequences were obtained from the National Center of Biotechnology Information (NCBI) database (https://www.ncbi.nlm.nih.gov/protein/, accessed on 1 July 2024). The conserved domains were identified using NCBI Conserved Domain Database (CDD) (https://www.ncbi.nlm.nih.gov/cdd/, accessed on 1 July 2024) and Eukaryotic Linear Motif (ELM) (http://elm.eu.org/, accessed on 1 July 2024). Multiple sequence alignment was performed using Clustal Omega (https://www.ebi.ac.uk/Tools/msa/clustalo/, accessed on 1 July 2024) to reveal the conservation of domains. The *Homo sapiens* Dynamin-1-like protein (DLP-1/Drp1) structure (PDB ID 4BEJ) was obtained from NCBI structure database (https://www.ncbi.nlm.nih.gov/structure/, accessed on 1 July 2024), and *Drosophila melanogaster Drp1* protein structure was developed using Phyre2 (http://www.sbg.bio.ic.ac.uk/phyre2/html/page.cgi?id=index, accessed on 1 July 2024) modelling tool. The final models were edited with the PyMOL version 2 software (https://pymol.org/2/, accessed on 1 July 2024) to highlight the N-terminus, C-terminus and consensus LC3-interacting region (LIR) regions.

2.2. *Drosophila Stocks and Media*

The stocks, *UAS-lacZ*$^{4\text{-}1\text{-}2}$; (BDSC_1776: w; P{UAS-lacZ.B}Bg4-1-2); *UAS-Drp1* (BDSC_51647: y w; P{UAS-Drp1.D}3); *UAS-Drp1-RNAi*^{JF02762} or *UAS-Drp1-RNAi1* (BDSC_27682: y v; P{TRiP.JF02762}attP2); *UAS-Drp1-RNAi*^{HMC03230} or *UAS-Drp1-RNAi2* (BDSC_51483: y v; P{TRiP.HMC03230}attP40); *UAS-Buffy* (BDSC_58358: w; P{UAS-Buffy.Q}2); *UAS-Buffy-RNAi* (BDSC_32060: w; P{UAS-Buffy.RNAi}3); *UAS-Debcl*^{EY05743} or *UAS-Debcl* (BDSC_20156: y w; P{EPgy}Debcl[EY05743]); and *Ddc-Gal4*$^{4.3D}$ (BDSC_7010: w; P{Ddc-Gal4}4.3D) were obtained from Bloomington Drosophila Stock Center (BDSC) at Indiana University, Bloomington, Indiana, USA. The *UAS-Debcl-RNAi*^{v47515} (VDRC_47515: P{GD1637}v47515) was obtained from Vienna Drosophila Resource Center (VDRC). The *Ddc-Gal4/CyO; UAS-Drp1/TM3, Ddc-Gal4/CyO; UAS-Drp1-RNAi/TM3* derivative lines were generated using standard recombination methods and used to overexpress or inhibit *Drp1* in the selected DA neurons using the *Ddc-Gal4*$^{4.3D}$ transgene. In brief, a standard "double balancer chromosome" line of the genotype w^{1118}; *L/CyO; Ki/TM3* was used to generate intermediate lines w^{1118}; *Ddc-Gal4/CyO; Ki/TM3*; w^{1118}; *L/CyO; UAS-Drp1/TM3* and w^{1118}; *L/CyO; UAS-Drp1-RNAi/TM3* which were then used to generate the w^{1118}; *Ddc-Gal4/CyO; Ki/TM3; UAS-Drp1/TM3* and w^{1118}; *Ddc-Gal4/CyO; UAS-Drp1-RNAi/TM3* lines.

Sources of validation of the transgenes from previous studies that are used in this study are as follows: *Ddc-Gal4*$^{4.3D}$ (Gal4 expression validated in [18]); *UAS-lacZ*$^{4\text{-}1\text{-}2}$ (directed expression verified in [19]); *UAS-Drp1* (directed expression validated in [20]); *UAS-Drp1-RNAi(1)* (RNA interference validated in [21]); *UAS-Drp1-RNAi2* (RNA interference validated in [22]); *UAS-Buffy* (directed expression validated in [12]); *UAS-Buffy-RNAi* (RNA interference validated in [23]); *UAS-Debcl*^{EY05743} (directed expression validated in [24]); and *UAS-Debcl-RNAi*^{v47515} (RNA interference validated in [25]).

All stocks and experiments were maintained on a standard cornmeal/molasses/yeast/agar medium treated with propionic acid and methylparaben to resist fungal growth. Aliquots of media were poured into plastic vials, allowed to solidify, and refrigerated at 4 °C until used. Stocks were kept at room temperature while crosses and experiments were carried out at 25 °C. In all experimental runs, control critical class individuals were generated and evaluated in parallel under conditions nearly identical to the experimental flies.

2.3. *Aging Assay*

Crosses of select virgin females and males were made, and a cohort of critical class males was collected upon eclosion. At least 250 flies were aged per genotype in the cohorts of 25 or less per vial on fresh media, replenished every two to five days to avoid crowding. Flies were observed and scored every two days for the presence of deceased adults. As a rule, flies were considered dead when they did not display movement upon agitation [26]. Longevity data were analyzed using GraphPad Prism version 8 statistical software (graphpad.com), and survival curves were compared by the Mantel–Cox test. Significance was determined at a 95% confidence level ($p \leq 0.05$) with Bonferroni correction.

2.4. Climbing Assay

A cohort of 70 critical-class male flies was collected within 24 h and maintained as a maximum of ten flies per vial at low density. The food was changed twice every week. Initially, every week, 50 males of each genotype were assayed, in groups of 10 or less, for their ability to climb a glass tube divided into five levels of 2 cm each according to standard protocol [26]. However, as flies died, the climbing assay was conducted on the survivors. The climbing index was calculated for each week using GraphPad prism version 8 statistical software. The climbing curve was fitted using non-linear regression and determined at a 95% confidence interval ($p \leq 0.05$).

3. Results

3.1. *Drp1* Is Highly Conserved between *Homo sapiens* and *Drosophila melanogaster*

The *D. melanogaster* Drp1 protein sequence was sourced from NCBI protein, and the conserved sequences were identified using NCBI CDD. NCBI protein Blast of Drp1 protein of *D. melanogaster* (NP_608694.2) with the *H. sapiens,* identified dynamin-1-like protein (isoform 4) (NP_001265392.1), it is 65% identical with a bit score of 957. The multiple sequence alignment of the two proteins derived by Clustal Omega (Figure 1a) shows a highly conserved dynamin-like protein family domain, a dynamin central domain, and a dynamin GTPase effector domain. Two well-documented phosphorylation sites are identified: S606 and S627 in *dynamin-1-like protein* isoform 4 of *H. sapiens;* and S616 and T637 in *Drp1* of *D. melanogaster*. A template-based modelling of *D. melanogaster* Drp1 protein by use of a combination of empirically derived energy functions and physics-based simulated folding was produced using Phyre2. The modelled *D. melanogaster* Drp1 protein (i) and the *H. sapiens* Dynamin-1-like protein (ii) from the NCBI database share a near identical structure (Figure 1b). The amino-terminus region of the Drp1 protein is highly conserved and has a consensus LC3-interacting region (LIR) sequence for binding to the ATG8/LC3 protein as determined by the Eukaryotic Linear Motif (ELM) resource. As this protein structure is so highly conserved, it seems very likely that the functions are highly conserved.

3.2. The Directed Overexpression and Knockdown of *Drp1* with *Ddc-Gal4$^{4.3D}$*

In this set of experiments, the control *Ddc-Gal4$^{4.3D}$*; *UAS-lacZ* critical class males were determined to have a median lifespan of 68 days (n = 340). The directed overexpression of *Drp1* by the *Ddc-Gal4* transgene results in a decreased lifespan of 56 days in 314 flies, much lower compared to the control as determined by log-rank (Mantel–Cox) test with a p-value at <0.0001 (Figure 2A). Inhibition of *Drp1* by two distinct RNAi transgenes via the *UAS-Drp1-RNAi1* and *UAS-Drp1-RNAi2* directed by the *Ddc-Gal4* transgene results in median lifespans of 70 (n = 377) and 72 days (n = 323), respectively; very similar to the control (Figure 2A) as determined by log-rank (Mantel–Cox) test with p-value 0.0566 and 0.0213. The non-linear fitting of the climbing curve shows that altering the *Drp1* expression in either direction compromises the climbing ability phenotype compared to control at 95% confidence interval (CI) (p-value < 0.0001) (Figure 2B) (n = 50).

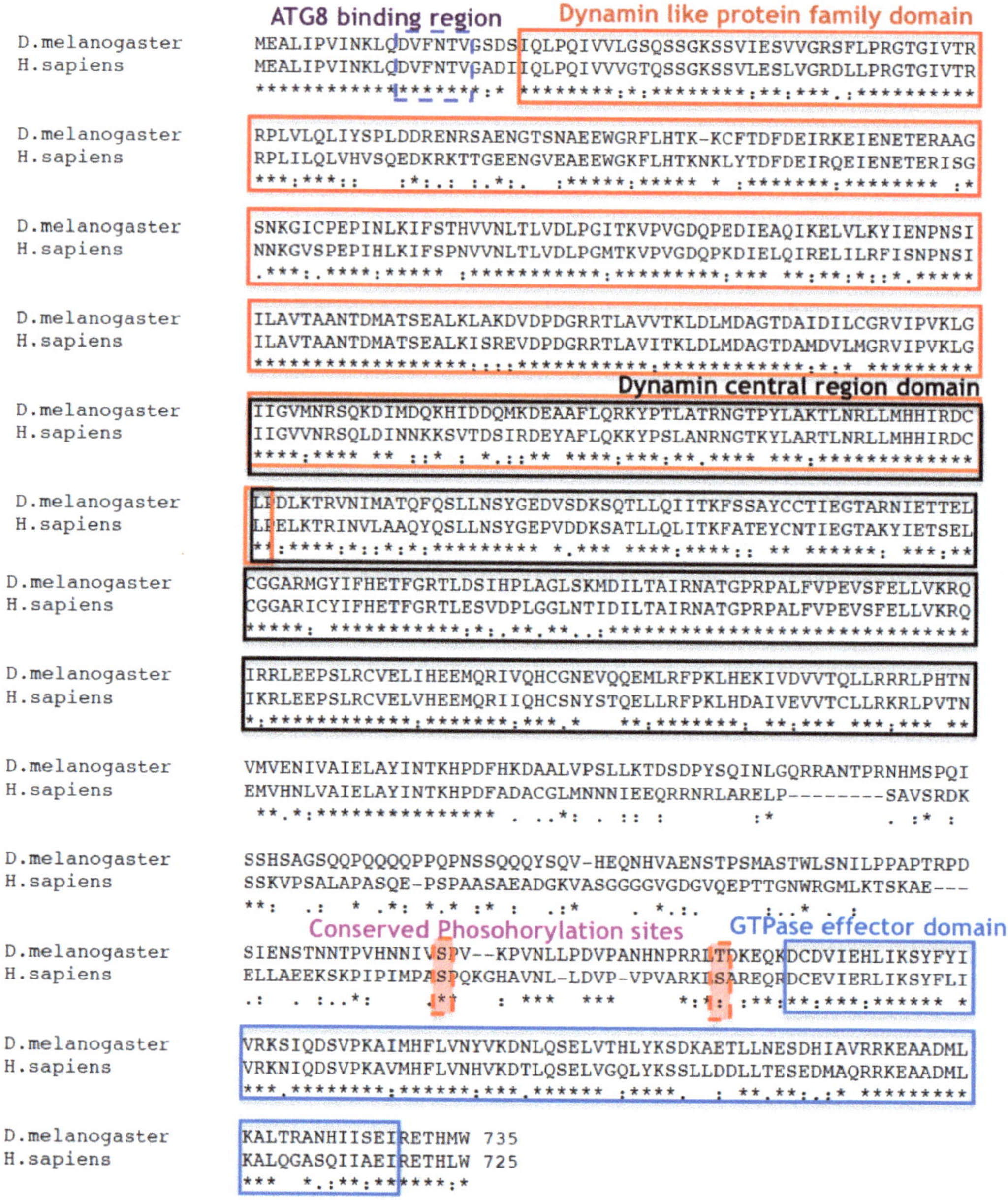

(a)

Figure 1. *Cont.*

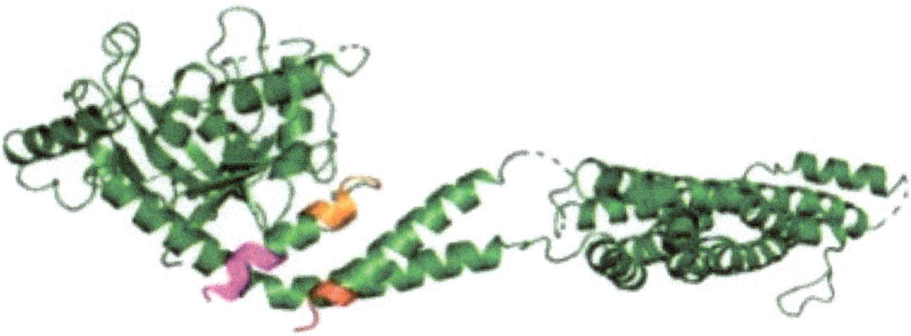

(**bi**)

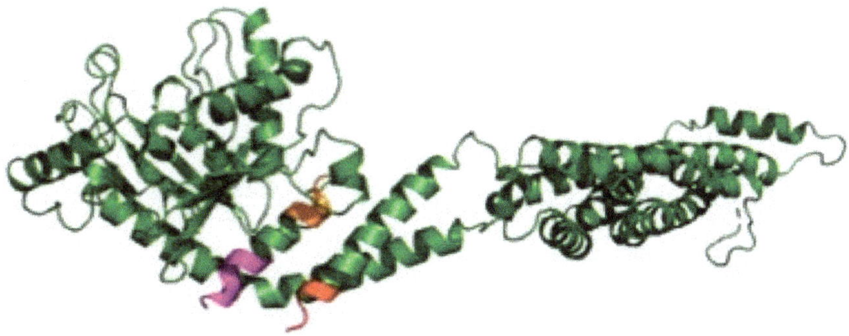

(**bii**)

Figure 1. *Drp1* is evolutionarily conserved between Drosophila and humans. (**a**) Clustal Omega multiple sequence alignment of *D. melanogaster* Drp1 (NP_608694.2) protein with the *H. sapiens* (NP_001265392.1) shows evolutionarily conserved domains identified using the NCBI Conserved Domain Database (CDD) and is further confirmed by the Eukaryotic Linear Motif (ELM) resource. The two well-documented phosphorylation sites are identified: S606 and S627 in dynamin-1-like protein (DLP-1) isoform 4 of *H. sapiens* and S616 and T637 in Drp1 of *D. melanogaster*. The asterisks indicate the residues that are identical, the colons indicate the conserved substitutions, and the dots indicates the semi-conserved substitutions. Colour differences indicate the chemical nature of amino acids: red indicates small hydrophobic (includes aromatic) residues; blue indicates acidic; magenta indicates basic; and green indicates basic with hydroxyl or amine groups. (**bi**) The original Dynamin-1-like protein (DLP-1) structure of H. sapiens (NP_001265392.1) from the NCBI structure database. (**bii**) The Phyre2 web portal for protein modelling, prediction, and analysis mediated the development of a model of the Drp1 protein of *D. melanogaster* (NP_608694.2) from a 76% identical protein with a confidence of 100%. The N terminus is coloured in Magenta; C terminus is coloured in Red, and a consensus ATG8 binding region at N terminus is coloured in orange.

(A)

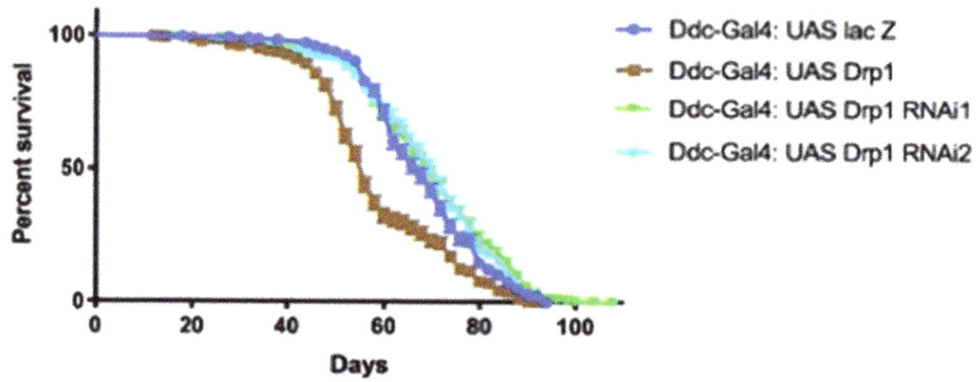

(B)

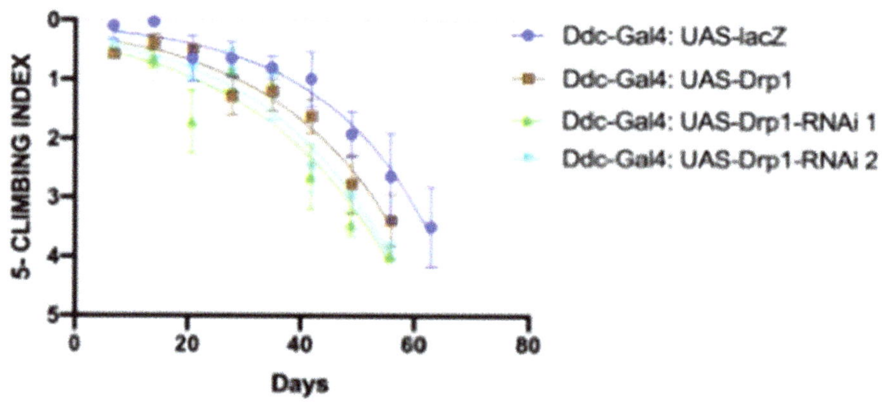

Figure 2. Altered *Drp1* expression under the control of *Ddc-Gal4*$^{4.3D}$ influences the survival and climbing ability of flies. (**A**) The GraphPad prism8 generated graph of the longevity assay for the expression of *Drp1*, *Drp1-RNAis* under the control of *Ddc-Gal4* transgene. The directed expression results in decreased median lifespan of 56 days compared to 68 days of control, as calculated by Log-rank Mantel–Cox test, with Bonferroni correction. The knockdown of *Drp1* under the control of the *Ddc-Gal4* transgene results in lifespan of 70 days with *UAS-Drp1-RNAi1* and 72 days with *UAS-Drp-RNAi2* compared to 68 days of control performed by Log-rank Mantel–Cox test, with Bonferroni correction. (**B**) The GraphPad prism8 generated graph of the climbing abilities of flies with overexpression of *Drp1*, the *Drp1-RNAis*, and the *lacZ* control. The climbing ability of *Drp1* overexpression and *Drp1-RNAi*s flies have decreased compared to control as determined in non-linear fitting of the climbing curve by 95% confidence interval (p-value < 0.0001).

3.3. *Phenotypic Rescue by Co-Expression of Drp1 and Drp1-RNAi Directed by Ddc-Gal4*$^{4.3D}$

In this set of experiments, we demonstrate that the phenotype caused by the directed expression of *Drp1* can be counteracted by knockdown via *Drp1-RNA*, and the phenotype caused by knockdown via *Drp1-RNA* can be counteracted by the directed expression of *Drp1*. Both are compared to *lacZ* controls. The control *Ddc-Gal4*$^{4.3D}$; *UAS-lacZ* critical class males were determined to have a median lifespan of 62 days (n = 308). The directed knockdown of *Drp1* by the *Ddc-Gal4* transgene results in a greater median lifespan of 70 days in 321 flies compared to the control as determined by log-rank (Mantel–Cox) test with *p*-value < 0.0001 (Figure 3). In contrast, the directed expression of *Drp1* by the *Ddc-Gal4* transgene results in a reduced median lifespan of 56 days in 255 flies compared to the control as determined by log-rank (Mantel–Cox) test with *p*-value < 0.0001 (Figure 3A). Furthermore, the *Ddc-Gal4 UAS-Drp1-RNAi UAS-lacZ* critical class males have a median lifespan of 70 days in 310 flies. The directed expression of *Drp1* along with *UAS-Drp1-RNAi* under the direction of the *Ddc-Gal4* transgene (*Ddc-Gal4; UAS-Drp1-RNAi; UAS-Drp1*) has a median lifespan of 64 days, similar to control (*Ddc-Gal4; UAS-lacZ*) with a *p* value of 0.0633 as determined by the Log-rank Mantel–Cox test with a Bonferroni correction (Figure 3A). The *Ddc-Gal4 UAS-Drp1 UAS-lacZ* critical class males have a median lifespan of 58 days in 294 flies. The inhibition of *Drp1* along with *UAS-Drp1* under the direction of the *Ddc-Gal4* transgene (*Ddc-Gal4; UAS-Drp1; UAS-Drp1-RNAi*) has a median lifespan of 64 days (n = 327), similar to control (*Ddc-Gal4; UAS-lacZ*) with a *p* value of 0.0582 as determined by the Log-rank Mantel–Cox test with a Bonferroni correction (Figure 3C). The non-linear fitting of the climbing ability curve shows the *Drp1* expression and inhibition both have compromised the climbing ability phenotype compared to control at 95% CI (*p* < 0.0001) (Figure 3B,D). The climbing ability curve of *Ddc-Gal4 UAS-Drp1-RNAi UAS-Drp1* and *Ddc-Gal4 UAS-Drp1 UAS-Drp1-RNAi* is very close to the control (*Ddc-Gal4; UAS-lacZ*) as determined by the non-linear fitting of the climbing curve at a 95% CI at *p*-value 0.2752 and 0.0589, respectively. The co-expression of *Drp1-RNAi* along with *Drp1* ectopic expression in flies has resulted in phenotypes that are similar to the control and suggests that these phenotypes are primarily due to the changes in the expression of *Drp1*.

In Figure 3A,B, the *Ddc-Gal4* and the *Drp1-RNAi* genes were maternally contributed. When compared with *lacZ*, *Drp1* acted to reduce the enhanced median lifespan and reduced climbing of *Drp1-RNAi* knockdown. In Figure 3C,D, the *Ddc-Gal4* and the *Drp1* genes were maternally contributed. When compared with *lacZ*, *Drp1-RNAi* knockdown resulted in the rescue of the climbing defect, the *Ddc-Gal4 UAS-lacZ* control, with little change to lifespan.

(A)

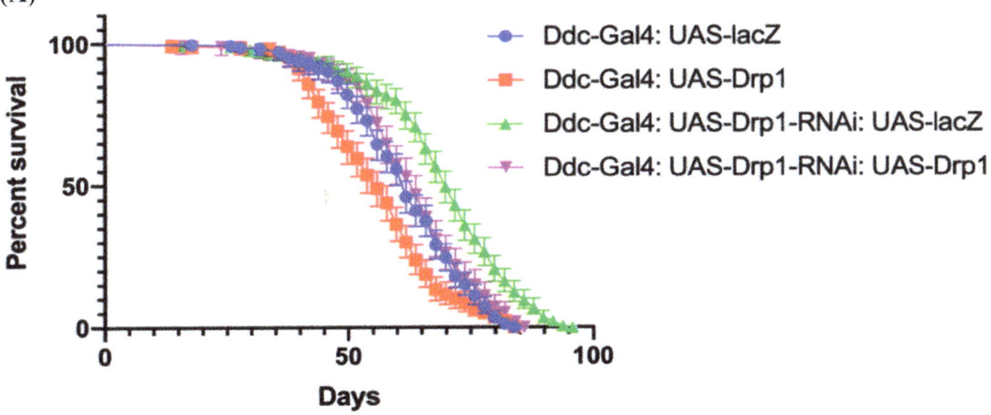

Figure 3. *Cont.*

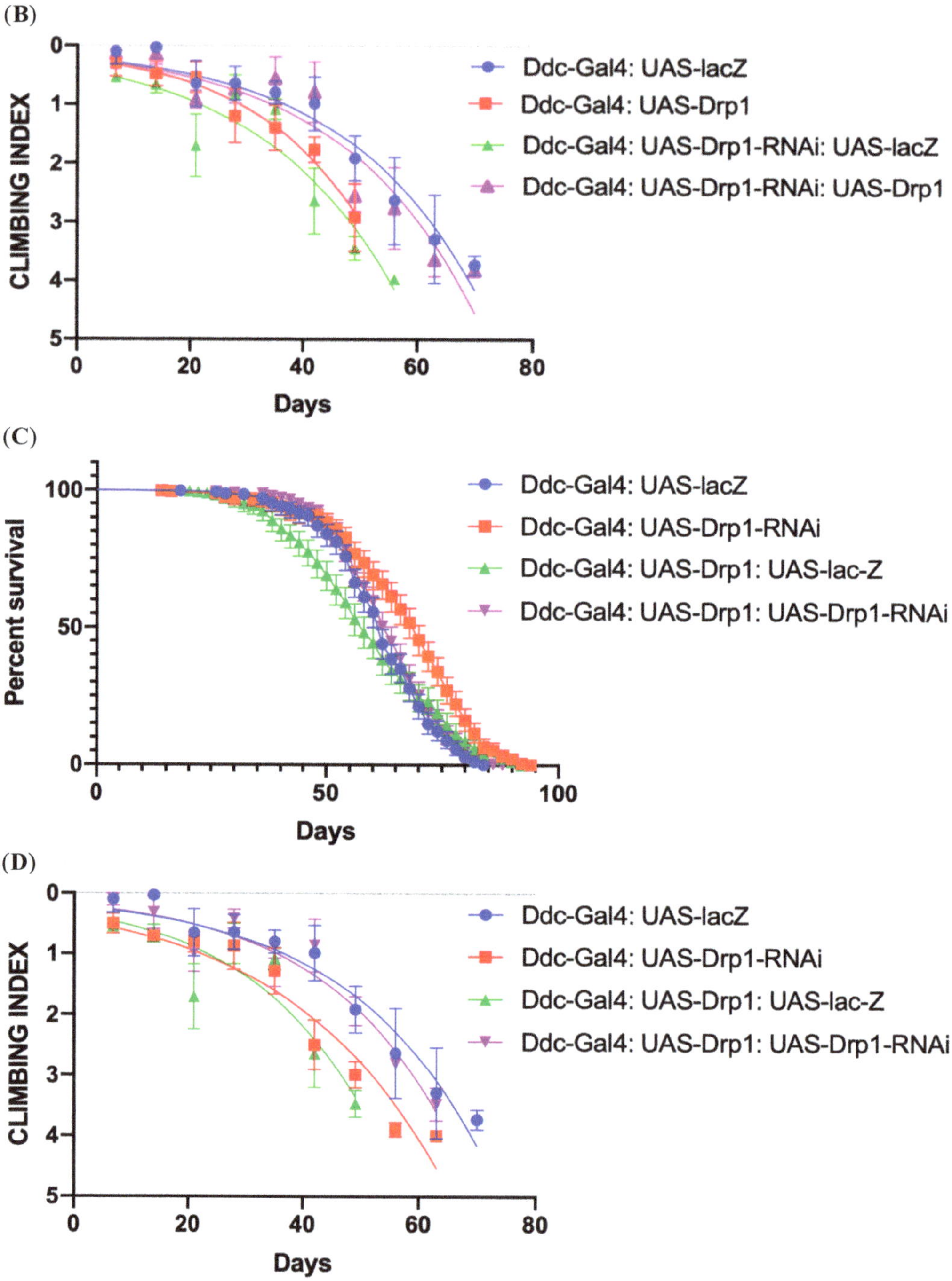

Figure 3. The ectopic expression of *Drp1-RNAi*, directed by *Ddc-Gal4*$^{4.3D}$, can increase median lifespan and decrease climbing. (**A**) In control, *Ddc-Gal4*$^{4.3D}$ *UAS-lacZ* critical class males resulted in

a median life span of 62 days (n = 308). Expression of *Drp1* in *Ddc-Gal4*$^{4.3D}$ resulted in a median life span of 56 days (n = 310), much lower than the *lacZ*-expressing control; expression of *Drp1* in *Ddc-Gal4*$^{4.3D}$ *UAS-Drp1-RNAi* resulted in a median life span of 64 days (n = 250) very similar to control (*Ddc/lacZ*) as determined by the Log-rank Mantel–Cox test (p value = 0.0633) with Bonferroni correction. The graph of the longevity assay was generated by GraphPad prism8. (**B**) The *Ddc-Gal4*$^{4.3D}$ flies express *UAS-lacZ* in control flies. The climbing abilities of *Ddc-Gal4*$^{4.3D}$ *UAS-Drp1* expressing flies have decreased compared to control as determined in the non-linear fitting of the climbing curve by a 95% confidence interval ($p < 0.0001$). The flies' climbing ability expressing *Drp1* in *Ddc-Gal4*$^{4.3D}$ *UAS-Drp1-RNAi* transgene is similar to control as determined in the non-linear fitting of the climbing curve by a 95% confidence interval at p value = 1.309. The graph of longevity assay was generated by GraphPad prism8 non-linear regression curve. (**C**) In control, *Ddc-Gal4*$^{4.3D}$ *UAS-lacZ* critical class males resulted in a median life span of 62 days (n = 308). Knockdown of *Drp1-RNAi* in *Ddc-Gal4*$^{4.3D}$ resulted in a median life span of 70 days (n = 321), much higher compared to the control; knockdown of *Drp1-RNAi* in *Ddc-Gal4*$^{4.3D}$ *UAS-Drp1* resulted in a median life span of 64 days (n = 327), very similar to control (*Ddc/lacZ*) as determined by the Log-rank Mantel–Cox test (p value = 0.0582) with Bonferroni correction. The graph of the longevity assay was generated by GraphPad prism8. (**D**) The *Ddc-Gal4*$^{4.3D}$ flies express *UAS-lacZ* in control flies. The climbing abilities of *Ddc-Gal4*$^{4.3D}$ *UAS-Drp1-RNAi* expressing flies have decreased compared to control as determined in the non-linear fitting of the climbing curve by a 95% confidence interval ($p < 0.0001$). The flies' climbing ability expressing *Drp1-RNAi* in *Ddc-Gal4*$^{4.3D}$ *UAS-Drp1* transgene is similar to control (*Ddc/lacZ*) as determined in the non-linear fitting of the climbing curve by a 95% confidence interval at p value = 0.0027. The graph of longevity assay was generated by GraphPad prism8 non-linear regression curve.

3.4. Alteration of the Expression of Buffy and Debcl in Combination with Drp1 Directed by the Ddc-Gal4$^{4.3D}$ Transgene

The control *Ddc-Gal4*$^{4.3D}$; *UAS-Drp1*; *UAS-lacZ* critical class males were determined to have a median lifespan of 58 days (n = 282). The overexpression of *Buffy* along with *UAS-Drp1* under the direction of the *Ddc-Gal4* transgene (*Ddc-Gal4; UAS-Drp1; UAS-Buffy*) has a median lifespan of 68 days (n = 255), much higher compared to control with a p value of <0.0001 as determined by log-rank (Mantel–Cox) test with a Bonferroni correction. The knockdown of *Buffy* along with *UAS-Drp1* under the direction of the *Ddc-Gal4* transgene (*Ddc-Gal4; UAS-Drp1; UAS-Buffy-RNAi*) has a median lifespan of 52 days (n = 274), much less compared to the control (Figure 4A) with a p value 0.0125 as determined by log-rank Mantel–Cox test with a Bonferroni correction. The overexpression of *Buffy* in neurons rescued the early onset of impairment in the climbing ability of *Ddc-Gal4; UAS-Drp1* flies. The non-linear fitting of the climbing curve shows *Buffy* overexpression has rescued the climbing ability defect compared to control at 95% CI ($p < 0.0001$) (Figure 4B). The knockdown of *Buffy* by *Ddc-Gal4 UAS-Drp1; UAS-Buffy-RNAi* further contributes to loss of the climbing ability throughout the life of critical class flies compared to control at 95% CI at a p-value 0.0125 (n = 50) (Figure 4B).

The overexpression of *Debcl* along with *UAS-Drp1* under the direction of *Ddc-Gal4* transgene (*Ddc-Gal4; UAS-Drp1; UAS-Debcl*^{EY05743}) has a median lifespan of 60 days (n = 331 flies), similar to the control (Figure 4A) with a p value at 0.0114 as determined by log-rank Mantel–Cox test with Bonferroni correction. The inhibition of *Debcl*, along with *UAS-Drp1* under the direction of the *Ddc-Gal4* transgene (*Ddc-Gal4; UAS-Drp1; UAS-Debcl-RNAi*^{v47515}) has a median lifespan of 66 days (n = 303), much higher compared to the control (Figure 4A) with a p value at 0.0004 as determined by log-rank Mantel–Cox test with Bonferroni correction. The non-linear fitting of the climbing curve shows that *Debcl* overexpression has no change in the climbing ability defect compared to control at 95% CI ($p = 0.3293$) (Figure 4B). The knockdown of *Debcl* by *Ddc-Gal4 UAS-Drp1; UAS-Debcl-RNAi*^{v47515} has rescued the climbing ability throughout the life of critical class flies compared to control at 95% CI at a p-value 0.0057 (n = 50) (Figure 4B).

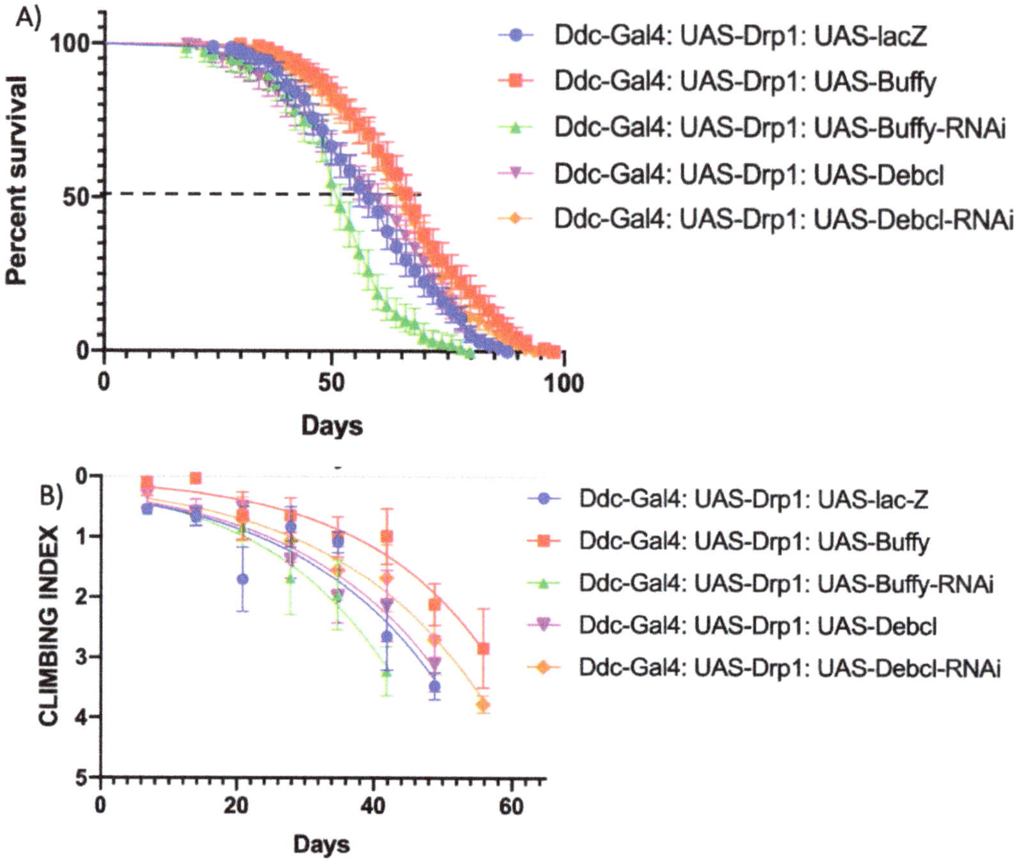

Figure 4. Altered expression of *Buffy* and *Debcl* can enhance and suppress climbing ability in *Drp1* over-expression flies. (**A**) In control, *Ddc-Gal4*$^{4.3D}$; *UAS-Drp1 UAS-lacZ* critical class males resulted in a median life span of 58 days (n = 282). The overexpression of *Buffy* results in a median lifespan of 68 days (n = 375) compares to 58 days of control (p value = 0.0002); the knockdown of *Buffy* directed by the *Ddc-Gal4*$^{4.3D}$ *UAS-Drp1* transgene results in the median lifespan of 52 (n = 274), much less compared to control, determined by Log-rank Mantel–Cox test at p-value < 0.0001, with Bonferroni correction. The overexpression of *Debcl*^{EY05743} results in a median lifespan of 60 days (n = 331) similar to 58 days of control determined by Log-rank Mantel–Cox test at p-value 0.3293; the inhibition of *Debcl* directed by the *Ddc-Gal4 UAS-Drp1* transgene result in the median lifespan of 66 (n = 303); much higher than control, determined by Log-rank Mantel–Cox test at p value 0.0057, with Bonferroni correction. (**B**) The GraphPad prism8 generated graph of the climbing abilities of *Ddc-Gal4 UAS-Drp1* flies with the expression of *Buffy*, *Buffy-RNAi*, *Debcl*^{EY05743}, *Debcl-RNAi*^{v47515} and control. The climbing abilities of flies overexpressing *Buffy* have rescued compared to control as determined in the climbing curve's non-linear fitting by a 95% confidence interval ($p < 0.0001$). The climbing ability of the flies was further weakened by the knockdown of *UAS-Buffy-RNAi* as determined in the non-linear fitting of the climbing curve by a 95% confidence interval at a p-value 0.0125 and 0.03293, respectively (n = 50). The climbing abilities of flies expressing *Debcl-RNAi*^{v47515} has been rescued compared to control as determined by the non-linear fitting of the climbing curve by a 95% confidence interval (p value = 0.0057). The graph of longevity assay was generated by GraphPad prism8 non-linear regression curve.

3.5. Altering the Expression of Buffy and Debcl along with Drp1-RNAi by *Ddc-Gal4*$^{4.3D}$ Transgene

The control *Ddc-Gal4*$^{4.3D}$; *UAS-Drp1-RNAi; UAS-lacZ* critical class males were determined to have a median lifespan of 70 days (n = 323). The overexpression of *Buffy* along with *UAS-Drp1-RNAi* under the direction of the *Ddc-Gal4* transgene (*Ddc-Gal4; UAS-Drp1-RNAi; UAS-Buffy*) has a median lifespan of 64 days (n = 308), much lower compared to control with a p value of <0.0001 as determined by log-rank (Mantel–Cox) test with a Bonferroni correction. The co-inhibition of *Buffy* and *Drp1* under the direction of the *Ddc-Gal4* transgene (*Ddc-Gal4; UAS-Drp1-RNAi; UAS-Buffy-RNAi*) has a median lifespan of 62 days (n = 273), much less compared to the control (Figure 5A) with a p value at <0.0001 as determined by log-rank Mantel–Cox test with a Bonferroni correction. The non-linear fitting of the climbing curve shows *Buffy* overexpression has rescued the climbing ability defect compared to control at 95% CI ($p < 0.0001$) (Figure 5B). The inhibition of *Buffy* by *Ddc-Gal4 UAS-Drp1-RNAi; UAS-Buffy-RNAi* further contributes to loss of the climbing ability throughout the life of critical class flies compared to control at 95% CI at a p-value < 0.0001 (n = 50) (Figure 5B).

The overexpression of *Debcl* along with *UAS-Drp1-RNAi* under the direction of *Ddc-Gal4* transgene (*Ddc-Gal4; UAS-Drp1-RNAi; UAS-Debc*^{EY05743}) has a median lifespan of 68 days (n = 156 flies), similar to the control (Figure 5A) with a p value at 0.0003 as determined by log-rank Mantel–Cox test with Bonferroni correction. The inhibition of *Debcl*, along with *UAS-Drp1-RNAi* under the direction of the *Ddc-Gal4* transgene (*Ddc-Gal4; UAS-Drp1-RNAi; UAS-Debcl-RNA*^{v47515}) results in a median lifespan of 72 days (n = 321), higher compared to the control (Figure 5A) with a p value at 0.0211 as determined by log-rank Mantel–Cox test with Bonferroni correction. The non-linear fitting of the climbing curve shows that *Debcl* overexpression has further increased the climbing ability defect compared to control at 95% CI ($p = 0.0004$) (Figure 5B). The inhibition of *Debcl* by *Ddc-Gal4 UAS-Drp1-RNAi; UAS-Debcl-RNAi*^{v47515} has rescued the climbing ability throughout the life of critical class flies compared to control at 95% CI at a p-value 0.0211 (n = 50) (Figure 5B).

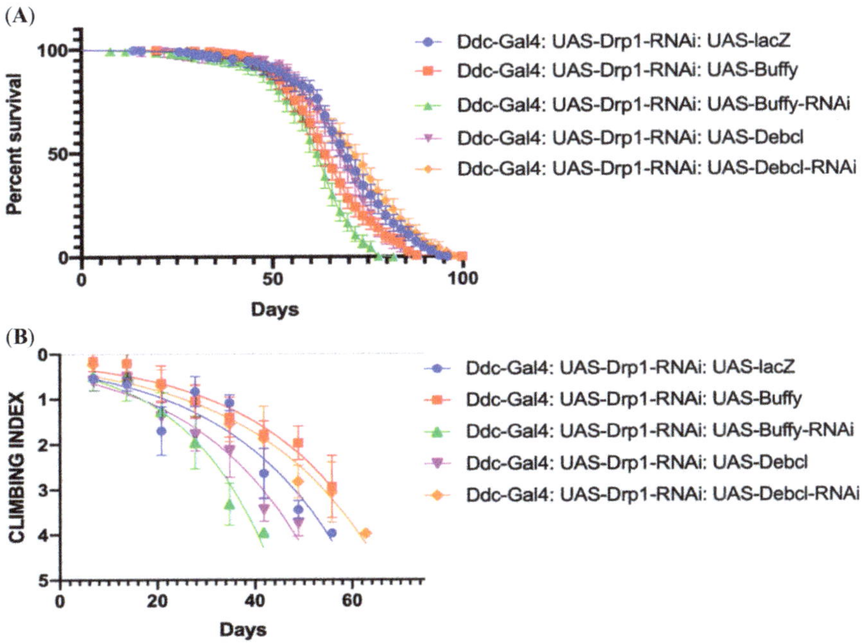

Figure 5. Altered expression of *Buffy* and *Debcl* can enhance and suppress climbing ability in *Drp1* knockdown flies. (**A**) In control, *Ddc-Gal4*$^{4.3D}$; *UAS-Drp1* transgene results in the median lifespan of

62 (n = 273), determined by Log-rank Mantel–Cox test at p-value < 0.0001, with Bonferroni correction. The overexpression of *Debcl*^{EY05743} results in a median lifespan of 68 days (n = 331), much higher compared to control as determined by Log-rank Mantel–Cox test at p-value 0.0003; the inhibition of *Debcl* directed by the *Ddc-Gal4; UAS-Drp1* transgene result in the median lifespan of 72 (n = 303); similar to control, determined by Log-rank Mantel–Cox test at p-value 0.021, with Bonferroni correction. (**B**) The GraphPad prism8 generated graph of the climbing abilities of *Ddc-Gal4 UAS-Drp1* flies with the expression of *Buffy*, *Buffy-RNAi*, *Debcl*^{EY05743}, *Debcl-RNAi*^{v47515} and control. The climbing abilities of flies overexpressing *Buffy* have rescued compared to control as determined in the climbing curve's non-linear fitting by a 95% confidence interval ($p < 0.0001$). The climbing ability of the flies has further diminished through the ectopic expression of *UAS-Buffy-RNAi* and *UAS-Debcl*^{EY05743} as determined in the non-linear fitting of the climbing curve by a 95% confidence interval at a p-value 0.0004 and 0.0002 respectively (n = 50). The climbing abilities of flies expressing *Debcl-RNAi*^{v47515} has been rescued compared to control as determined by the non-linear fitting of the climbing curve by a 95% confidence interval (p value < 0.0001). The graph of longevity assay was generated by GraphPad prism8 non-linear regression curve.

4. Discussion

The protein product of the *Drp1* gene, along with the participation of other mitochondrial protection proteins, is involved in the processes of mitochondrial fission, apoptosis, and mitophagy. Excessive mitochondrial fragmentation can be associated with dysfunctional metabolic diseases, whereas a "hyper-fused" mitochondrial network can serve to protect cells from metabolic insult and autophagy [27]. In the skeletal muscle of mice, *Drp1* overexpression can cause a severe impairment of post-natal muscle growth as the production of protein may become attenuated and growth hormone pathways may be downregulated [28]. Conditions of high fat and/or high glucose levels can cause excessive oxidative stress along with mitochondrial fragmentation as mediated by the Drp1 protein [29,30]. These phenotypes are similar to the increased activity of *Drp1* as observed with Cos and PC12 cells [31]. In humans, protein kinase A (PKA) can phosphorylate and inactivate the pro-apoptotic Bcl-2 family member protein Bad [32] and the Drp1 protein [33] in a complex effort to promote cell survival. The effect of *Drp1* overexpression and, consequently, excessive mitochondrial fragmentation can be toxic to many physiological processes.

The Bcl-2 family proteins assist the pro-fission activity of the Drp1 protein during apoptosis in nematodes and mammals [34]. However, in non-apoptotic cells of mammals, the Bcl-2 family proteins have both pro-fission and pro-fusion activities. The overexpression of *Drp1* in selected neurons, along with the overexpression of *Buffy* or the inhibition of *Debcl*, has resulted in an increase in median lifespan and the ability to climb over the increased lifespan. In complementary experiments, *Buffy* knockdown and *Debcl* overexpression have resulted in reduced lifespans accompanied with impaired climbing abilities, consistent with the conclusion that *Buffy* can function as the antithesis of *Debcl* [12]. The rescue of the *Drp1* expression phenotype is in accordance with the role of the Buffy protein as "the guardian of mitochondria". As proteins, Buffy can interact with Debcl to inhibit *Debcl*-induced cell death. As a mechanism, this process could be due to decreased activity of the Debcl protein, which influences cooperation with the Drp1 protein in the promotion of cell death [13]. The pro-apoptotic Debcl protein acts to induce apoptosis through a caspase-independent mechanism that triggers the release of Cytochrome C in an activity that resembles the loss of *Drp1* [6]. The overexpression of *Debcl* and *Drp1* together in selected neurons does not alter the phenotype generated by the overexpression of *Drp1* without *Debcl*. This may not be surprising as Drp1 and Debcl proteins seem to cooperate to promote apoptosis. Indeed, this and earlier studies have demonstrated that *Drp1* can play various roles in mitochondrial fragmentation and apoptosis to act in concert with anti- and pro-survival proteins, dependent upon the stimuli.

The directed knockdown in *Drosophila melanogaster* of *Drp1*, via *Drp1-RNAi*, in a subset of neurons results in an accelerated age-dependent loss in climbing ability, a phenotype strongly associated with the modelling of age-related disease in flies. The overexpression of *Buffy* in neurons that co-express *Drp1-RNAi* led to a decrease in the median lifespan accompanied with, or perhaps balanced by, a rescue of the impaired locomotor ability. The recovery in age-dependent climbing ability over time may be evidence of a complicated regulatory relationship. A study shows *Drp1* inhibition reduces the total accumulation of pro-apoptotic Bcl-2 protein, Bax, on the mitochondria outer membrane in HeLa cell lines [8]. This intermediate phenotype was not expected but may be important in the determination of the pathology of neurological diseases. The knockdown of anti-apoptotic *Buffy* or overexpression of pro-apoptotic *Debcl* enhanced the phenotype induced through the knockdown of *Drp1* in *Ddc-Gal4* expressing neurons. The interaction of *Bax* with *Drp1* in mammals seems to be conserved with this relationship. Drp1 protein interacts directly and indirectly with the Bcl-2 family protein to facilitate the permeability of the mitochondrial outer membrane in apoptotic cells [34]. Overall, we believe that we have established that *Buffy* confers a survival advantage to flies overexpressing *Drp1* and provides a partially rescued intermediate phenotype in flies with a knockdown of *Drp1* function.

5. Conclusions

Closely associated with cell death pathways in neurons, *Drp1* has recently been associated with ALS, Parkinson's disease, and age-related disease. Our studies demonstrate that the directed overexpression and knockdown of *Drp1* activity in selected neurons can phenocopy some neurodegenerative-like symptoms in Drosophila and, therefore, may represent a novel model of disease. Importantly, the decrease in lifespan and age-dependent loss in climbing ability observed with overexpression of *Drp1* in flies is "rescued" to near controls either by overexpression of *Buffy* or by *Debcl* knockdown. The age-dependent loss of climbing ability in flies expressing *Drp1-RNAi* can be rescued by *Buffy* overexpression or *Debcl-RNAi*-directed knockdown. Future studies of these interactions will be required to chart out a pathway for *Drp1* and interactions with *Buffy* and *Debcl* in Drosophila and, importantly, the molecular changes associated with the loss-of-function of these genes in the development, function, and aging of the organism.

Author Contributions: All experiments were initiated, conceived, designed, and presented by A.H. with experimental design, consultation, critical review, and revision provided by B.E.S. All authors have read and agreed to the published version of the manuscript.

Funding: This research is funded by the Natural Sciences and Engineering Research Council of Canada (NSERC) Discovery Grant Program, grant number 2016-04828, and the Aging Research Centre of Newfoundland and Labrador (ARC-NL) Research Grant Program to Brian E. Staveley, Memorial University of Newfoundland School of Graduate Studies Baseline Funding and Memorial University of Newfoundland Department of Biology Teaching Assistantships.

Institutional Review Board Statement: The animal study protocol was approved by Animal Care Committee (ACC) of Memorial University of Newfoundland (ACC 21-01-BS) and by the Institutional Biosafety Committee of Memorial University of Newfoundland (Biosafety Certificate S-78).

Informed Consent Statement: Not applicable.

Data Availability Statement: Data are contained within the article.

Conflicts of Interest: The authors declare no conflicts of interest.

References

1. Galluzzi, L.; Kepp, O.; Kroemer, G. Mitochondria: Master regulators of danger signalling. *Nat. Rev. Mol. Cell Biol.* **2012**, *13*, 780–788. [CrossRef]
2. Adhikary, A.; Mukherjee, A.; Banerjee, R.R.; Nagotu, S. DRP1: At the crossroads of dysregulated mitochondrial dynamics and altered cell signaling in cancer cells. *ACS Omega* **2023**, *8*, 45208–45223. [CrossRef] [PubMed]

3. Zerihun, M.; Sukumaran, S.; Qvit, N. The Drp1-mediated mitochondrial fission protein interactome as an emerging core player in mitochondrial dynamics and cardiovascular disease therapy. *Int. J. Mol. Sci.* **2023**, *24*, 5785. [CrossRef]
4. Ikeda, Y.; Shirakabe, A.; Maejima, Y.; Zhai, P.; Sciarretta, S.; Toli, J.; Nomura, M.; Mihara, K.; Egashira, K.; Ohishi, M.; et al. Endogenous Drp1 mediates mitochondrial autophagy and protects the heart against energy stress. *Circ. Res.* **2015**, *116*, 264–278. [CrossRef]
5. Schmitt, K.; Grimm, A.; Dallmann, R.; Oettinghaus, B.; Restelli, L.M.; Witzig, M.; Ishihara, N.; Mihara, K.; Ripperger, J.A.; Albrecht, U.; et al. Circadian control of DRP1 activity regulates mitochondrial dynamics and bioenergetics. *Cell Metab.* **2018**, *27*, 657–666.e5. [CrossRef] [PubMed]
6. Oettinghaus, B.; D'Alonzo, D.; Barbieri, E.; Restelli, L.M.; Savoia, C.; Licci, M.; Tolnay, M.; Frank, S.; Scorrano, L. DRP1-dependent apoptotic mitochondrial fission occurs independently of BAX, BAK and APAF1 to amplify cell death by BID and oxidative stress. *Biochim. Biophys. Acta* **2016**, *1857*, 1267–1276. [CrossRef] [PubMed]
7. Hoitzing, H.; Johnston, I.G.; Jones, N.S. What is the function of mitochondrial networks? A theoretical assessment of hypotheses and proposal for future research. *BioEssays* **2015**, *37*, 687–700. [CrossRef] [PubMed]
8. Maes, M.E.; Grosser, J.A.; Fehrman, R.L.; Schlamp, C.L.; Nickells, R.W. Completion of BAX recruitment correlates with mitochondrial fission during apoptosis. *Sci. Rep.* **2019**, *9*, 16565. [CrossRef]
9. Scott, I.; Youle, R.J. Mitochondrial fission and fusion. *Essays Biochem.* **2010**, *47*, 85–98. [CrossRef]
10. Chan, D.C. Fusion and fission: Interlinked processes critical for mitochondrial health. *Annu. Rev. Genet.* **2012**, *46*, 265–287. [CrossRef]
11. Sebastián, D.; Palacín, M.; Zorzano, A. Mitochondrial dynamics: Coupling mitochondrial fitness with healthy aging. *Trends Mol. Med.* **2017**, *23*, 201–215. [CrossRef]
12. Quinn, L.; Coombe, M.; Mills, K.; Daish, T.; Colussi, P.; Kumar, S.; Richardson, H. Buffy, a Drosophila Bcl-2 protein, has anti-apoptotic and cell cycle inhibitory functions. *EMBO J.* **2003**, *22*, 3568–3579. [CrossRef]
13. Clavier, A.; Ruby, V.; Rincheval-Arnold, A.; Mignotte, B.; Guénal, I. The Drosophila retinoblastoma protein, Rbf1, induces a Debcl- and Drp1-dependent mitochondrial apoptosis. *J. Cell Sci.* **2015**, *128*, 3239–3249. [CrossRef] [PubMed]
14. Pradeep, H.; Sharma, B.; Rajanikant, G.K. Drp1 in ischemic neuronal death: An unusual suspect. *Curr. Med. Chem.* **2014**, *21*, 2183–2189. [CrossRef] [PubMed]
15. Choi, S.Y.; Lee, J.-H.; Chung, A.-Y.; Jo, Y.; Shin, J.-H.; Park, H.-C.; Kim, H.; Lopez-Gonzalez, J.R.; Ryu, R.; Sun, W. Prevention of mitochondrial impairment by inhibition of protein phosphatase 1 activity in amyotrophic lateral sclerosis. *Cell Death Dis.* **2020**, *11*, 888. [CrossRef] [PubMed]
16. Zhang, X.; Huang, W.; Shao, Q.; Yang, Y.; Xu, Z.; Chen, J.; Zhang, X.; Xiaoqun, G. Drp1, a potential therapeutic target for Parkinson's disease, is involved in olfactory bulb pathological alteration in the Rotenone-induced rat model. *Toxicol. Lett.* **2020**, *325*, 1–13. [CrossRef]
17. Xiong, Y.; Yu, J. Modeling Parkinson's disease in Drosophila: What have we learned for dominant traits? *Front. Neurol.* **2018**, *9*, 228. [CrossRef]
18. Li, H.; Chaney, S.; Forte, M.; Hirsh, J. Ectopic g-protein expression in dopamine and serotonin neurons blocks cocaine sensitization in *Drosophila melanogaster*. *Curr. Biol.* **2000**, *10*, 211–214. [CrossRef]
19. Brand, A.H.; Perrimon, N. Targeted gene expression as a means of altering cell fates and generating dominant phenotypes. *Development* **1993**, *118*, 401–415. [CrossRef]
20. Hwang, R.-D.; Wiemerslage, L.; LaBreck, C.J.; Khan, M.; Kannan, K.; Wang, X.; Zhu, X.; Lee, D.; Fridell, Y.-W.C. The neuroprotective effect of human uncoupling protein 2 (hUCP2) requires cAMP-dependent protein kinase in a toxin model of Parkinson's disease. *Neurobiol. Dis.* **2014**, *69*, 180–191. [CrossRef]
21. Wang, Z.-H.; Clark, C.; Geisbrecht, E.R. *Drosophila clueless* is involved in Parkin-dependent mitophagy by promoting VCP-mediated Marf degradation. *Hum. Mol. Genet.* **2016**, *25*, 1946–1964. [CrossRef]
22. Zhou, J.; Xu, L.; Duan, X.; Liu, W.; Zhao, X.; Wang, X.; Shang, W.; Fang, X.; Yang, H.; Jia, L.; et al. Large-scale RNAi screen identified Dhpr as a regulator of mitochondrial morphology and tissue homeostasis. *Sci. Adv.* **2019**, *5*, eaax0365. [CrossRef] [PubMed]
23. M'Angale, P.G.; Staveley, B.E. The *Bcl-2* homologue *Buffy* rescues *α-synuclein*-induced Parkinson disease-like phenotypes in Drosophila. *BMC Neurosci.* **2016**, *17*, 24. [CrossRef] [PubMed]
24. Terhzaz, S.; Finlayson, A.J.; Stirrat, L.; Yang, J.; Tricoire, H.; Woods, D.J.; Dow, J.A.T.; Davies, S.-A. Cell-specific inositol 1,4,5 trisphosphate 3-kinase mediates epithelial cell apoptosis in response to oxidative stress in Drosophila. *Cell Signal.* **2010**, *22*, 737–748. [CrossRef]
25. Shan, Z.; Li, S.; Gao, Y.; Jian, C.; Ti, X.; Zuo, H.; Wang, Y.; Zhao, G.; Wang, Y.; Zhang, Q. mtDNA extramitochondrial replication mediates mitochondrial defect effects. *iScience* **2024**, *27*, 108970. [CrossRef]
26. Todd, A.; Staveley, B. Expression of Pink1 with α-synuclein in the dopaminergic neurons of Drosophila leads to increases in both lifespan and healthspan. *Genet. Mol. Res.* **2012**, *11*, 1497–1502. [CrossRef]
27. Wai, T.; Langer, T. Mitochondrial dynamics and metabolic regulation. *Trends Endocrinol. Metab.* **2016**, *27*, 105–117. [CrossRef]
28. Touvier, T.; De Palma, C.; Rigamonti, E.; Scagliola, A.; Incerti, E.; Mazelin, L.; Thomas, J.-L.; D'Antonio, M.; Politi, L.; Schaeffer, L.; et al. Muscle-specific Drp1 overexpression impairs skeletal muscle growth via translational attenuation. *Cell Death Dis.* **2015**, *6*, e1663. [CrossRef] [PubMed]

29. Yu, T.; Robotham, J.L.; Yoon, Y. Increased production of reactive oxygen species in hyperglycemic conditions requires dynamic change of mitochondrial morphology. *Proc. Natl. Acad. Sci. USA* **2006**, *103*, 2653–2658. [CrossRef]
30. Sun, Y.; Ge, X.; Li, X.; He, J.; Wei, X.; Du, J.; Sun, J.; Li, X.; Xun, Z.; Liu, W.; et al. High-fat diet promotes renal injury by inducing oxidative stress and mitochondrial dysfunction. *Cell Death Dis.* **2020**, *11*, 914. [CrossRef]
31. Cribbs, J.T.; Strack, S. Reversible phosphorylation of Drp1 by cyclic AMP-dependent protein kinase and calcineurin regulates mitochondrial fission and cell death. *EMBO Rep.* **2007**, *8*, 939–944. [CrossRef] [PubMed]
32. Virdee, K.; Parone, P.A.; Tolkovsky, A.M. Phosphorylation of the pro-apoptotic protein BAD on serine 155, a novel site, contributes to cell survival. *Curr. Biol.* **2000**, *10*, 1151–1154. [CrossRef] [PubMed]
33. Cereghetti, G.M.; Stangherlin, A.; Martins De Brito, O.; Chang, C.R.; Blackstone, C.; Bernardi, P.; Scorrano, L. Dephosphorylation by calcineurin regulates translocation of Drp1 to mitochondria. *Proc. Natl. Acad. Sci. USA* **2008**, *105*, 15803–15808. [CrossRef] [PubMed]
34. Rolland, S.G.; Conradt, B. New role of the BCL2 family of proteins in the regulation of mitochondrial dynamics. *Curr. Opin. Cell Biol.* **2010**, *22*, 852–858. [CrossRef]

biomolecules

MDPI

Article

3-Hydroxyanthranilic Acid Delays Paralysis in *Caenorhabditis elegans* Models of Amyloid-Beta and Polyglutamine Proteotoxicity

Bradford T. Hull [1], Kayla M. Miller [2], Caroline Corban [3], Grant Backer [3], Susan Sheehan [3], Ron Korstanje [3] and George L. Sutphin [1,2,*]

[1] Molecular and Cellular Biology Department, University of Arizona, Tucson, AZ 85721, USA
[2] Cancer Biology Graduate Interdisciplinary Program, University of Arizona, Tucson, AZ 85721, USA
[3] The Jackson Laboratory, Bar Harbor, ME 04609, USA
* Correspondence: sutphin@arizona.edu

Abstract: Age is the primary risk factor for neurodegenerative diseases such as Alzheimer's and Huntington's disease. Alzheimer's disease is the most common form of dementia and a leading cause of death in the elderly population of the United States. No effective treatments for these diseases currently exist. Identifying effective treatments for Alzheimer's, Huntington's, and other neurodegenerative diseases is a major current focus of national scientific resources, and there is a critical need for novel therapeutic strategies. Here, we investigate the potential for targeting the kynurenine pathway metabolite 3-hydroxyanthranilic acid (3HAA) using *Caenorhabditis elegans* expressing amyloid-beta or a polyglutamine peptide in body wall muscle, modeling the proteotoxicity in Alzheimer's and Huntington's disease, respectively. We show that knocking down the enzyme that degrades 3HAA, 3HAA dioxygenase (HAAO), delays the age-associated paralysis in both models. This effect on paralysis was independent of the protein aggregation in the polyglutamine model. We also show that the mechanism of protection against proteotoxicity from HAAO knockdown is mimicked by 3HAA supplementation, supporting elevated 3HAA as the mediating event linking HAAO knockdown to delayed paralysis. This work demonstrates the potential for 3HAA as a targeted therapeutic in neurodegenerative disease, though the mechanism is yet to be explored.

Keywords: metabolism; tryptophan; kynurenine; Alzheimer's disease; Huntington's disease; neurodegeneration; proteotoxicity; *Caenorhabditis elegans*

1. Introduction

Neurodegenerative diseases have become increasingly prevalent as our population has shifted to an older demographic over the last 30 years. Alzheimer's disease (AD) is now the seventh leading cause of the death in the United States and the most common form of neurodegeneration in the elderly [1]. As our cells age, a functional decline in the machinery that maintains protein homeostasis results in accumulation of damaged, misfolded, and aggregate-prone proteins [2–4]. The disruption of proteostasis is considered a hallmark of aging [5,6]. The aggregation of specific protein species is a defining feature—and is thought to be involved in the pathogenesis—of age-associated neurodegenerative diseases including AD and Huntington's disease (HD) [7–12]. Current treatments for these diseases generally target symptoms (cholinergic and glutamatergic signaling in neurons in AD; huntingtin production in HD), but do not address the underlying pathology [13–15]. Hundreds of clinical trials targeting different aspects of AD, HD, and other neurodegenerative disease neuropathology—amyloid toxicity, tau toxicity, huntingtin production, oxidative stress, and neuroinflammation—have largely failed to demonstrate efficacy, with some limited success in delaying symptomatic progression [16–18]. Identifying effective treatments for AD and HD is a major current focus of scientific resources, and there is a critical need for novel therapeutic strategies.

The kynurenine pathway (Figure 1) is a highly conserved metabolic pathway that converts ingested tryptophan to nicotinamide adenine dinucleotide (NAD+) [19]. Several studies have shown promise in delaying the aging and age-associated decline in protein homeostasis through the inhibition of tryptophan degradation via the kynurenine pathway. In the roundworm *Caenorhabditis elegans*, proteotoxicity is commonly modeled by expressing aggregate-prone proteins in body wall muscle, which causes the accumulation of muscle protein aggregates and accelerated paralysis with age [20,21]. van der Goot et al. [22] reported that knockdown of the gene *tdo-2*, encoding the kynurenine pathway enzyme tryptophan 2,3-dioxygenase (TDO), delayed the age-associated pathology in several of these models, including worms expressing human amyloid-beta (Aβ) or a 35-unit polyglutamine tract (Q35), modeling proteotoxicity in AD and HD, respectively. We validated this observation and further found that knockdown of *kynu-1* encoding the kynurenine pathway enzyme kynureninase (KYNU) similarly delays paralysis in these models [23]. Both interventions extend the lifespan in wild-type animals, supporting the broader link between the maintenance of protein homeostasis and aging [22,23]. We recently reported that knockdown of *haao-1*, a third kynurenine pathway gene encoding the enzyme 3-hydroxyanthranilate 3,4-dioxygenase (HAAO), increases the lifespan in *C. elegans* [24]. Worms lacking *haao-1* accumulate the HAAO metabolic substrate, 3-hydroxyanthranilic acid (3HAA), and supplementing 3HAA in the growth media similarly extends the lifespan [24].

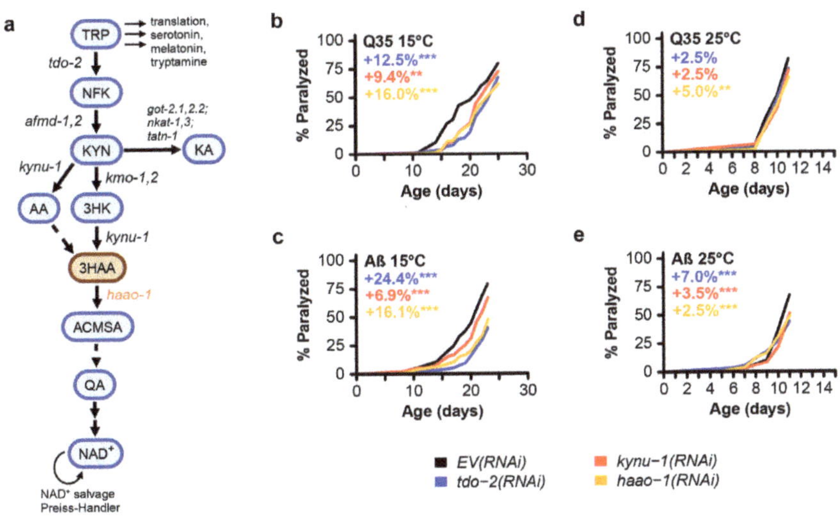

Figure 1. Knockdown of the kynurenine pathway gene *haao-1* delays paralysis in *C. elegans* models of polyglutamine and amyloid-beta proteotoxicity. (**a**) Schematic of the kynurenine pathway in *C. elegans*. Knockdown of *haao-1* substantially delayed age-associated pathology in *C. elegans* expressing either (**b**) Aβ or (**c**) a 35-unit polyglutamine tract (Q35) in body wall muscle at 15 °C to a similar degree as previously reported knockdown of *tdo-2* or *kynu-1* [23]. Knockdown of *haao-1* slightly delayed age-associated pathology in *C. elegans* expressing either (**d**) Aβ or (**e**) a Q35 in body wall muscle at 25 °C to a similar degree as previously reported knockdown of *tdo-2* or *kynu-1* [23]. Colored numbers indicate the percent change in mean age of paralysis vs. *EV(RNAi)*. ** $p < 0.01$, *** $p < 0.001$ for log-rank test vs. *EV(RNAi)*.

Here, we examine the impact of *haao-1* inhibition on Aβ and Q35 proteotoxicity in *C. elegans*. We find that *haao-1* knockdown delays paralysis in both contexts but does not prevent Q35 protein aggregation. Supplementing 3HAA similarly delays paralysis. The effectiveness at delaying paralysis was also dependent on environmental temperature. We include previously published data on knockdown of *tdo-2* or *kynu-1* in these models for comparison. These results suggest that the inhibition of HAAO or supplementation with 3HAA may hold potential as a clinical intervention for neurodegenerative disease.

2. Materials and Methods

2.1. *C. elegans* Strains and Maintenance

The following strains were obtained from the *Caenorhabditis* Genetic Center (CGC) in the College of Biological Sciences at the University of Minnesota: *dvIs2[pCL12(unc-54p::Aβ_{1-42}) + rol-6(su1006)]* (CL2006) and *rmIs132[unc-54p::Q35::YFP]* (AM140). We maintained worms on solid nematode growth media (NGM) seeded with *Escherichia coli* strain HT115 bacteria at 20 °C as previously described [25]. Worms were age-synchronized via bleach prep and transferred to plates containing 50 µM 5-fluorodeoxyuridine (FUdR) to prevent reproduction at the L4 larval stage [25].

2.2. RNA Interference

We conducted RNA interference (RNAi) feeding assays using standard media preparation procedures [25]. RNAi feeding bacteria were obtained from the Ahringer *C. elegans* RNAi feeding library [26,27]. All RNAi plasmids were sequenced to verify the correct target sequence. All RNAi experiments were conducted on NGM containing 1 mM Isopropyl β-D-1 thiogalactopyranoside (IPTG) to activate production of RNAi transcripts and 25 µg/mL carbenicillin to select RNAi plasmids and seeded with live *E. coli* (HT115) containing RNAi feeding plasmids. Worms were raised on egg plates seeded with RNAi bacteria until the L4 larval stage, and then transferred to FUdR-containing plates seeded with RNAi bacteria as described above.

2.3. 3HAA Supplementation

3HAA supplementation was achieved by adding the appropriate mass of solid 97% pure 3HAA (Sigma-Aldrich, St. Louis, MO, USA; catalog number 148776) to NGM media during preparation prior to autoclaving. Worms were raised on egg plates containing 1 mM 3HAA until the L4 larval stage, and then transferred to FUdR-containing plates containing 1 mM 3HAA as described above.

2.4. Paralysis Analysis

Paralysis experiments were conducted as previously described [23,25]. Briefly, adult worms were maintained on NGM RNAi or 3HAA plates with FUdR for the duration of their lives. Animals were examined every 1–2 days (25 °C) or 2–3 days (15 °C) by nose- and tail-prodding with a platinum wire pick. Worms were scored as paralyzed if they were unable to move relative to the plate surface. Worms showing vulva rupture were included in all analyses and worms that fled the surface of the plate were excluded. *p*-values for statistical comparison of paralysis between test groups were calculated using the log-rank test via the *survdiff* function in the R (version 4.3.2) "survival" package (version 3.5-7). For paralysis experiments, raw data are provided in Supplementary Appendix and summary statistics are provided in Table S1.

2.5. Polyglutamine Aggregate Imaging and Analysis

Polyglutamine aggregate imaging and quantification was completed as previously described [23]. Briefly, we quantified polyglutamine aggregate number and volume using 3D fluorescence microscopy of transgenic worms expressing fluorescently tagged polyglutamine repeats (Q35::YFP) at 7, 12, and 17 days of age at 15 °C, and 7 and 11 days of age at 25 °C. At least 8 animals were examined for each RNAi at each age. Worms were immobilized using 25 nM sodium azide in M9 buffer on microscope slides with 6% agarose pads. Worms were imaged using a Leica TCS SP8 Laser Scanning Microscope equipped with a 10× objective (Leica Microsystems Inc., Deerfield, IL, USA). Z-stack images were collected for whole worms with stack spacing selected based on the Nyquist sampling theorem (minimum 2.3 images per minimum aggregate diameter). The microscope was set to capture images with resonance scanning and line averaging of 16. Images for worms that did not fit a single field of view were merged with XuvStitch version 1.8.099x64. Protein aggregates in each worm were detected and analyzed using Imaris x64 version 8.1.2. The surface tool was used to identify aggregates with thresholds for minimum fluorescence intensity and aggregate volume to ensure the analyzed surfaces were aggregates and not discrete proteins: surface = 3.0, absolute intensity = 75, seed points = 5:15, intensity > 150, volume > 45, and number of voxels > 500. Aggregate volume and number were exported to R for analysis. Two-sided Welch's *t*-tests were used to determine significance in observed differences between target and EV RNAi at each age. For aggregate quantification experiments, raw data are provided in Supplementary Appendix and summary statistics are provided in Table S2.

3. Results

To test the efficacy of *haao-1* disruption at improving proteotoxicity linked to neurodegeneration, we used RNAi to knockdown *haao-1* in *C. elegans* transgenically expressing polyglutamine fused to a yellow fluorescence protein (Q35::YFP) or human Aβ in body wall muscle, modeling the proteotoxicity in HD and AD, respectively. We include data for both *tdo-2* and *kynu-1* knockdown for comparison [23]. Similar to previous reports for knockdown of *tdo-2* or *kynu-1* [22,23], *haao-1* knockdown delayed the age-associated paralysis in worms at 15 °C (Figure 1b,c). At 25 °C, the effect was present but much smaller in both models (Figure 1d,e). These results show that *haao-1* disruption may be protective against AD- and HD-associated proteotoxicity.

We next investigated whether the mechanism of delayed paralysis in Q35::YFP worms might be a consequence of the inhibition of Q35 protein aggregate formation, a hallmark of HD. We used confocal microscopy to quantify the number and size of Q35::YFP aggregates at different ages for worms cultured at both 15 °C and 25 °C. *haao-1* knockdown did not affect either the number or volume of protein aggregates at either 15 °C (Figure 2) or 25 °C (Figure 3), with the exception of a small but significant reduction in aggregate volume at 25 °C in 11-day-old worms (Figure 3b). This is in contrast to our previously reported observation that *tdo-2(RNAi)* can reduce the number of aggregates at 25 °C (Figure 3a), and aggregate volume at both 15 °C (Figure 2b) and 25 °C (Figure 3b), and that *kynu-1(RNAi)* can reduced the mid-life aggregate volume at 15 °C (Figure 2b) [23]. This suggests that the protective effects of *haao-1* knockdown against polyglutamine toxicity are independent of protein aggregation.

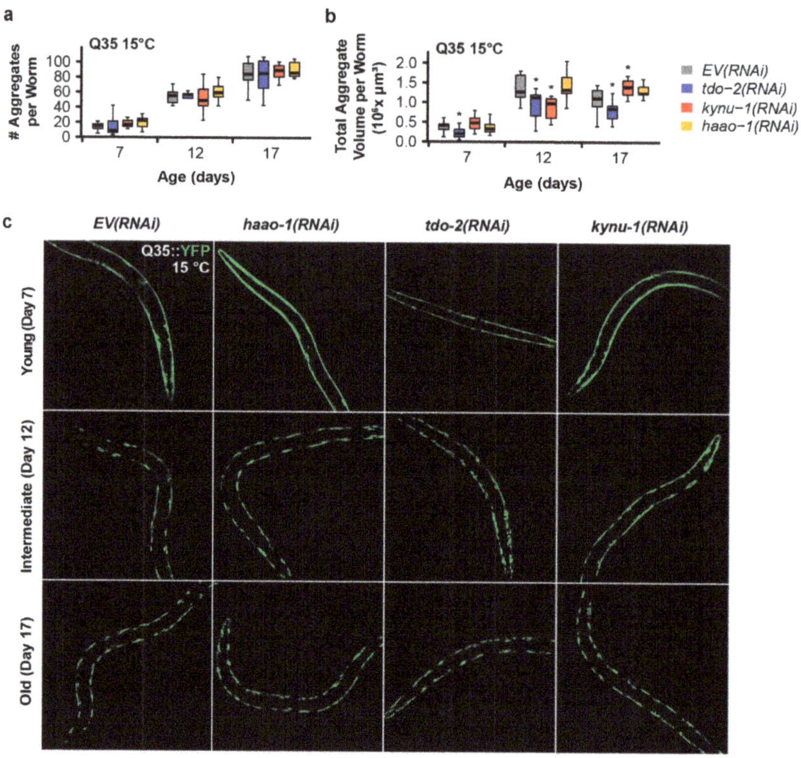

Figure 2. Knockdown of *haao-1* does not prevent polyglutamine aggregation at 15 °C. (**a**) *haao-1* knockdown did not affect the number of YFP-labeled muscle polyglutamine (Q35::YFP) protein aggregates at 15 °C, similar to our previous report for knockdown of *tdo-2* or *kynu-1* [23]. (**b**) *haao-1* knockdown did not affect the volume of Q35::YFP protein aggregates at 15 °C, unlike our previous report for knockdown of *tdo-2*, which reduced Q35::YFP aggregate volume at all ages examined, or *kynu-1*, which reduced Q35::YFP volume in mid-life [23]. (**c**) Representative images of Q35::YFP aggregation in each test group and age at 15 °C. * $p < 0.05$ for two-sided Welch's t-test vs. age-matched *EV(RNAi)*.

In an earlier work, we found that the beneficial effects of *haao-1* knockdown on the lifespan are mediated by increased physiological levels of the HAAO substrate metabolite 3HAA (Figure 1). Animals lacking functional *haao-1* accumulate 3HAA with age, and a 1 mM 3HAA supplementation in the growth media is sufficient to extend the mean *C. elegans* lifespan by 22%, a similar degree as knockout (26.2%) or RNAi knockdown (21.9%) of *haao-1* [24]. Meek et al. [28] predicted, in silico, that 3HAA should bind human Aβ at a critical region for protein misfolding, and found, in vitro, that 3HAA prevented protein Aβ aggregation. To investigate whether 3HAA supplementation can similarly recapitulate the protective effects of *haao-1(RNAi)* against proteotoxicity in *C. elegans*, we supplemented worms expressing muscle Q35 and Aβ with 1 mM 3HAA. The 1 mM 3HAA supplementation delayed the age-associated paralysis to a nearly identical degree as experiment-matched worms subjected to *haao-1(RNAi)* in both Q35 (Figure 4a) and Aβ (Figure 4b) animals. This observation supports a model in which elevated physiological 3HAA mediates the protection against proteotoxicity by *haao-1* disruption.

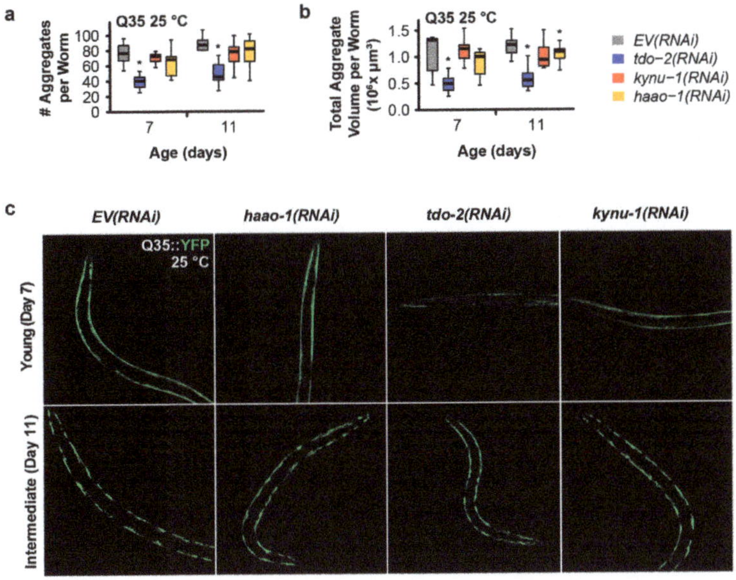

Figure 3. Knockdown of *haao-1* does not prevent polyglutamine aggregation at 25 °C. (**a**) *haao-1* knockdown did not affect the number of YFP-labeled muscle polyglutamine (Q35::YFP) protein aggregates at 25 °C, similar to our previous report for knockdown of *kynu-1* and unlike knockdown of *tdo-2*, which reduced Q35::YFP aggregate number at all ages examined [23]. (**b**) *haao-1* knockdown slightly reduced the volume of Q35::YFP protein aggregates at 25 °C, unlike our previous report for knockdown of *tdo-2*, which substantially reduced Q35::YFP aggregate volume both at all ages examined, or *kynu-1*, which did not reduce Q35::YFP volume [23]. (**c**) Representative images of Q35::YFP aggregation in each test group and age at 25 °C. * $p < 0.05$ for two-sided Welch's *t*-test vs. age-matched *EV(RNAi)*.

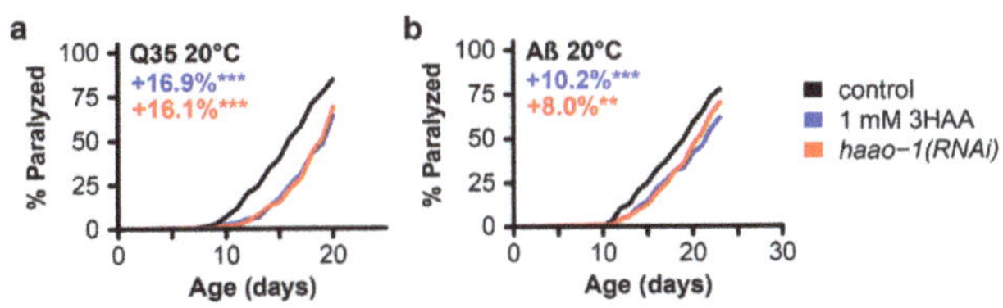

Figure 4. 3HAA supplementation recapitulates the protective effects of *haao-1* knockdown against polyglutamine and Aβ proteotoxicity in *C. elegans*. Supplementing the growth media with 1 mM 3HAA in both Huntington's (**a**) and Alzheimer's (**b**) models, supplementation with 3HAA closely mimicked the effects of haao-1 knockdown. Colored numbers indicate the percent change in mean age of paralysis vs. control. ** $p < 0.01$, *** $p < 0.001$ for log-rank test vs. *EV(RNAi)*.

4. Discussion

Here, we describe a small study examining the impact of *haao-1* disruption or 3HAA supplementation on the pathology in *C. elegans* models of AD- and HD-associated proteotoxicity. We first showed that *haao-1* knockdown delayed the age-associated paralysis in worms expressing Aβ or Q35::YFP, suggesting that *haao-1* disruption may be protective against proteotoxicity in AD and HD, respectively. We next investigated the mechanism of delayed paralysis and found that *haao-1* knockdown did not meaningfully affect either the number or volume of Q35::YFP protein aggregates. This suggests that the protective effect of *haao-1* knockdown against polyglutamine proteotoxicity in the *C. elegans* muscle is independent of protein aggregation. Because the transgenic Aβ construct is not labeled, we did not evaluate the Aβ aggregation, leaving open the possibility that 3HAA may improve proteotoxicity by preventing aggregation in this model. Finally, we showed that media supplementation with 1 mM 3HAA closely mimicked the benefit of *haao-1* knockdown on the age-associated paralysis, supporting a model in which the protection against proteotoxicity by *haao-1* disruption is mediated by elevated physiological 3HAA.

These observations suggest that *haao-1* disruption likely acts to reduce proteotoxicity via a mechanism that is at least partially distinct from both *tdo-2* and *kynu-1* for two reasons. First, *haao-1* disruption appears to be acting by elevating physiological levels of 3HAA. 3HAA levels are not elevated by the disruption of either *tdo-2* or *kynu-1* as we and others previously reported [22–24]. Specifically, *haao-1* knockdown resulted in an age-dependent increase in 3HAA that was 40-fold or higher relative to wild-type ones by 14 days of age, while *tdo-2* or *kynu-1* knockdown resulted in 3HAA levels at or below wild-type ones [22–24]. In contrast, *tdo-2* knockdown results in elevated tryptophan (TRP) while *kynu-1* knockdown results in elevated kynurenine (KYN) and 3-hydroxykynurenine (3HK) [22–24], as suggested by the structure of the kynurenine pathway (Figure 1a). Second, *haao-1* disruption does not impact the Q35::YFP aggregate size or volume, while the disruption of both *tdo-2* and *kynu-1* affects the aggregate number or volume in at least one context (Figures 2 and 3). This leaves open the possibility that these interventions act via both common mechanisms (such as reducing production downstream metabolites) and distinct mechanisms (with *haao-1* knockdown elevating 3HAA and *tdo-2* and *kynu-1* acting via a separate mechanism). The lifespan extension from the disruption of *tdo-2*, *kynu-1*, and *haao-1* was partially dependent on different combinations of established aging pathways [23,24]. Evidence presented by van der Goot et al. [22] suggests that the protection of *tdo-2* knockdown against pathology in a *C. elegans* model of alpha-synuclein proteotoxicity (modeling Parkinson's disease) may be mediated by elevated physiological levels of tryptophan. These observations provide guidance for future detailed mechanistic studies of the broader mechanisms mediating the protection of kynurenine pathway interventions against proteotoxicity.

The role of 3HAA in neurodegenerative disease has only been explored in a few small studies. Meek et al. [28] showed, in silico, that 3HAA was capable of binding to and preventing the misfolding of Aβ in an energetically favorable reaction. Other groups have found evidence linking 3HAA to an improved prognosis and survival from spinal cord injury [29,30] and the repression of pro-inflammatory signaling and behavior of innate immune cells [31–36], including both astrocytes and microglia [37]. Parrott et al. [38] found that HAAO knockout mice were protected against behavioral despair induced by lipopolysaccharide (LPS)-induced inflammation. Together, these studies suggest that elevating 3HAA may have specific benefits in preventing Aβ aggregation in AD and broader benefits in neurodegenerative disease by reducing neuroinflammation.

Quinolinic acid (QA), another metabolite of the kynurenine pathway which is created from 3HAA by HAAO (Figure 1a), has been described as an endogenous neurotoxin produced in active microglia and implicated in several human neurological diseases, including AD and HD [39–41]. Tau, an endogenous microtubule-associated protein whose phosphorylation and subsequent aggregation is considered a pathological hallmark of AD, has been shown to have elevated phosphorylation levels in the presence of QA [42]. QA has also been proposed to modulate excitotoxic neuronal death in HD [43]. Combined, this evidence suggests that HAAO disruption may be doubly beneficial for the treatment of neurodegenerative disease pathogenesis in both increasing 3HAA and decreasing QA. Reduced QA is also a candidate for a shared mechanism of action for the benefits of knocking down *tdo-2*, *kynu-1*, and *haao-1* in *C. elegans* Aβ and Q35 proteotoxicity models (Figure 1b–e) [23]. While the pathway structure suggests that QA should be reduced by all three interventions (Figure 1a), we have found QA challenging to measure via mass spectrometry relative to other metabolites in the kynurenine pathway [23,24] and have not yet been successful in developing a consistent protocol for QA quantification.

C. elegans are poikilothermic and temperature is a major determinant of lifespan, with worms cultured at 15 °C living almost twice as long on average as worms cultured at 25 °C [44,45]. In this work, we initially examined each intervention at 15 °C and 25 °C, temperatures near the extremes of the viable range (~12 to 26 °C). The expression of both Aβ in strain CL2006 [46] and Q35::YFP in strain AM140 [21] are driven by the constitutive muscle-specific *unc-54* promoter which is not known to be temperature-sensitive, and we observe age-dependent paralysis in untreated animals across the temperature spectrum (Figures 1b–e and 2a,b). In an earlier work, we found that the lifespan extension from *tdo-2*, but not *kynu-1* or *haao-1*, was dependent on environmental temperature [24]. Surprisingly, all three interventions were more effective at delaying paralysis in both Aβ and Q35 animals at 15 °C relative to 25 °C (Figure 1b–e). Interestingly, the benefits of *haao-1* knockdown were similar for Q35, but reduced for Aβ (Figure 3a,b), at 20 °C relative to 15 °C (Figure 1b,c), suggesting that these models may have a differential sensitivity to environmental temperature. Unlike *tdo-2(RNAi)* and *kynu-1(RNAi)*, which reduced both the number and volume of protein aggregates, particularly at higher temperatures, *haao-1* knockdown did not significantly alter protein aggregation, except for a slight reduction in aggregate volume in 11-day-old worms at 25 °C (Figure 3b). This differential effect suggests that the protective mechanisms of *haao-1* disruption against polyglutamine toxicity are independent of the alterations in protein aggregation, highlighting the complexity of the kynurenine pathway in modulating proteotoxicity in neurodegenerative diseases. The temperature-dependent differences observed in our study shed light on the nuanced effects of *haao-1* disruption on proteotoxicity associated with neurodegenerative diseases in *C. elegans* models. The implications for this temperature dependence for proteotoxicity in mammals are unclear, as, unlike *C. elegans*, human tissues maintain their temperature within a narrow range. However, because interventions that influence the *C. elegans* lifespan are often differentially effective at different temperatures [44], pathways that modulate the lifespan at 15 °C may be worth examining as potential mediators of 3HAA-dependent improvements in proteotoxicity.

Related to the environmental temperature, our practice is to not remove animals with age-associated vulva integrity defects (Avid [45], aka vulva rupture, as noted in Materials and Methods), which may alter interpretation. While many studies make a practice of removing Avid worms from analysis, many do not and there is not a clear standard in the field. A detailed examination in previous work suggests that Avid is a form of age-associated pathology in *C. elegans* that is influenced by pathways that also impact the lifespan [45]. As a pathology of aging, we cannot see a clear justification from excluding these animals from analysis. We acknowledge that the rupture may be confounded with paralysis if both are influenced by the intervention under study. However, removing Avid worms does not fix this issue, but simply shifts the bias in a different direction; removing ruptured worms may preferentially remove worms that will experience earlier or later paralysis, depending on the relationship between the intervention, Avid, and paralysis. Based on our experience, paralysis in the Aβ and Q35 models tends to occur prior to a significant onset of Avid in most cases (Figures 1 and 3) [45], so we believe that the potential impact of choosing to either remove or retail ruptured animals in the analysis in this case is minimal.

Finally, we note that, while our results are positive, there are inherent limitations with these *C. elegans* proteotoxicity models. First, these models express Q35 or Aβ in body wall muscle instead of neurons, a distinct cellular context from the proteotoxicity that occurs in neurodegenerative disease. Second, while *C. elegans* do have cells that carry out analogous roles to the mammalian innate immune system and share many conserved immune-signaling pathways with mammals, they lack many elements of mammalian immunity, including an analogous adaptive immune system. This may be relevant to the present study because the kynurenine pathway is active in immune cells and responsive to inflammatory signaling, and kynurenine metabolites produced by glial cells have been implicated in neurodegenerative pathology [42,43]. Third, the Q35 polyglutamine repeat model of HD disease mimics the polyglutamine expansion in the huntingtin gene that causes human HD but does not contain the full-length huntingtin gene. These considerations limit a strong interpretation of the presented data with respect to the impact of 3HAA and HAAO in human AD and HD. However, this study and the related literature discussed above motivates future detailed mechanistic studies of 3HAA supplementation and HAAO inhibition in worm models of neuronal Aβ and polyglutamine, which have a more subtle pathology than the muscle transgenic strains used in this study, and in preclinical rodent models.

5. Conclusions

In this study, we show that HAAO disruption improves the pathology in *C. elegans* models of AD- and HD-associated proteotoxicity. Additionally, we show that 3HAA supplementation is sufficient to mimic the effects of *haao-1* knockdown, suggesting that these beneficial effects are a result of 3HAA accumulation, consistent with our prior work demonstrating that the lifespan benefits of *haao-1* knockdown are also mediated by 3HAA in wild-type animals [24]. While our findings provide valuable insights into potential mechanisms underlying proteotoxicity and therapeutic interventions, *C. elegans* lack physiological features relevant to the pathophysiology of both AD and HD. Even considering these limitations, our study highlights 3HAA as a potential therapeutic target for both AD and HD. Further mechanistic studies are warranted to elucidate the precise pathways through which *haao-1* disruption and 3HAA supplementation exert their protective effects. Additionally, investigating these interventions in more clinically relevant models, such as neuronal Aβ and polyglutamine *C. elegans* models and preclinical rodent models, will be important steps for validating the therapeutic potential of HAAO and 3HAA.

Supplementary Materials: The following supporting information can be downloaded at: https://www.mdpi.com/article/10.3390/biom14050599/s1, Table S1: Summary statistics for paralysis experiments in *C. elegans* Alzheimer's (Aβ) and Huntington's (Q35) disease models; Table S2: Summary statistics for Q35::YFP aggregate quantification experiments in *C. elegans*. Supplementary Appendix: Contains the raw data for paralysis, aggregate number, and aggregate volume.

Author Contributions: Conceptualization, S.S., R.K., and G.L.S.; methodology, G.B., S.S., and G.L.S.; software, B.T.H., K.M.M., C.C., S.S., and G.L.S.; formal analysis, B.T.H., K.M.M., C.C., S.S., and G.L.S.; investigation, B.T.H., K.M.M., C.C., G.B., S.S., R.K., and G.L.S.; data curation, S.S., R.K., and G.L.S.; writing—original draft preparation, B.T.H., K.M.M., and G.L.S.; writing—review and editing, B.T.H., K.M.M., S.S., R.K., and G.L.S.; visualization, B.T.H., K.M.M., C.C., S.S., and G.L.S.; supervision, R.K. and G.L.S.; project administration, R.K. and G.L.S.; funding acquisition, R.K. and G.L.S. All authors have read and agreed to the published version of the manuscript.

Funding: This research was funded by the National Institute on Aging grant P30AG038070, and The Jackson Laboratory and National Institute on Aging grant R35GM133588 to G.L.S.

Institutional Review Board Statement: Not applicable.

Informed Consent Statement: Not applicable.

Data Availability Statement: The raw data for paralysis and aggregate quantification are provided in Supplementary Appendix. The raw images for Q35::YFP aggregation are available for download at: https://omero.jax.org/webclient/?show=project-1601 (accessed on 15 April 2024).

Conflicts of Interest: G.L.S. is the co-founder, owner, and one of the Managing Members of Senfina Biosystems LLC. All other authors declare that they have no competing interests.

References

1. Xu, J.; Murphy, S.L.; Kochanek, K.D.; Arias, E. *Mortality in the United States, 2021*; NCHS Data Brief, No. 456; CDC: Atlanta, GA, USA, 2022.
2. Hipp, M.S.; Kasturi, P.; Hartl, F.U. The Proteostasis Network and Its Decline in Ageing. *Nat. Rev. Mol. Cell Biol.* **2019**, *20*, 421–435. [CrossRef]

3. Labbadia, J.; Morimoto, R.I. The Biology of Proteostasis in Aging and Disease. *Annu. Rev. Biochem.* **2015**, *84*, 435–464. [CrossRef] [PubMed]
4. Taylor, R.C.; Dillin, A. Aging as an Event of Proteostasis Collapse. *Cold Spring Harb. Perspect. Biol.* **2011**, *3*, a004440. [CrossRef]
5. López-Otín, C.; Blasco, M.A.; Partridge, L.; Serrano, M.; Kroemer, G. The Hallmarks of Aging. *Cell* **2013**, *153*, 1194–1217. [CrossRef]
6. López-Otín, C.; Blasco, M.A.; Partridge, L.; Serrano, M.; Kroemer, G. Hallmarks of Aging: An Expanding Universe. *Cell* **2023**, *186*, 243–278. [CrossRef] [PubMed]
7. Bates, G.P.; Dorsey, R.; Gusella, J.F.; Hayden, M.R.; Kay, C.; Leavitt, B.R.; Nance, M.; Ross, C.A.; Scahill, R.I.; Wetzel, R.; et al. Huntington Disease. *Nat. Rev. Dis. Primer* **2015**, *1*, 15005. [CrossRef]
8. Irvine, G.B.; El-Agnaf, O.M.; Shankar, G.M.; Walsh, D.M. Protein Aggregation in the Brain: The Molecular Basis for Alzheimer's and Parkinson's Diseases. *Mol. Med.* **2008**, *14*, 451–464. [CrossRef] [PubMed]
9. Knopman, D.S.; Amieva, H.; Petersen, R.C.; Chételat, G.; Holtzman, D.M.; Hyman, B.T.; Nixon, R.A.; Jones, D.T. Alzheimer Disease. *Nat. Rev. Dis. Primer* **2021**, *7*, 33. [CrossRef]
10. Masters, C.L.; Simms, G.; Weinman, N.A.; Multhaup, G.; McDonald, B.L.; Beyreuther, K. Amyloid Plaque Core Protein in Alzheimer Disease and Down Syndrome. *Proc. Natl. Acad. Sci. USA* **1985**, *82*, 4245–4249. [CrossRef]
11. Masters, C.L.; Bateman, R.; Blennow, K.; Rowe, C.C.; Sperling, R.A.; Cummings, J.L. Alzheimer's Disease. *Nat. Rev. Dis. Primer* **2015**, *1*, 15056. [CrossRef]
12. Scherzinger, E.; Lurz, R.; Turmaine, M.; Mangiarini, L.; Hollenbach, B.; Hasenbank, R.; Bates, G.P.; Davies, S.W.; Lehrach, H.; Wanker, E.E. Huntingtin-Encoded Polyglutamine Expansions Form Amyloid-like Protein Aggregates in Vitro and in Vivo. *Cell* **1997**, *90*, 549–558. [CrossRef] [PubMed]
13. Andhale, R.; Shrivastava, D. Huntington's Disease: A Clinical Review. *Cureus* **2022**, *14*, e28484. [CrossRef] [PubMed]
14. Huang, L.-K.; Kuan, Y.-C.; Lin, H.-W.; Hu, C.-J. Clinical Trials of New Drugs for Alzheimer Disease: A 2020–2023 Update. *J. Biomed. Sci.* **2023**, *30*, 83. [CrossRef] [PubMed]
15. Wild, E.J.; Tabrizi, S.J. Therapies Targeting DNA and RNA in Huntington's Disease. *Lancet Neurol.* **2017**, *16*, 837–847. [CrossRef] [PubMed]
16. Kim, C.K.; Lee, Y.R.; Ong, L.; Gold, M.; Kalali, A.; Sarkar, J. Alzheimer's Disease: Key Insights from Two Decades of Clinical Trial Failures. *J. Alzheimers Dis.* **2022**, *87*, 83–100. [CrossRef] [PubMed]
17. Travessa, A.M.; Rodrigues, F.B.; Mestre, T.A.; Ferreira, J.J. Fifteen Years of Clinical Trials in Huntington's Disease: A Very Low Clinical Drug Development Success Rate. *J. Huntingt. Dis.* **2017**, *6*, 157–163. [CrossRef]
18. Prasad, E.M.; Hung, S.-Y. Current Therapies in Clinical Trials of Parkinson's Disease: A 2021 Update. *Pharmaceuticals* **2021**, *14*, 717. [CrossRef]
19. Castro-Portuguez, R.; Sutphin, G.L. Kynurenine Pathway, NAD+ Synthesis, and Mitochondrial Function: Targeting Tryptophan Metabolism to Promote Longevity and Healthspan. *Exp. Gerontol.* **2020**, *132*, 110841. [CrossRef]
20. Link, C.D. Expression of Human Beta-Amyloid Peptide in Transgenic Caenorhabditis Elegans. *Proc. Natl. Acad. Sci. USA* **1995**, *92*, 9368–9372. [CrossRef]
21. Morley, J.F.; Brignull, H.R.; Weyers, J.J.; Morimoto, R.I. The Threshold for Polyglutamine-Expansion Protein Aggregation and Cellular Toxicity Is Dynamic and Influenced by Aging in Caenorhabditis Elegans. *Proc. Natl. Acad. Sci. USA* **2002**, *99*, 10417–10422. [CrossRef]
22. van der Goot, A.T.; Zhu, W.; Vazquez-Manrique, R.P.; Seinstra, R.I.; Dettmer, K.; Michels, H.; Farina, F.; Krijnen, J.; Melki, R.; Buijsman, R.C.; et al. Delaying Aging and the Aging-Associated Decline in Protein Homeostasis by Inhibition of Tryptophan Degradation. *Proc. Natl. Acad. Sci. USA* **2012**, *109*, 14912–14917. [CrossRef]
23. Sutphin, G.L.; Backer, G.; Sheehan, S.; Bean, S.; Corban, C.; Liu, T.; Peters, M.J.; van Meurs, J.B.J.; Murabito, J.M.; Johnson, A.D.; et al. Caenorhabditis Elegans Orthologs of Human Genes Differentially Expressed with Age Are Enriched for Determinants of Longevity. *Aging Cell* **2017**, *16*, 672–682. [CrossRef] [PubMed]
24. Dang, H.; Castro-Portuguez, R.; Espejo, L.; Backer, G.; Freitas, S.; Spence, E.; Meyers, J.; Shuck, K.; Gardea, E.A.; Chang, L.M.; et al. On the Benefits of the Tryptophan Metabolite 3-Hydroxyanthranilic Acid in Caenorhabditis Elegans and Mouse Aging. *Nat. Commun.* **2023**, *14*, 8338. [CrossRef]
25. Sutphin, G.L.; Kaeberlein, M. Measuring Caenorhabditis Elegans Life Span on Solid Media. *JoVE J. Vis. Exp.* **2009**, *27*, e1152. [CrossRef]
26. Fraser, A.G.; Kamath, R.S.; Zipperlen, P.; Martinez-Campos, M.; Sohrmann, M.; Ahringer, J. Functional Genomic Analysis of *C. elegans* Chromosome I by Systematic RNA Interference. *Nature* **2000**, *408*, 325–330. [CrossRef] [PubMed]
27. Kamath, R.S.; Fraser, A.G.; Dong, Y.; Poulin, G.; Durbin, R.; Gotta, M.; Kanapin, A.; Le Bot, N.; Moreno, S.; Sohrmann, M.; et al. Systematic Functional Analysis of the Caenorhabditis Elegans Genome Using RNAi. *Nature* **2003**, *421*, 231–237. [CrossRef] [PubMed]
28. Meek, A.; Simms, G.; Weaver, D. Searching for an Endogenous Anti-Alzheimer Molecule: Identifying Small Molecules in the Brain That Slow Alzheimer Disease Progression by Inhibition of β-Amyloid Aggregation. *J. Psychiatry Neurosci.* **2013**, *38*, 269–275. [CrossRef]
29. Jacobs, K.R.; Lovejoy, D.B. Inhibiting the Kynurenine Pathway in Spinal Cord Injury: Multiple Therapeutic Potentials? *Neural Regen. Res.* **2018**, *13*, 2073–2076. [CrossRef]

30. Yates, J.R.; Gay, E.A.; Heyes, M.P.; Blight, A.R. Effects of Methylprednisolone and 4-Chloro-3-Hydroxyanthranilic Acid in Experimental Spinal Cord Injury in the Guinea Pig Appear to Be Mediated by Different and Potentially Complementary Mechanisms. *Spinal Cord.* **2014**, *52*, 662–666. [CrossRef]
31. Berg, M.; Stenuit, B.; Ho, J.; Wang, A.; Parke, C.; Knight, M.; Alvarez-Cohen, L.; Shapira, M. Assembly of the Caenorhabditis Elegans Gut Microbiota from Diverse Soil Microbial Environments. *ISME J.* **2016**, *10*, 1998–2009. [CrossRef]
32. Zhang, L.; Ovchinnikova, O.; Jönsson, A.; Lundberg, A.M.; Berg, M.; Hansson, G.K.; Ketelhuth, D.F.J. The Tryptophan Metabolite 3-Hydroxyanthranilic Acid Lowers Plasma Lipids and Decreases Atherosclerosis in Hypercholesterolaemic Mice. *Eur. Heart J.* **2012**, *33*, 2025–2034. [CrossRef] [PubMed]
33. Fallarino, F.; Grohmann, U.; Vacca, C.; Bianchi, R.; Orabona, C.; Spreca, A.; Fioretti, M.C.; Puccetti, P. T Cell Apoptosis by Tryptophan Catabolism. *Cell Death Differ.* **2002**, *9*, 1069–1077. [CrossRef] [PubMed]
34. Morita, T.; Saito, K.; Takemura, M.; Maekawa, N.; Fujigaki, S.; Fujii, H.; Wada, H.; Takeuchi, S.; Noma, A.; Seishima, M. 3-Hydroxyanthranilic Acid, an L-Tryptophan Metabolite, Induces Apoptosis in Monocyte-Derived Cells Stimulated by Interferon-Gamma. *Ann. Clin. Biochem.* **2001**, *38*, 242–251. [CrossRef] [PubMed]
35. Lee, W.-S.; Lee, S.-M.; Kim, M.-K.; Park, S.-G.; Choi, I.-W.; Choi, I.; Joo, Y.-D.; Park, S.-J.; Kang, S.-W.; Seo, S.-K. The Tryptophan Metabolite 3-Hydroxyanthranilic Acid Suppresses T Cell Responses by Inhibiting Dendritic Cell Activation. *Int. Immunopharmacol.* **2013**, *17*, 721–726. [CrossRef] [PubMed]
36. Gargaro, M.; Vacca, C.; Massari, S.; Scalisi, G.; Manni, G.; Mondanelli, G.; Mazza, E.M.C.; Bicciato, S.; Pallotta, M.T.; Orabona, C.; et al. Engagement of Nuclear Coactivator 7 by 3-Hydroxyanthranilic Acid Enhances Activation of Aryl Hydrocarbon Receptor in Immunoregulatory Dendritic Cells. *Front. Immunol.* **2019**, *10*, 1973. [CrossRef] [PubMed]
37. Krause, D.; Suh, H.-S.; Tarassishin, L.; Cui, Q.L.; Durafourt, B.A.; Choi, N.; Bauman, A.; Cosenza-Nashat, M.; Antel, J.P.; Zhao, M.-L.; et al. The Tryptophan Metabolite 3-Hydroxyanthranilic Acid Plays Anti-Inflammatory and Neuroprotective Roles During Inflammation. *Am. J. Pathol.* **2011**, *179*, 1360–1372. [CrossRef] [PubMed]
38. Parrott, J.M.; Redus, L.; Santana-Coelho, D.; Morales, J.; Gao, X.; O'Connor, J.C. Neurotoxic Kynurenine Metabolism Is Increased in the Dorsal Hippocampus and Drives Distinct Depressive Behaviors during Inflammation. *Transl. Psychiatry* **2016**, *6*, e918. [CrossRef] [PubMed]
39. Espey, M.G.; Chernyshev, O.N.; Reinhard, J.F.; Namboodiri, M.A.; Colton, C.A. Activated Human Microglia Produce the Excitotoxin Quinolinic Acid. *Neuroreport* **1997**, *8*, 431–434. [CrossRef] [PubMed]
40. Guillemin, G.J.; Brew, B.J.; Noonan, C.E.; Takikawa, O.; Cullen, K.M. Indoleamine 2,3 Dioxygenase and Quinolinic Acid Immunoreactivity in Alzheimer's Disease Hippocampus. *Neuropathol. Appl. Neurobiol.* **2005**, *31*, 395–404. [CrossRef]
41. Lugo-Huitrón, R.; Ugalde Muñiz, P.; Pineda, B.; Pedraza-Chaverrí, J.; Ríos, C.; Pérez-de la Cruz, V. Quinolinic Acid: An Endogenous Neurotoxin with Multiple Targets. *Oxid. Med. Cell. Longev.* **2013**, *2013*, 104024. [CrossRef]
42. Rahman, A.; Ting, K.; Cullen, K.M.; Braidy, N.; Brew, B.J.; Guillemin, G.J. The Excitotoxin Quinolinic Acid Induces Tau Phosphorylation in Human Neurons. *PLoS ONE* **2009**, *4*, e6344. [CrossRef] [PubMed]
43. Guidetti, P.; Luthi-Carter, R.E.; Augood, S.J.; Schwarcz, R. Neostriatal and Cortical Quinolinate Levels Are Increased in Early Grade Huntington's Disease. *Neurobiol. Dis.* **2004**, *17*, 455–461. [CrossRef] [PubMed]
44. Miller, H.; Fletcher, M.; Primitivo, M.; Leonard, A.; Sutphin, G.L.; Rintala, N.; Kaeberlein, M.; Leiser, S.F. Genetic Interaction with Temperature Is an Important Determinant of Nematode Longevity. *Aging Cell* **2017**, *16*, 1425–1429. [CrossRef] [PubMed]
45. Leiser, S.F.; Jafari, G.; Primitivo, M.; Sutphin, G.L.; Dong, J.; Leonard, A.; Fletcher, M.; Kaeberlein, M. Age-Associated Vulval Integrity Is an Important Marker of Nematode Healthspan. *AGE* **2016**, *38*, 419–431. [CrossRef]
46. Wu, Y.; Wu, Z.; Butko, P.; Christen, Y.; Lambert, M.P.; Klein, W.L.; Link, C.D.; Luo, Y. Amyloid-Beta-Induced Pathological Behaviors Are Suppressed by Ginkgo Biloba Extract EGb 761 and Ginkgolides in Transgenic Caenorhabditis Elegans. *J. Neurosci. Off. J. Soc. Neurosci.* **2006**, *26*, 13102–13113. [CrossRef]

MDPI AG
Grosspeteranlage 5
4052 Basel
Switzerland
Tel.: +41 61 683 77 34

Biomolecules Editorial Office
E-mail: biomolecules@mdpi.com
www.mdpi.com/journal/biomolecules

www.ingramcontent.com/pod-product-compliance
Lightning Source LLC
Chambersburg PA
CBHW061631210926
51850CB00009B/2508